PROCEEDINGS
of the
STEKLOV INSTITUTE
of
MATHEMATICS

edited by

I. G. Petrovskiĭ and S. M. Nikol'skiĭ

number 81 (1966)

TOPOLOGICAL IMBEDDINGS IN EUCLIDEAN SPACE

by

L. V. KELDYS

AMERICAN MATHEMATICAL SOCIETY

Providence, Rhode Island

1968

Академия Наук
Союза Советских Социалистических Республик

ТРУДЫ МАТЕМАТИЧЕСКОГО ИНСТИТУТА
имени В. А. СТЕКЛОВА
LXXXI

Ответственный редактор академик И. Г. ПЕТРОВСКИЙ
Заместитель ответственного редактора профессор С. М. НИКОЛЬСКИЙ

ТОПОЛОГИЧЕСКИЕ ВЛОЖЕНИЯ В ЕВКЛИДОВО ПРОСТРАНСТВО

Л. В. Келдыш

Издательство „Наука"
Москва—1966

Translated from the Russian by J. Zilber

PROCEEDINGS OF THE STEKLOV INSTITUTE OF MATHEMATICS
IN THE ACADEMY OF SCIENCES OF THE USSR

(Труды математического института им. В. А. Стеклова, т. LXXXI, 1966)

TABLE OF CONTENTS

Page

INTRODUCTION

This monograph is devoted to a presentation of the foundations of the set-theoretical theory of topological imbeddings in Euclidean space E^n and of a number of new results in this area.

To be considered are questions concerning the topological imbedding in E^n of cells (topological cubes), spheres and zero-dimensional compacta. This area of investigation began with the study of imbeddings in the plane, which were completely dealt with in the work of Schoenflies [69] and Antoine [20]. In the twenties the first results were obtained concerning imbeddings in three-dimensional Euclidean space E^3. Alexander constructed an example of a sphere S^2 in E^3 (the so-called *horned sphere*), bounding a domain which is not homeomorphic to E^3 [18] (see Figure 4.7), and Antoine [20] and Uryson [11] constructed zero-dimensional compacta in E^3 whose complements are not simply-connected (see Chapter IV). On the other hand, Alexander proved that every polyhedral two-dimensional sphere in E^3 bounds a domain whose closure is a three-dimensional cell.

At the end of the forties began the detailed study of the topology of three-dimensional Euclidean space E^3. Artin and Fox constructed a number of new examples of "wild" topological imbeddings in E^3 of simple arcs, disks (topological squares) and two-dimensional spheres [21].

Next, at the end of the forties and in the fifties Moise [53-60] and Bing [23-30] carried out an extensive study of the polyhedral structure of E^3 and topological imbeddings in E^3. These papers also contain some important theorems on three-dimensional manifolds; in particular, the proof of the Principal Conjecture (Hauptvermutung) and the theorem on triangulability of three-dimensional manifolds [29, 57] We shall present in this monograph only the examples of wild imbeddings in E^3 (see Chapter IV), since the study of the polyhedral structure of E^3 and of topological imbeddings in E^3 depends on a specialized, very cumbersome polyhedral and set-theoretical technique.

As for imbeddings in E^n and S^n for $n > 3$, their study was until recently represented only by some examples of wild imbeddings [31, 3] obtained by starting from wild imbeddings in E^3.

Only in 1959 did the systematic study begin of topological imbeddings in E^n for $n > 3$. Two papers by Mazur [49, 50] appeared in 1959, and one by Brown [32] in 1960, giving a sufficiency criterion for an imbedding of a sphere S^{n-1} into a sphere S^n to be "nice," i.e., for the closures of both complementary domains of S^{n-1} in S^n to be n-cells—the latter being equivalent to the existence of a homeomorphism of S^n onto itself taking S^{n-1} into the equator.

The theorem proved by Mazur and Brown bears the name of the "generalized Schoenflies theorem" (see Chapter V). It received wide applications immediately. A number of theorems were proved by means of it concerning the topology of manifolds (see Chapter V). One of the corollaries of the generalized Schoenflies theorem (Theorem 2.1 of Chapter V) was used by Stallings to prove the generalized Poincaré conjecture [71]. Brown and Michael showed that the nature of the imbedding of an $(n-1)$-sphere into an n-sphere depends on local conditions [34, 35] (see Chapter V, § 4).

It was proved by Newman that a combinatorial sphere (see Chapter V, § 5) Σ^{n-1} which is a subpolyhedron of a combinatorial sphere Σ^n satisfies the conditions of the Schoenflies theorem, and consequently bounds topological balls [61] (see Chapter V, Theorem 5.3). However, the question concerning the nature of the imbedding in E^n of an arbitrary polyhedral sphere remains open for $n > 5$. For $n = 4$ and 5, it has been solved by Rosen on the basis of Newman's theorem and the properties of three-dimensional manifolds [65]. At the same time, there have been added to the classical, as yet unsolved problems of the topology of manifolds some newly formulated ones. An important place among them is occupied by the so-called annulus problem. Suppose S_1 and S_2 are two disjoint $(n-1)$-spheres in S^n (or E^n) satisfying the conditions of the Schoenflies theorem. Is the closed region bounded by $S_1 \cup S_2$ a topological spherical annulus? This problem is related to a number of others, in particular to the problem of the structure of the group of homeomorphisms of the n-sphere. The efforts to find an approach to these problems have recently given rise to a number of papers devoted to the study of the homeomorphisms of the n-sphere S^n. Brown and Gluck [36] have investigated the so-called stable homeomorphisms (see Chapter V, § 3). An interesting theorem on the structure of the homeomorphisms of a sphere has been proved by Černavskiĭ [13]. Connected with Newman's theorem is the so-called combinatorial Schoenflies problem: is a topological ball bounded by a combinatorial sphere Σ^{n-1} in Σ^n a combinatorial ball? Any method based on the Schoenflies theorem offers no possibility of deciding this question.

In very recent years, a number of important results have been obtained concerning imbeddings of a k-sphere into S^n (or E^n), where $k < n-1$. For $k \leq n-2$, $n \geq 5$, Stallings has described sufficient conditions for the imbedding in S^n to be "nice," i.e., for the existence of a homeomorphism H of S^n onto itself taking S^k into the boundary of a spherical $(k+1)$-simplex [73] (see Chapter V, Conclusion). For $n = 4$, $k = 1$, the problem has been solved by Gluck [41]. For $n = 4$, $k = 2$, the question is still open. Zeeman has studied the imbedding in E^n of combinatorial spheres Σ^k of dimension $k \leq n-3$ [77]. He formulates the problem differently, however. Stallings gives, for $n \geq 5$, $k \leq n-3$, local topological conditions for the existence of the homeomorphism H; while Zeeman proves, for a combinatorial sphere Σ^k, $k \leq n - 3$, the existence of a semilinear isotopy $F_t\colon E^n \to E^n$, i.e., of a continuous family of semilinear homeomorphisms (see Chapter V, Conclusion), where $0 \leq t \leq 1$, F_0 is the identity mapping, and $F_1(\Sigma^k)$ is the boundary of a $(k+1)$-simplex.

In the Stallings case, of a topological imbedding of S^k into S^n for $k \leq n - 3$, $n \geq 5$, as well as in the case of a topological imbedding of S^{n-1} into S^n, the question of the existence of an isotopy from the identity mapping of S^n onto itself, taking a "nice" sphere S^k into the boundary of a $(k+1)$-simplex, is in its general form still open. Very recently, a partial solution has been obtained for sufficiently small k. Gluck and Greathouse solved it for $k \leq n/2 - 1$ [42–43]. Černavskiĭ has improved the bound to $k < 2n/3 - 1$, and recently to $k \leq n - 3$ [13, 81].

Černavskiĭ's result follows from his theorem on the structure of the homeomorphisms of S^n [13] (see the Conclusion). The general problem of the existence of an isotopy for an imbedding of S^k into S^n, $k \neq n - 2$, is closely related to the problem of the structure of the group of homeomorphisms of a sphere, and in particular to the problem of the existence of an isotopy joining an arbitrary orientation-preserving homeomorphism of S^n with the identity mapping (see Chapter V, § 3).

From what has been said, it is evident that the theory of topological imbeddings in S^n is an area of topology which is at present undergoing intensive developement. Besides the problems mentioned, it also includes those concerning a bound to the dimension of the space E^n in which a manifold M^k, of dimension $k < n$, can be topologically imbedded, as a function of the homotopy properties of M^k. A number of results have also been recently obtained in this direction [63]. These latter questions do not, however, enter into the present monograph. The purpose of the monograph is to acquaint the reader with the current state of the theory of topological imbeddings and to prepare him to read the literature in this area. It is divided into five chapters.

Chapter I describes the basic properties of the simplest kinds of homotopy, namely isotopy and pseudo-isotopy, and gives examples of the application of these notions to the study of topological imbeddings. Included here are results of the author on imbedding of zero-dimensional compacta in E^n.

Chapter II contains the classical theorems on imbedding in the plane and on planar isotopies.

Chapter III contains the information needed in Chapter IV concerning the fundamental group of a space, particularly information not previously available in Russian. We prove the van Kampen theorem and apply it to computing fundamental groups, and in particular the fundamental group of a three-dimensional knot. The last section describes a procedure for computing the fundamental group of any domain in E^n.

In Chapter IV, we give examples of "wild" imbeddings in three-dimensional Euclidean space E^3. Included here are a number of examples of wild simple arcs, 2-spheres and disks described in the paper of Artin and Fox [21], the horned sphere of Alexander [18] and some others. The last section concerns the wild zero-dimensional compactum of Antoine [20].

Chapter V is devoted to a study of the new results. It begins with the proof of the generalized Schoenflies theorem given by Brown [32] (see § 1). A number of applications of the theorem are then given to the topology of manifolds and to topological imbeddings (see § 2). In § 3 we study the homeomorphisms of the n-sphere. Here we prove the theorem of Brown and Gluck on stable homeomorphisms [36] and give some of its applications. We give also Černavskiĭ's sufficient condition for regularity of an imbedding of finitely many $(n-1)$-spheres [12].

Along with the new results, this section contains the statements of a number of problems in the topology of manifolds, primarily those relating to the structure of the homeomorphisms of the n-sphere.

In § 4 we prove the Brown-Michael theorem concerning the dependence of an imbedding on local conditions and give some applications [34, 35]. § 5 concerns the imbedding of combinatorial spheres. We give there the basic definitions of the topology of combinatorial manifolds and prove Newman's theorem concerning the imbedding of a star sphere (in particular, a combinatorial sphere) Σ^{n-1} into E^n. In addition, some problems are formulated in that section concerning the topology of combinatorial manifolds.

The concluding § 6 contains the statements, without proofs, of some important recent results. Included are the results of Stallings on the imbedding of S^k into S^n for $n \geq 5$, $k \leq n-3$; the results of Zeeman on the imbedding, into E^n, of combinatorial n-cells and spheres of dimension $k \leq n-3$; and the theorem of Černavskiĭ on the structure of the homeomorphisms of the n-sphere, which implies, for a "nice" imbedding into S^n of the sphere S^k of dimension $k < 2n/3 - 1$, the existence of an isotopy F_t from the identity mapping, taking S^k into the boundary of a spherical $(k+1)$-simplex.

The full proofs of the results of Stallings [73] and of Zeeman [77] were published in 1963. Černavskiĭ's theorem was published in 1965 in Matematičeskiĭ Sbornik [13, 81]. Let us merely remark here that whereas all the results presented in detail in Chapter V are relatively easy to obtain from the Schoenflies theorem, these new results are considerably more intricate and required for their proof the creation of new methods. They are based on the so-called engulfing lemma proved by Stallings in [72].

The question of refining Černavskiĭ's result, i.e. of proving the existence of a similar isotopy for tamely imbedded spheres S^k of dimension $k \neq n - 2$, is related to some very hard problems, in particular to the annulus problem.

CHAPTER I

ISOTOPY AND PSEUDO-ISOTOPY

§ 1. Homotopy

1. Let X and Y be two topological spaces. The points of X will be denoted by x, those of Y by y.

Definition 1.1. Two continuous mappings f and ϕ of X into Y are said to be homotopic if there exists a continuous mapping $F(x, t)$ of the topological product $X \times I^1$ (of the space X by the line segment $I^1 = [0, 1]$) into Y such that

$$F(x, 0) = f(x); \quad F(x, 1) = \varphi(x). \tag{1}$$

For each fixed $t = t_0$, $F(x, t_0)$ is a continuous mapping of X into Y. Thus $F(x, t)$ can be regarded as a family of continuous mappings of X into Y which depends continuously on the linear parameter t. We shall denote this family of continuous mappings by $F_t(x)$ or F_t, and write

$$F_t : X \to Y, \quad 0 \leqslant t \leqslant 1.$$[1)]

F_t is called a homotopy taking the continuous mapping f into the continuous mapping ϕ, or joining f and ϕ.

A continuous mapping $f: X \to Y$ is called constant if every point $x \in X$ maps into the same point $y_0 \in Y$:

$$f(x) = y_0, \qquad x \in X.$$

In this case we write $f \equiv \text{const}$.

A continuous mapping T of X into itself is called the identity if $T(x) = x$ for every $x \in X$. In this case we write

$$T(x) = 1. \tag{2}$$

If a homotopy F_t takes the identity mapping of X into a mapping $\phi: X \to X$, we shall say that F_t is a *homotopy from the identity mapping*. We shall often be considering such homotopies of Euclidean space E^n or of the sphere S^n.

Examples.

Example 1.1. Suppose Y is arcwise connected, i.e. any two points of Y can be joined by a simple arc. Then any two constant mappings of an arbitrary

---

1) One often writes $F_t: X \times [0, 1] \to Y$, but we find the shorter notation convenient here.

space X into Y are homotopic.

Indeed, for $x \in X$ let $f(x) = y_1$ and $\phi(x) = y_2$, where y_1 and y_2 are points of Y. There exists a simple arc $l \subset Y$ joining y_1 and y_2. Let $y = l(t)$ be a homeomorphic mapping of the segment $[0 \leq t \leq 1]$ into Y defining the arc l. Set

$$F_t(x) = l(t), \quad x \in X, \quad 0 \leqslant t \leqslant 1.$$

Then $F_0(x) = y_1$ and $F_1(x) = y_2$, so that F_t is a homotopy joining the constant mappings f and ϕ.

Example 1.2. Every continuous mapping f of the cube I^n or of Euclidean space E^n into an arbitrary space Y is homotopic to a constant mapping.

Indeed, the points of the cube can be regarded as vectors with origin at the center of the cube. Then the homotopy

$$F_t(x) = f[(1-t)x]$$

takes the mapping f into the constant mapping $\phi(x) = f(0)$.

Example 1.3. Every continuous mapping of an arbitrary space X into the cube I^n or into E^n is homotopic to a constant mapping. The homotopy is defined by

$$F_t(x) = (1-t)f(x).$$

Example 1.4. Let π be the orthogonal projection of the unit cube onto its face I^{n-1}. Then π is a continuous mapping of the cube into itself, which is homotopic to the identity mapping. If we specify the points $x \in I^n$ by two coordinates:

$$\xi \in I^{n-1}, \quad z = \rho(x, I^{n-1})$$

we may define a homotopy F_t by

$$F_t(\xi, z) = [\xi, (1-t)z].$$

Then F_0 is the identity mapping, and $F_1 = \pi$.

2. Elementary properties of homotopy.

Property 1.1. *Let f, ϕ and ψ be three continuous mappings of X into Y, with f homotopic to ϕ and ϕ homotopic to ψ. Then f is homotopic to ψ.*

Let F_t be a homotopy taking f into ϕ, and Φ_t a homotopy taking ϕ into ψ. Changing variables in the homotopies F_t and Φ_t, let $\tau = t/2$ for F_t and $\tau = (1+t)/2$ for Φ_t, and set

$$\begin{cases} \Psi_\tau = F_{2\tau}, & \text{for} \quad 0 \leqslant \tau \leqslant \frac{1}{2}, \\ \Psi_\tau = \Phi_{2\tau-1}, & \text{for} \quad \frac{1}{2} \leqslant \tau \leqslant 1. \end{cases} \tag{3}$$

Then $\Psi_0 = F_0 = f$; $\Psi_{1/2} = F_1 = \Phi_0 = \phi$; $\Psi_1 = \Phi_1 = \psi$.

Definition 1.2. The homotopy Ψ_τ is called the *product* of the homotopies F_t and Φ_t, and is denoted by

$$\Psi_\tau = F_t \cdot \Phi_t.^{1)}$$

If $f_0, f_1, \cdots, f_k$ are finitely many continuous mappings of X into Y, and F_t^i, $1 \le i \le k$, is a homotopy taking f_{i-1} into f_i, we may define a homotopy Φ_τ which is the product of the homotopies F_t^i:

$$\Phi_\tau = F_t^1 F_t^2 \ldots F_t^k.$$

The homotopy Φ_τ is given on the interval $[(i-1)/k, i/k]$ by transforming the variable t in the homotopy F_t^i:

$$\tau = \frac{1}{k}(t + i - 1);$$

$$\Phi_\tau = F^i_{k\tau - i + 1} \qquad \text{for} \quad \frac{i-1}{k} \leqslant \tau \leqslant \frac{i}{k}.$$

Property 1.2. *If $f: X \to Y$ and $\phi: X \to Y$ are homotopic mappings of X into Y, and $\psi: Y \to Z$ is a continuous mapping of Y into Z then ψf and $\psi\phi$ are homotopic mappings of X into Z.*

Indeed, if F_t is a homotopy taking f into ϕ, then ψF_t is a homotopy taking ψf into $\psi\phi$.

Property 1.3. *If $f: X \to Y$ and $\phi: X \to Y$ are homotopic mappings of X into Y, and $\psi: Z \to X$, then $f\psi$ and $\phi\psi$ are homotopic mappings of Z into Y.*

Indeed, if F_t takes f into ϕ, then $F_t\psi$ takes $f\psi$ into $\phi\psi$.

By an n-cell we shall mean a topological n-cube, i.e. a compactum homeomorphic to the n-cube I^n.

Corollary. *Any continuous mapping of an arbitrary space X into an n-cell Q^n, or into a space $\mathfrak{E}^n$ homeomorphic to E^n, is homotopic to a constant mapping; and likewise for any continuous mapping of an n-cell Q^n (or of $\mathfrak{E}^n$) into X.*

Let $f: X \to Q^n$ and $Q^n = h(I^n)$, where h is a homeomorphism. Then $h^{-1}f: X \to I^n$; by Example 1.3 there exists a homotopy F_t taking $h^{-1}f$ into a constant mapping:

$$F_0 = h^{-1}f; \qquad F_1 \equiv \text{const.}$$

By Property 1.2 the mapping $hF_0 = f$ of X into the n-cell Q^n is homotopic to the mapping hF_1, which is constant since $F_1 = \text{const.}$

Now let $f: Q^n \to X$ be any continuous mapping of an n-cell Q^n into X. Then $fh: I^n \to X$, and by Example 1.2 there exists a homotopy $\Phi_t: I^n \to X$ such that $\Phi_0 = fh$ and $\Phi_1 \equiv \text{const.}$

---

1) *Translator's note.* The dot is later omitted or replaced by a small circle.

By Property 1.3, $\Phi_0 h^{-1}$ and $\Phi_1 h^{-1}$ are homotopic mappings of Q^n into X. But $\Phi_0 h^{-1} = f$ and $\Phi_1 h^{-1} \equiv \text{const}$, since $\Phi_1 \equiv \text{const}$, and f is consequently homotopic to a constant mapping.

The proof is similar if Q^n is replaced by $\mathfrak{S}^n$.

Example 1.5. A continuous mapping f of a space X into a topological sphere S^n such that $f(X)$ does not cover all of S^n is homotopic to a constant mapping.

Let p be a point of $S^n \setminus f(X)$. Then our assertion follows from the Corollary, since $S^n \setminus \{p\}$ is homeomorphic to E^n.

Property 1.4. *If X and Y are homeomorphic spaces: $Y = h(X)$, and f and ϕ are homotopic mappings of X into itself, then hfh^{-1} and $h\phi h^{-1}$ are homotopic mappings of Y into itself.*

If F_t is a homotopy taking f into ϕ, then $hF_t h^{-1}$ is a homotopy taking hfh^{-1} into $h\phi h^{-1}$. Note that if f is the identity mapping, then so is hfh^{-1}, and if f is a constant mapping, so is hfh^{-1}.

If A is the set of fixed points for F_t (or for f), i.e. $F_t(x) = x$ for $x \in A$ and $0 \le t \le 1$, then $h(A)$ is the set of fixed points for $hF_t h^{-1}$ (or for hfh^{-1}).

3. We shall study certain types of homotopy which have application to questions of imbedding in E^n and S^n, as well as to the study of certain continuous mappings of E^n and S^n.

Definition 1.3. By the distance $\rho[f, \phi]$ between two continuous mappings of a space X into a metric space Y we mean the least upper bound of the distances between the images $f(x)$ and $\phi(x)$ of points of X:

$$\rho[f, \varphi] = \sup_{x \in X} \rho\,[f(x), \varphi(x)].$$

Definition 1.4. Two continuous mappings f and ϕ of a metric space X into a metric space Y are called *ϵ-homotopic* if for every t the homotopy F_t, $0 \le t \le 1$, taking f into ϕ satisfies the condition

$$\rho\,[F_0, F_t] < \varepsilon, \quad 0 \leqslant t \leqslant 1.$$

Theorem 1.1. *Let*

$$f_1, f_2, \ldots, f_n, \ldots$$

be a sequence of continuous mappings of a metric compactum X onto itself, such that f_n is homotopic to f_{n+1} for each n, and for any $\epsilon > 0$ there exists an integer N such that f_p and f_q are ϵ-homotopic for $p \ge N$ and $q \ge N$. Then the sequence $\{f_n\}$ converges uniformly to a continuous mapping f of X onto itself which is homotopic to f_1.

Since f_p and f_q are ϵ-homotopic for $p \ge N$, $q \ge N$, it follows that for each point $x_0 \in X$ the sequence of points $f_n(x_0)$ satisfies the Cauchy criterion, and consequently, X being compact, converges to a point $f(x_0) \in X$. Since N is independent of x, the sequence $\{f_n\}$ converges uniformly. Consequently the limit mapping is a continuous mapping of the compactum X into itself. We assert that f maps X onto itself. Suppose the contrary: there exists a point $x_0 \in X \backslash f(X)$. Since $f(X)$ is a compact subset of X, there exists a spherical neighborhood of x_0: $\delta_\epsilon(x_0) = \{x : \rho(x, x_0) < \epsilon\}$ whose intersection with $f(X)$ is void. Pick an integer N so large that

$$\rho\,[f_N, f_p] < \frac{\varepsilon}{2}, \qquad p > N,$$

for any $p > N$. Then obviously

$$\rho\,[f_N, f] \leqslant \frac{\varepsilon}{2} < \varepsilon.$$

Since $f_N(X) = X$, there exists a point $x_1 \in X$ such that $f_N(x_1) = x_0$. But then, since $\rho[f_N, f] \leq \epsilon/2$, we have $f(x_1) \in \delta_\epsilon(x_0)$, contradicting the choice of $\delta_\epsilon(x_0)$. Thus the assumption that $f(X) \neq X$ has led to a contradiction, and therefore $f(X) = X$.

We now construct a homotopy Φ_τ taking the mapping f_1 into the mapping f. Let F_t^n, $0 \leq t \leq 1$, be the homotopy taking f_n into f_{n+1}:

$$F_0^n = f_n; \qquad F_1^n = f_{n+1} = F_0^{n+1}.$$

By hypothesis, the F_t^n may be so chosen that for any $\epsilon > 0$ there exists an N such that F_t^n is an ϵ-homotopy for $n > N$.

For each homotopy F_t^n we make a change of variable, replacing t by a new variable τ such that the interval $[0, 1]$ of the t-axis goes into the interval $[(n-1)/n, n/(n+1)]$ of the τ-axis. The change is given by the relation

$$\tau = \frac{n-1}{n}(1-t) + \frac{n}{n+1}t.$$

Now set

$$\Phi_\tau(x) = F_t^n(x) \qquad \text{for} \qquad \frac{n-1}{n} \leqslant \tau \leqslant \frac{n}{n+1}.$$

Φ_τ is thus defined for all $\tau < 1$ and is a mapping of the space $X \times [0, 1)$ onto X, where $[0, 1)$ is the half-open interval omitting the right-hand endpoint. This mapping is continuous. Indeed, it is certainly continuous on each compactum $X \times [(n-1)/n, n/(n+1)]$, since it is obtained from F_t^n by a continuous transformation of the variable t. Its continuity for the values $\tau = (n-1)/n$ follows from the fact that

$$\Phi_{\frac{n-1}{n}} = F_1^{n-1} = f_n = F_0^n.$$

It remains to show that if we set $\Phi_1(x) = f(x)$, the mapping Φ_τ becomes a continuous mapping of the compactum $X \times [0, 1]$ onto X. Since the sequence $\{f_n\}$ converges uniformly to f, there exists an integer N_1 such that

$$\rho\,[f, f_n] < \frac{\varepsilon}{4}, \qquad n > N_1.$$

On the other hand, the homotopies F_t^n have been chosen so that there exists an $N \geq N_1$ such that

$$\rho\,[f_n, F_t^n] < \frac{\varepsilon}{4}, \quad 0 \leqslant t \leqslant 1, \quad n \geqslant N.$$

Replacing F_t^n by Φ_τ, $(n-1)/n \leq \tau \leq n/(n+1)$, we have

$$\rho\,[f_n, \Phi_\tau] < \frac{\varepsilon}{4}, \quad \frac{n-1}{n} \leqslant \tau \leqslant \frac{n}{n+1}, \quad n \geqslant N.$$

Combining the inequalities, we obtain

$$\rho\,[f, \Phi_\tau] < \frac{\varepsilon}{2} \qquad \text{for} \quad \tau > \frac{N-1}{N}.$$

Since f is uniformly continuous, there exists a number $\eta > 0$ such that $\rho[f(x), f(x')] < \epsilon/2$ whenever $\rho(x, x') < \eta$.

We find, therefore, setting $f(x) = \Phi_1(x)$, that

$$\rho\,[\Phi_\tau(x'), \Phi_1(x)] < \rho\,[\Phi_\tau(x'), \Phi_1(x')] + \rho\,[\Phi_1(x'), \Phi_1(x)] < \varepsilon$$
$$\text{for} \quad \delta(x, x') < \eta, \quad \tau > \frac{N-1}{N},$$

i.e. the mapping Φ_τ: $X \times [0, 1] \to X$ is continuous at the points $(x, 1)$. Thus Φ_τ, $0 \leq \tau \leq 1$, is a homotopy taking the mapping f_1 into the mapping f.

Corollary. *Let $f_1, f_2, \cdots, f_k, \cdots$ be a sequence of continuous mappings of Euclidean space E^n onto itself, satisfying the conditions of Theorem 1.1, such that each homotopy F_t^k taking f_k into f_{k+1} is stationary outside a certain bounded domain U independent of k:*

$$F_t^k(x) = x, \quad x \in E^n \setminus U, \quad 0 \leqslant t \leqslant 1, \quad k = 1, 2, \ldots,$$

and, in addition, $F_t^k(\overline{U}) = \overline{U}$. Then the sequence $\{f_k\}$ converges uniformly to a continuous mapping f of E^n onto itself, which is homotopic to f_1, and the homotopy Φ_τ taking f_1 into f is stationary outside U.

This follows from the hypothesis that $f_k(x) = x$ for $x \in E^n \setminus U$, $k = 1, 2, \cdots$.

Since U is bounded, $\overline{U}$ is a compactum and $F_t^k(x) = x$ for $x \in \operatorname{Fr} U$, where $\operatorname{Fr} U$ is the boundary of U and $\overline{U}$ is the closure of U. Applying Theorem 1.1 to the compactum $\overline{U}$, we obtain a homotopy Φ_τ defined on $\overline{U}$ and taking the mapping $f_1|_U$ into a mapping $f|_U$. Extend Φ_τ to E^n by setting

$$\Phi_\tau(x) = x, \quad x \in E^n \setminus U, \quad 0 \leqslant \tau \leqslant 1.$$

Then the homotopy Φ_τ satisfies the required conditions.

§2. Isotopy

1. Definition and examples.

Definition 2.1. Two homeomorphisms f and g of a space X into a space Y are said to be *isotopic* if there exists a homotopy joining f and g which for every fixed t is a homeomorphism. Such a homotopy F_t joining two homeomorphisms is called an isotopy.

Two homeomorphic subsets A and B of a space X are said to be *isotopic* if there exists an isotopy F_t of X onto itself such that F_0 is the identity mapping and $F_1(A) = B$.

We shall then say that the isotopy F_t takes A into B.

Examples of isotopies.

Example 2.1. Any two points in E^n (or in S^n) are isotopic. Suppose the points of E^n are determined by vectors x issuing from the origin of coordinates, and let x_0 and y_0 be two given vectors. Then the isotopy $F_t\colon E^n \to E^n$ taking x_0 into y_0 is given by

$$F_t(x) = x + t(y_0 - x_0). \tag{4}$$

Example 2.2. Any two concentric balls in E^n are isotopic. Let O be the common center of the balls, c the ratio of their radii, and x the vector from the center to any point of E^n. Then the isotopy is given by $F_t = x[1 + (c - 1)t]$,

Example 2.3. An n-simplex $\Delta^n \subset E^n$ is isotopic to a ball Q^n containing it. Let O be an interior point of Δ^n. For x a vector issuing from O, the isotopy F_t is given by

$$F_t(x) = x\left[1 + \left(\frac{R_2(x)}{R_1(x)} - 1\right)t\right],$$

where $R_1(x)$ and $R_2(x)$ are the distances from O to the points of intersection of the line of the vector x with the boundaries of the ball and the simplex, respectively.

F_t takes the ball Q^n into the simplex Δ^n.

2. The construction of an isotopy taking one set into another is frequently carried out in consecutive steps, based on the following theorem.

Theorem 2.1. *Let $A_1, \ldots, A_m$ be homeomorphic subsets of the space X, with A_k isotopic to A_{k+1} for each $k < m$. Then A_1 is isotopic to A_m.*

Indeed, let F_t^k be the isotopy from the identity mapping of X onto itself which takes A_k into A_{k+1}, i.e.

$$F_0^k \equiv 1; \quad F_t^k(X) = X, \quad 0 \leqslant t \leqslant 1,$$
$$F_1^k(A_k) = A_{k+1}.$$

Then the isotopy Φ_τ taking A_1 into A_k is the product of the isotopies

$F_t^k F_1^{k-1} \cdots F_1^1$, $1 \le k \le m-1$ (see Definition 1.2).

That Φ_τ is continuous at the points $(x, k/(m-1))$ follows from the fact that F_0^k is the identity mapping, so that

$$F_0^k F_1^{k-1} \cdots F_1^1 = F_1^{k-1} \cdots F_1^1.$$

The transformation of A_1 into A_m via Φ_τ proceeds according to the schema

$$\Phi_\tau : A_1 \xrightarrow{0 \leqslant \tau \leqslant \frac{1}{m-1}} A_2 \xrightarrow{\frac{1}{m-1} \leqslant \tau \leqslant \frac{2}{m-1}} A_3 \to \cdots \to A_{m-1} \xrightarrow{\frac{m-2}{m-1} \leqslant \tau \leqslant 1} A_m. \tag{5}$$

The inverse isotopy. Let F_t be an isotopy of the space X taking a set A into a set B.

For each fixed t there corresponds to the homeomorphism $F_t(x) = x'$ the inverse homeomorphism $F_t^{-1}(x') = x$. The mapping $F(x, t) = (F_t(x), t)$ is a continuous one-to-one mapping of the space $X \times I^1$ onto itself. But in general $F^{-1}(x', t) = (F_t^{-1}(x'), t)$ need not be a continuous mapping, so that $F(x, t)$ may not be a homeomorphism.

If $F(x, t)$ is in fact a homeomorphism of $X \times I^1$ onto itself, then, for $0 \le t \le 1$, F_t^{-1} is an isotopy of X onto itself. In particular, *F_t^{-1} is an isotopy if X is a Euclidean space*, since for each point $x \in X$ there then exists a neighborhood u_x such that $\bar{u}_x$, and so also $\bar{u}_x \times I^1$, is compact, and each point (x, t), $0 \le t \le 1$, goes into an interior point of the image of $\bar{u}_x \times I^1$ by the theorem on invariance of domain, so that F is a homeomorphism.

We call F_t^{-1} the *inverse isotopy*. Obviously F_t^{-1} takes the set B into the set A if F_t takes A into B.

Using Theorem 2.1 and inverse isotopy, we may now construct some further isotopies.

Example 2.4. Any two n-simplices in E^n are isotopic.

Let us show first that any two n-balls Q_1^n and Q_2^n in E^n are isotopic. Let O_1 and O_2 be the centers of Q_1^n and Q_2^n, and R_1 and R_2 their radii. By Example 2.2 there exists an isotopy F_t^2 taking Q_1^n into the ball $\tilde{Q}^n$ concentric with Q_1^n and of radius R_2; and by Example 2.1 there exists an isotopy F_t^3 taking the point O_1 into the point O_2, given by (4). For every t this isotopy is a parallel displacement of all points of E^n by the vector $t(O_2 - O_1)$. Hence for $t = 1$ the ball $\tilde{Q}^n$ goes into the ball Q_2^n. Applying Theorem 2.1, we have then an isotopy taking Q_1^n into Q_2^n.

Now let Δ_1^n and Δ_2^n be two n-simplices in E^n, and Q_1^n and Q_2^n balls containing them. By Example 2.3 there exists an isotopy F_t^1 taking Q_1^n into Δ_1^n, and an isotopy F_t^4 taking Q_2^n into Δ_2^n.

By Theorem 2.1 we may construct an isotopy Φ_τ: $E^n \to E^n$ taking the simplex Δ_1^n into the simplex Δ_2^n, by means of the four consecutive isotopies:

$$(F_t^1)^{-1}:\Delta_1^n \to Q_1^n; \qquad F_t^2: Q_1^n \to \widetilde{Q}^n;$$
$$F_t^3:\widetilde{Q}^n \to Q_2^n; \qquad F_t^4: Q_2^n \to \Delta_2^n.$$

Φ_τ takes Δ_1^n into Δ_2^n via the schema

$$\Phi_\tau:\Delta_1^n \to Q_1^n \to \widetilde{Q}^n \to Q_2^n \to \Delta_2^n.$$

Example 2.5. Any two line segments in E^n are isotopic.

Let $\overline{x_0x_1}$ and $\overline{y_0y_1}$ be the segments. An isotopy F_t^1 as described in Example 2.1 takes the point x_0 into the point y_0 and the segment $\overline{x_0x_1}$ into a segment $\overline{y_0y_2'}$. An isotopy F_t^2 like that in Example 2.2 takes the segment $\overline{y_0y_2'}$ into a segment $\overline{y_0y_2}$ of length equal to that of $\overline{y_0y_1}$.

Now choose a coordinate system in E^n such that the segments $\overline{y_0y_1}$ and $\overline{y_0y_2}$ lie in the same coordinate plane and the origin coincides with the point y_0. Let α be the angle between the vectors $\overrightarrow{y_0y_1}$ and $\overrightarrow{y_0y_2}$. Then an isotopy F_t^3 taking the segment $\overline{y_0y_2}$ into the segment $\overline{y_0y_1}$ is defined for points $\xi\in E^n$ with coordinates $(\rho, \phi, x_3, \cdots, x_n)$, where (ρ, ϕ) are polar coordinates in the plane (y_0, y_1, y_2) with origin y_0, by

$$\rho(t, \xi) = \rho(\xi); \ \phi(t, \xi) = \phi(\xi) - \alpha t;$$
$$x_i(t, \xi) = x_i(\xi), \quad i = 3, \cdots, n.$$

By Theorem 2.1 we can now use the isotopies F_t^1, F_t^2, F_t^3 to construct an isotopy Φ_τ taking the segment $\overline{x_0x_1}$ into the segment $\overline{y_0y_1}$.

3. An important role is played by the isotopies of E^n (or of S^n) which are stationary outside some bounded domain U, i.e. such that

$$F_t(x) = x \quad \text{for} \quad x\in E^n\setminus U, \quad 0\leqslant t\leqslant 1.$$

Example 2.6. A ball $Q^n\subset E^n$ of radius R can be taken into a concentric ball q^n of radius r by an isotopy which is stationary outside a ball Q'^n containing $Q^n\cup q^n$.

Let the center of the balls O be at the origin, and let x, y, z be the points of intersection of the ray l issuing from O with the boundaries of the balls q^n, Q^n, Q'^n, respectively. Then the isotopy is given on the ray l by

$$\begin{gathered} F_t(\alpha y) = \alpha\,[y(1-t)+xt]; \\ F_t\,[z(1-\alpha)+\alpha y] = z(1-\alpha)+\alpha\,[y(1-t)+tx], \\ 0\leqslant\alpha\leqslant 1,\ 0\leqslant t\leqslant 1. \end{gathered} \tag{6}$$

Theorem 2.2. *Let Q_1^n and Q_2^n be two balls in E^n, and U a domain containing them. Then there exists an isotopy of E^n onto itself taking Q_1^n into Q_2^n, taking the center of Q_1^n into the center of Q_2^n, and stationary on $E^n\setminus U$.*

Consider first the special case that the radii of Q_1^n and Q_2^n are the same, and the domain U contains a closed spherical cylinder π' which has the balls Q_1^n and Q_2^n in its interior and whose axis $\overline{AB}$ contains the segment $\overline{O_1O_2}$ joining the centers of Q_1^n and Q_2^n (Figure 1.1):

$$U \supset \pi' \supset \operatorname{int} \pi' \supset Q_1^n \cup Q_2^n.$$

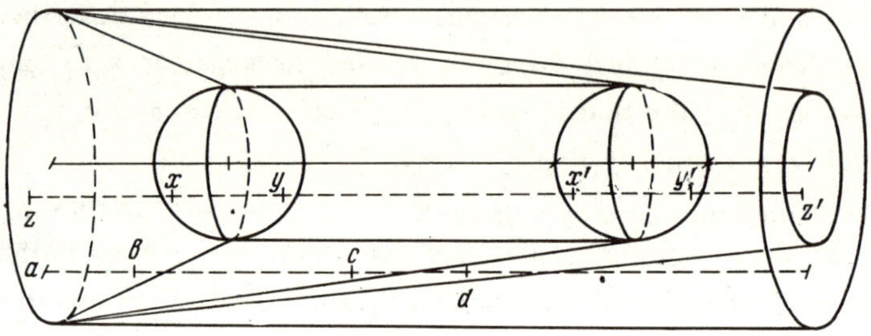

Figure 1.1

The base κ of the cylinder π' is an $(n-1)$-ball lying in the plane perpendicular to the axis $\overline{AB}$. Let r be the radius of the balls Q_1^n and Q_2^n, and π the cylinder whose bases are the $(n-1)$-balls κ_1 and κ_2 which are the intersections of Q_1^n and Q_2^n with the $(n-1)$-hyperplanes through O_1 and O_2 perpendicular to $\overline{AB}$. Obviously,

$$\pi \cup Q_1^n \cup Q_2^n \subset \operatorname{int} \pi'.$$

Let $\Gamma_1, \Gamma_2, \Gamma_3$ be three truncated cones whose axes lie on the segment $\overline{AB}$ and whose bases are: the balls (κ, κ_1) for Γ_1; (κ, κ_2) for Γ_2; and for Γ_3, the ball κ and the ball κ_3 of radius r and center at B which lies in the hyperplane perpendicular to AB (see Figure 1.1).

It is easily seen that $\pi \subset \Gamma_2 \subset \Gamma_3 \subset \pi' \subset U$. Let l be a line parallel to O_1O_2 and intersecting π; let z, x, y be its points of intersection with κ and the boundary of Q_1^n, and x', y', z' its points of intersection with the boundary of Q_2^n and κ_3.

The isotopy on the segment zz' is given by

$$\begin{cases} F_t[(1-\alpha)z+\alpha x] = z(1-\alpha)+\alpha[(1-t)x+tx']; \\ F_t[\alpha x+(1-\alpha)y] = \alpha[(1-t)x+tx']+(1-\alpha)[(1-t)y+ty']; \\ F_t[(1-\alpha)y+\alpha z'] = (1-\alpha)[(1-t)y+ty']+\alpha z', \end{cases} \tag{7}$$
$$0 \leqslant \alpha \leqslant 1,\ 0 \leqslant t \leqslant 1.$$

Let l' be a line parallel to O_1O_2 and intersecting Γ_3 but not π, and let a, b, c, d be the points of intersection of l with κ and the lateral sides of Γ_1, Γ_2, and Γ_3. The isotopy for points of the segment $\overline{ad}$ is given by

$$F_t\,[(1-\alpha)a+\alpha b]=(1-\alpha)a+\alpha\,[(1-t)b+tc],$$
$$F_t\,[\alpha b+(1-\alpha)d]=\alpha\,[(1-t)b+tc]+(1-\alpha)d, \tag{8}$$
$$0\leqslant\alpha\leqslant 1,\quad 0\leqslant t\leqslant 1.$$

The isotopies (7) and (8) define an isotopy of the truncated cone Γ_3 onto itself which takes the ball Q_1^n into the ball Q_2^n and is stationary on the boundary of Γ_3, since

$$F_t(z)\equiv z,\quad F_t(z')\equiv z',$$
$$F_t(a)\equiv a,\quad F_t(d)\equiv d,$$
$$F_1(\overline{xy})=\overline{x'y'}.$$

Completing the definition of F_t at the points of $E^n\backslash\Gamma_3$ by setting

$$F_t(x)\equiv x,\quad x\in E^n\diagdown\Gamma_3,$$

we obtain an isotopy taking Q_1^n into Q_2^n and stationary outside Γ_3, and therefore outside U.

General case. Let Q_1^n and Q_2^n be two closed balls in E^n with centers O_1 and O_2, and U a domain containing them. Join O_1 and O_2 by a polygonal line L in U:

$$L=\bigcup_{i=0}^{k-1}\overline{a_ia_{i+1}}\subset U,\quad a_0=O_1,\quad a_k=O_2.$$

Each segment $\overline{a_ia_{i+1}}$ is contained in a neighborhood $u_i\subset U$ such that

$$\overline{a_ia_{i+1}}\subset u_i;\quad u_i\cap u_j=\Lambda\quad\text{for}\quad |i-j|>1.$$

Choose r so small that for each $i=0,1,\cdots,k$ the ball δ_i with center a_i and radius r is contained in $u_{i-1}\cap u_i$, and the cylinder π_i with axis a_ia_{i+1} and base of radius r is contained in u_i. For each i, by the special case, there exists an isotopy F_t^i taking δ_i into δ_{i+1} and stationary on $E^n\backslash u_i$.

By Theorem 2.1 and Example 2.6, there exists an isotopy Φ_τ taking Q_1^n into Q_2^n via the schema

$$\Phi_\tau: Q_1^n\to\delta_0^n\to\delta_1^n\to\ldots\to\delta_k^n\to Q_2^n.$$

Since the isotopies taking Q_1^n into δ_0^n and δ_k^n into Q_2^n may be chosen so that each is stationary outside a sphere contained in U, and since each F_t^i is stationary on $E^n\backslash u_i$ and therefore on $E^n\backslash U$, the isotopy Φ_τ is also stationary on $E^n\backslash U$:

$$\Phi_\tau(x)\equiv x\quad\text{for}\quad x\in E^n\diagdown U,\ 0\leqslant\tau\leqslant 1.$$

By construction, the center of each δ_i goes into the center of δ_{i+1}, and consequently the center of Q_1^n goes into the center of Q_2^n.

4. Isotopy of polyhedral simple arcs in E^n. By a *polyhedral simple arc* (or a

polygonal line) is meant a simple arc which is the union of finitely many line segments. We shall now show that in a Euclidean space of arbitrary dimension n any two *polyhedral* simple arcs are isotopic. It is unknown whether the same is true for polyhedral topological cubes of dimension k in E^n, $n \geq k > 1$, if $n > 3$. For $n = 3$ and $k = 2$, i.e. for two polyhedral disks (topological squares) Q_1^2 and Q_2^2 in E^3, there exists an isotopy of E^3 taking Q_1^2 into Q_2^2 [66]. The proof is cumbersome and will not be given in the present monograph.

Two nonpolyhedral simple arcs may be nonisotopic in E^n, and so may two topological squares. We shall give examples in Chapter IV of nonisotopic simple arcs and topological squares in E^3. In the case of the plane, $n = 2$, we shall show in Chapter II that any two topological squares or simple arcs are isotopic.

Lemma 2.1. *Let* $\overline{pqr}$ *be a closed two-dimensional simplex in* E^n. *Then there exists an isotopy of* E^n *which is stationary outside an arbitrary* n*-simplex* Δ^n *containing* $\overline{pqr}\backslash\{q\}$ *and having* q *as a vertex, and which takes the segment* $\overline{qp}$ *into the segment* $\overline{qr}$.

In the intersection of Δ^n with the plane pqr pick three points p', q', r' in the following way. The points p' and r' are to be on the line pr outside the segment $\overline{pr}$ and near the points p and r, respectively, so that $\overline{(pqr)} \subset \overline{(p'qr')} \subset \Delta^n$. The point q' is to be near the midpoint O of the segment pr, outside the simplex $\overline{pqr}$, and on the line qO, so that

$$\overline{(pq'r)} \subset (p'q'r') \subset \Delta^n.$$

Pass through O the $(n-2)$-hyperplane perpendicular to the plane pqr, and in it take a closed $(n-2)$-simplex Δ^{n-2} such that $\overline{\Delta}^{n-2} \subset \Delta^n$, $O \in \Delta^{n-2}$.

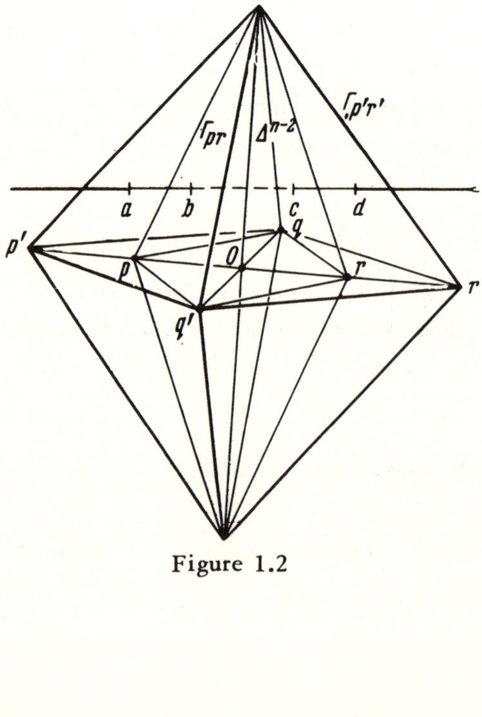

Figure 1.2

Let γ_{pr} be the double cone (= suspension) over Δ^{n-2} with vertices p and r, i.e. the union of the two $(n-1)$-cones with base Δ^{n-2} and vertices p and r, respectively (Figure 1.2). Similarly, $\gamma_{p'r'}$ is the double cone with base Δ^{n-2} and vertices p' and r'.

Next, construct the double n-cone Γ_{pr} over γ_{pr} with vertices q and q', and the double n-cone $\Gamma_{p'r'}$ over $\gamma_{p'r'}$ with vertices q and q'. It is easily seen that γ_{pr} (resp. $\gamma_{p'r'}$) is the union of two closed $(n-1)$-simplices and Γ_{pr}

(resp. $\Gamma_{p'r'}$) is the union of four closed n-simplices.

Furthermore

$$\Gamma_{pr} \setminus \{q\} \subset \Gamma_{p'r'} \setminus \{q\} \subset \Delta^n,$$

$$\gamma_{pr} \subset \gamma_{p'r'} \subset \Delta^n,$$

since $\overline{\Delta}^{n-2} \subset \Delta^n$ and, because Δ^n is convex, the cone over a set in Δ^n with vertex in Δ^n is likewise in Δ^n. Also, $\Gamma_{p'r'}$ is obviously a neighborhood of $\overline{pqr} \setminus \{q\}$.

On any line parallel to pr and intersecting $\Gamma_{p'r'}$, consider the following four consecutive points:

$$a \in \Delta_a \subset \operatorname{Fr} \Gamma_{p'r'};\ b \in \Delta_b \subset \operatorname{Fr} \Gamma_{pr};\ c \in \Delta_c \subset \operatorname{Fr} \Gamma_{pr};\ d \in \Delta_d \subset \operatorname{Fr} \Gamma_{p'r'}.$$

The simplices Δ_a and Δ_d (resp. Δ_b and Δ_c) of the boundary of $\Gamma_{p'r'}$ (Γ_{pr}) differ in only one vertex, namely the point p' for Δ_a (p for Δ_b) and r' for Δ_d (r for Δ_c).

Define the isotopy F_t on $\Gamma_{p'r'}$ by equations (8), and set $F_t(x) = x$ for $x \in E^n \setminus \Gamma_{p'r'}$. The isotopy is stationary outside Δ^n and takes $\overline{qp}$ into $\overline{qr}$, since each simplex Δ_b goes into the corresponding simplex Δ_c.

Theorem 2.3. *For any polygonal line $L \subset E^n$, there exists an isotopy of E^n onto itself which is stationary outside an arbitrary neighborhood of L and takes L into a line segment.*

We prove the theorem by induction on the number r of segments of the polygonal line

$$L = \overline{a_0a_1} \cup \overline{a_1a_2} \cup \cdots \cup \overline{a_{r-1}a_r}.$$

The theorem is true if $r = 1$. Suppose $r > 1$ and the theorem proved for $k < r$, and let U be an arbitrary neighborhood of L. We shall construct an isotopy which is stationary outside U and takes L into $L' = \overline{L \setminus \overline{a_{r-1}a_r}}$.

First step. Choose a ball $\delta^n \subset U$ with center at the point a_{r-1}, and an $(n-1)$-ball $\delta^{n-1} \subset S^n$ with center also at a_{r-1}, such that

$$\delta^n \cap L \subset \overline{a_{r-2}a_{r-1}} \cup \overline{a_{r-1}a_r};$$

$$\delta^{n-1} \cap L = \{a_{r-1}\};$$

$$\Gamma_{a_r} \cap L = \overline{a_{r-1}a_r}; \quad \Gamma_{a_r} \subset U,$$

where Γ_{a_r} is the cone over δ^{n-1} with vertex a_r.

On the ray $\overrightarrow{a_{r-1}a_r}$ choose two points: $a'_r \in \overline{a_{r-1}a_r} \cap \delta^n$, and $x \in U \setminus \overline{a_{r-1}a_r}$ such that the cone Γ_x over δ^{n-1} with vertex x is contained in U (Figure 1.3).

On lines parallel to $a_{r-1}a_r$, take the points

$$a \in \operatorname{Fr}\Gamma_x, \quad b \in \operatorname{Fr}\Gamma_{a_r},$$

$$c \in \operatorname{Fr}\Gamma_{a'_r}, \quad d \in \delta^{n-1},$$

where $\Gamma_{a'_r}$ is the cone over δ^{n-1} with vertex a'_r, and define an isotopy F^1_t by equations (8) on Γ_x and by the equation $F^1_t(x) = x$ for $x \in E^n \setminus \Gamma_x$.

The isotopy F^1_t takes the segment $\overline{a_{r-1}a_r}$ into the segment $\overline{a_{r-1}a'_r}$, and is stationary outside the cone Γ_x and consequently on $E^n \setminus U$.

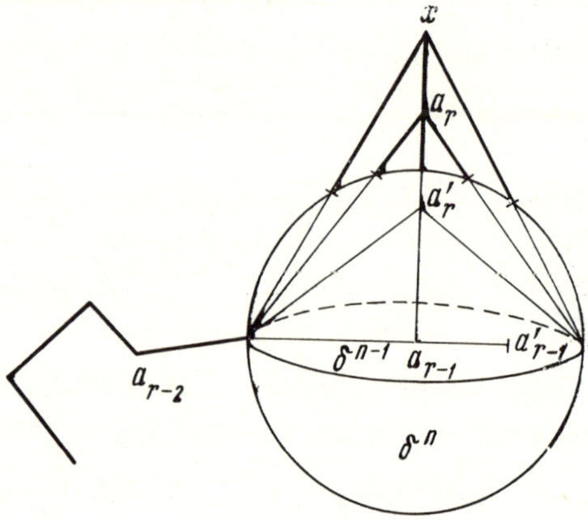

Figure 1.3

Second step. Choose a point $a'_{r-1} \subset \delta^n$ on the extension of the segment $\overline{a_{r-2}a_{r-1}}$.

Applying Lemma 2.1 to the simplex $a'_{r-1}a_{r-1}a'_r$, we obtain an isotopy F^2_t, which is stationary outside some simplex $\Delta^n \subset \delta^n$ such that $\Delta^n \cap L' = \{a_{r-1}\}$, and consequently outside δ^n and on L', and which takes the segment $\overline{a_{r-1}a'_r}$ into the segment $\overline{a_{r-1}a'_{r-1}}$ on the extension of $\overline{a_{r-2}a_{r-1}}$. Next, applying again the argument of the first step, construct an isotopy F^3_t taking the segment $\overline{a_{r-2}a'_{r-1}}$ into the segment $\overline{a_{r-2}a_{r-1}}$ and stationary on $L' \setminus \overline{a_{r-2}a_{r-1}}$ and outside U.

The polygonal line L' consists of $n-1$ segments.

By the induction assumption, there exists an isotopy F^4_t of E^n onto itself which is stationary on $E^n \setminus U$ and takes L' into a line segment l. Using the isotopies F^1_t, F^2_t, F^3_t, F^4_t, we may now construct, by Theorem 2.1, an isotopy Φ_τ which is stationary on $E^n \setminus U$ and takes L into the line segment l via the schema

$$\Phi_\tau : L \to L' \cup \overline{a_{r-1}a'_r} \to L' \cup \overline{a_{r-1}a'_{r-1}} \to L' \to l.$$

Corollary. *Any two polyhedral simple arcs L_1 and L_2 in E^n are isotopic.*

Let U be a bounded domain containing $L_1 \cup L_2$. By Theorem 2.3, there exist for L_1 and L_2 isotopies F^1_t and F^3_t, respectively, taking them into line segments l_1 and l_2, and stationary outside U. By Example 2.5 there exists an isotopy F^2_t taking l_1 into l_2. By Theorem 2.1, using the isotopies F^1_t, F^2_t, $(F^3_t)^{-1}$, we may now construct an isotopy Φ_τ taking L_1 into L_2.

Remark 2.1. If L_1 and L_2 have a common endpoint x_0, the isotopy can be supposed stationary at the point x_0.

Indeed, this is the case in Theorem 2.3 and in Example 2.5, if l_1 and l_2 have a common endpoint. In particular, L can be taken into its initial segment by an isotopy which is stationary outside an arbitrary neighborhood U of $L \setminus \{x_0\}$, since all the steps in the proof of Theorem 2.3 are of this description.

§3. Neighborhoods and coverings of sets in E^n

1. Open coverings. The question of the imbedding of a set A in E^n is closely related to the question of the structure of the neighborhoods of A in E^n and the coverings of A by open sets in E^n: such coverings will be called *open coverings of the set A*. For any $\epsilon > 0$, the set A can be written as a finite or countable union

$$A = \bigcup A_i, \quad i = 1, 2, \ldots,$$

where each A_i is the closure in A of a set open in A, and the diameter of each A_i is less than ϵ.

If for each A_i we take an open set U_i containing it and also of diameter $< \epsilon$, we obtain an open ϵ-covering of A:

$$A \subset \bigcup U_i; \quad A_i \subset U_i; \quad \text{diam } U_i < \varepsilon.$$

Theorem 3.1. *If a compactum K in E^n (or S^n) is a union of finitely many compacta K_i, then there exists an open covering $\{U_i\}$ of K such that $K_i \subset U_i$ and any intersection $\overline{U}_{i_1} \cap \overline{U}_{i_2} \cap \cdots \cap \overline{U}_{i_r}$ is nonvoid if and only if the intersection $K_{i_1} \cap K_{i_2} \cap \cdots \cap K_{i_r}$ is nonvoid.*

We prove the theorem by induction on the multiplicity of the covering $\{K_i\}$, i.e. the largest r for which there exists a nonvoid intersection $K_{i_1} \cap \cdots \cap K_{i_r}$. For $r = 1$ the theorem is obvious, since a finite number of disjoint compacta can be covered by open sets with disjoint closures.

Suppose the theorem is true for all $r < N$; we prove it for $r = N$. Let $K = \bigcup_{i=1}^p K_i$ be a covering of the compactum of multiplicity N. Consider all possible nonvoid intersections

$$\pi_j = K_{i_1} \cap K_{i_2} \cap \cdots \cap K_{i_N}.$$

Obviously $\pi_j \cap \pi_{j'} = \Lambda$ for $j \neq j'$. Thus there exists an open covering $\{O_j\}$ of the set $\bigcup \pi_j$ such that

$$\pi_j \subset O_j; \quad \overline{O}_j \cap \overline{O}_{j'} = \Lambda \quad \text{for} \quad j \neq j';$$
$$\overline{O}_j \cap K_i = \Lambda \quad \text{for} \quad \pi_j \cap K_i = \Lambda.$$

Let $T_i = K_i \setminus \bigcup O_j$. Then $\bigcup_{i=1}^p T_i$ is covered by finitely many compacta with multiplicity $\leq N - 1$. By the induction assumption there exists an open covering $\{U_i\}$ of this union, $T_i \subset U_i$, which satisfies the conditions of the theorem. Moreover, U_i may be chosen so that $\overline{U}_i \cap K_s = \Lambda$ if $T_i \cap K_s = \Lambda$, and $\overline{U}_i \cap \overline{O}_j = \Lambda$ if $K_i \cap \pi_j = \Lambda$, i.e. if $i \neq j_\lambda$, where $\pi_j = K_{j_1} \cap K_{j_2} \cap \cdots \cap K_{j_N}$.

Let V_i be the union of U_i together with all O_j intersecting K_i:

$$V_i = U_i \cup \bigcup_{O_j \cap K_i \neq \Lambda} O_j.$$

Then $K_i \subset V_i$, so that $\{V_i\}$ is a covering of K. We assert that it satisfies the conditions of the theorem.

That $K_{i_1} \cap \cdots \cap K_{i_r} \neq \Lambda$ implies $\overline{V}_{i_1} \cap \cdots \cap \overline{V}_{i_r} \neq \Lambda$, $r \leq N$, is obvious. We must show that the converse holds. Suppose $\overline{V}_{i_1} \cap \cdots \cap \overline{V}_{i_r} \neq \Lambda$. If there is a point in this intersection which belongs to $\overline{U}_{i_1} \cap \cdots \cap \overline{U}_{i_r}$, then $T_{i_1} \cap \cdots \cap T_{i_r} \neq \Lambda$ by the induction assumption, so that, since $T_i \subset K_i$, we have $K_{i_1} \cap \cdots \cap K_{i_r} \neq \Lambda$. If there is no such point, then every point of the intersection $\overline{V}_{i_1} \cap \cdots \cap \overline{V}_{i_r}$ belongs to some O_j, and, indeed, only to one.

Let x be such a point of the intersection $\overline{V}_{i_1} \cap \cdots \cap \overline{V}_{i_r}$, with

$$x \in \overline{U}_{i_1} \cap \ldots \cap \overline{U}_{i_s}, \quad s < r;$$

$$O_j \supset \pi_j = K_{\lambda_1} \cap \ldots \cap K_{\lambda_N}.$$

Since $\overline{O}_j \cap K_i = \Lambda$ for $i \neq \lambda_t$, we have $\overline{U}_i \cap \overline{O}_j = \Lambda$ for $i \neq \lambda_t$, so that each index i_k, $k = 1, \cdots, s$, must be among the indices λ_t; and the same is true for $k > s$, since O_j belongs only to those V_i for which $\overline{O}_j \cap K_i \neq \Lambda$, i.e. for which $i = \lambda_t$. Since $K_{\lambda_1} \cap \cdots \cap K_{\lambda_N} \neq \Lambda$, this proves the theorem.

A covering $\{V_i\}$ of the compactum K which satisfies the conditions of Theorem 3.1 will be called an open covering of K *similar* to the covering $\{K_i\}$.

If I^p, $p < n$, is a p-dimensional cube in E^n, then for any $\epsilon > 0$ there exists a covering of I^p by finitely many p-dimensional cubes δ_i^p: $I^p = \bigcup_{i=1}^N \delta_i^p$ such that the diameter of every δ_i^p is less than ϵ and the multiplicity of the covering is $p + 1$. Hence there exists a covering of I^p by open cubes of diameter $< \epsilon$ and dimension n which is similar to the covering $\{\delta_i^p\}$:

$$I^p \subset \bigcup_{i=1}^N u_i, \quad u_i \supset \delta_i^p.$$

Indeed, we may take for u_i the interior of an n-cube $\overline{u}_i$ concentric with δ_i^p such that u_i is a sufficiently small neighborhood of δ_i^p.

Now let Q^p be an arbitrary p-cell in E^n. Then it can likewise be represented in the form $Q^p = \bigcup_{i=1}^N q_i^p$, where the q_i^p are p-cells of diameter $< \epsilon$. To do so we can take a sufficiently fine cubical subdivision of the cube I^p and set $q_i^p = H(\delta_i^p)$, where H is the homeomorphism of I^p onto Q^p. By Theorem 3.1 Q^p has an open covering similar to the covering $\{q_i^p\}$:

$$Q^p \subset \bigcup_{i=1}^N v_i, \quad v_i \supset q_i^p.$$

But it is not always possible to choose the v_i so that they are homeomorphic to E^n or, all the more, so that their closures are n-cells. In Chapter IV we shall give an example of a simple arc l in E^3 for which the closures of the elements v_i of a covering similar to the linear covering of l cannot be 3-cells.

The union of all the elements v_i of an open ϵ-covering of a p-cell Q^p is an ϵ-neighborhood of Q^p. We shall see in Chapter IV that there exist p-cells in E^n for which an ϵ-neighborhood, for sufficiently small ϵ, cannot be homeomorphic to E^n or have its closure homeomorphic to an n-cell. It is obvious that if there exists a homeomorphism $H\colon E^n \xrightarrow{\text{onto}} E^n$ such that $Q^p = H(I^p)$, then for any $\epsilon > 0$ there exists an ϵ-neighborhood v of Q^p whose closure is an n-cell, and there exists an ϵ-covering $\{v_i\}$ of Q^p, similar to a covering $\{q_i^p\}$ of multiplicity $p+1$, for which every $\bar{v}_i$ is an n-cell. It suffices for the latter to take as the q_i^p the p-cells $H(\delta_i^p)$ for a sufficiently fine cubical subdivision of the cube I^p, and as the v_i the sets $H(u_i)$.

A zero-dimensional compactum A in E^n has for any $\epsilon > 0$ an ϵ-covering by finitely many domains whose closures are disjoint:

$$A \subset \bigcup_{i=1}^{N} u_i; \quad \bar{u}_i \cap \bar{u}_j = \Lambda, \ i \neq j.$$

$A \cap u_i = a_i$ is an open-closed subset of A. If A lies on a line $E^1 \subset E^n$ (e.g. if A is the Cantor perfect set), the u_i may be chosen so that their closures are n-balls. But for $n \geq 3$ it is not the case that every zero-dimensional compactum in E^n has an arbitrarily fine covering by disjoint domains whose closures are n-cells. Chapter IV contains an example of a zero-dimensional compactum, constructed by Antoine in 1921 [20] (see also Uryson [11]), which does not have an arbitrarily fine covering by disjoint n-cells.

If we consider a decreasing sequence of coverings $\gamma_k = \{X_i^k\}$ of a compactum X by closed subsets, we can represent X in the form

$$X = \bigcap_{k=1}^{\infty} \bigcup_{i=1}^{n_k} X_i^k. \tag{9}$$

We may, in addition, require that the diameters of the X_i^k be less than $1/k$. By Theorem 3.1, for each covering $\{X_i^k\}$, if X lies in E^n, there exists a similar open covering $\{v_i^k\}$, with $X_i^k \subset v_i^k$; the coverings $\{v_i^k\}$ may be so chosen that they too constitute a decreasing sequence, i.e. so that the closure of each v_i^k is contained in some v_j^{k-1}, and also so that the diameter of each v_i^k is less than $1/k$. Then X can be represented in the form

$$X = \bigcap_{k=1}^{\infty} \bigcup_{i=1}^{n_k} v_i^k; \qquad \bar{v}_i^k \subset v_j^{k-1}; \ \operatorname{diam} v_i^k < \frac{1}{k}. \tag{10}$$

We shall need one further representation of a compactum in E^n as an intersection of open sets.

Lemma 3.1. *An arbitrary compactum A in a manifold M can be represented in the form*

$$A = \bigcap_{k=1}^{\infty} \bigcup_{i=1}^{N_k} u_i^k, \tag{11}$$

where the u_i^k are domains in M, the closure $\bar{u}_i^k$ of each u_i^k is contained in some u_j^{k-1}, and $\bar{u}_i^k \cap \bar{u}_j^k = \Lambda$ for $i \neq j$.

Pick a sequence of positive numbers decreasing to zero:

$$\varepsilon_1 > \varepsilon_2 > \ldots > \varepsilon_k > \ldots, \quad \lim_{k\to\infty} \varepsilon_k = 0.$$

Let V be any ϵ_1-neighborhood of A. V has only finitely many components V_i intersecting A, and each intersection $A \cap V_i$ is open-closed in A. In each V_i pick a domain u_i^1 such that

$$A \cap V_i \subset u_i^1 \subset \bar{u}_i^1 \subset V_i.$$

Then obviously

$$\bar{u}_i^1 \cap \bar{u}_j^1 = \Lambda, \quad i \neq j$$

Suppose the sets $U_r = \bigcup_{i=1}^{N_r} u_i^r$ have been constructed for all $r \leq k$ so as to satisfy the conditions of the lemma, with each U_r, $r \leq k$, lying in an ϵ_r-neighborhood of A.

In u_k pick an arbitrary ϵ_{k+1}-neighborhood of A, and construct in it, as for $k = 1$, a set $U_{k+1} = \bigcup_j u_j^{k+1}$ such that

$$A \cap u_j^{k+1} \neq \Lambda; \quad \bar{u}_i^{k+1} \cap \bar{u}_j^{k+1} = \Lambda, \quad i \neq j.$$

Equality (11) follows from the facts that U_k lies in an ϵ_k-neighborhood of A, the numbers ϵ_k approach zero with increasing k, and $U_k \subset U_{k-1}$.

The condition $\bar{u}_i^k \cap \bar{u}_j^k = \Lambda$, $i \neq j$, implies that each component ξ of the set A is *entirely contained* in some u_i^k. Since each $\bar{u}_i^k$ is contained in some u_j^{k-1}, there exists a descending chain of sets $u_{i_k}^k$ containing ξ:

$$u_{i_1}^1 \supset u_{i_2}^2 \supset \ldots \supset u_{i_k}^k \supset \ldots, \quad \xi \subset u_{i_k}^k, \ k = 1, 2, \ldots$$

Since $\bigcap_{k=1}^{\infty} u_{i_k}^k = \bigcap_{k=1}^{\infty} \bar{u}_{i_k}^k$ because $\bar{u}_{i_k}^k \subset u_{i_{k-1}}^{k-1}$, it follows that $\bigcap_{k=1}^{\infty} u_{i_k}^k$ is a continuum; hence it coincides with ξ, since $\bigcap_{k=1}^{\infty} u_{i_k}^k \subset A$. It is easily seen that the intersection of the elements of any descending chain $\{u_{i_k}^k\}$ is a component of A: the intersection is nonvoid, since $\bar{u}_{i_k}^k \subset u_{i_{k-1}}^{k-1}$; it is a continuum, since each $\bar{u}_{i_k}^k$ is connected; and it coincides with a component of A because of the condition $\bar{u}_i^k \cap \bar{u}_j^k = \Lambda$, $i \neq j$.

2. Cellularly imbedded and cellularly separated sets in E^n.

Definition 3.1 (**M. Brown** [32]). A set A in a space X of dimension n is said to be *cellularly imbedded* in X if in every neighborhood U of A there is another neighborhood V, contained in U together with its closure, $A \subset V \subset \overline{V} \subset U$, such that $\overline{V}$ is an n-cell and V its interior.

The neighborhood V is called a cellular neighborhood of A.

A line segment, a square, and in general a k-dimensional cube, $k \leq n$, in E^n is cellularly imbedded. The topological structure of a cellularly imbedded continuum can be quite complicated.

The graph of the function $y = \sin(1/x)$ on $[0, 1]$ together with the limit segment $[-1, +1]$ on the axis OY is cellularly imbedded in the plane.

The indecomposable snakelike continua of Uryson [11] and of Knaster [47] are cellularly imbedded in the plane, as is any acyclic planar continuum (see Chapter II, §2). On the other hand, a k-cell in E^n need not be cellularly imbedded.

Examples will be given in Chapter IV of simple arcs and disks (topological squares) in E^3 which are not cellularly imbedded.

Thus the notion of cellular imbedding is not a characterization of the structure of a set, but rather of its imbedding in the containing space.

Theorem 3.2. *A set A is cellularly imbedded in an n-manifold M^n if and only if A is a continuum which is the intersection of a sequence of n-cells $\overline{Q}_k$ such that $\overline{Q}_k$ is contained in the interior of $\overline{Q}_{k-1}$:*

$$A = \bigcap_{k=1}^{\infty} \overline{Q}_k; \quad \overline{Q}_k \subset Q_{k-1} = \operatorname{int} \overline{Q}_{k-1} \tag{12}$$

Necessity. Let U be any ½-neighborhood of a cellularly imbedded set A; it contains a cellular neighborhood Q_1 of A. Suppose we have constructed k cellular neighborhoods $Q_1, \dots, Q_k$ of A such that $\overline{Q}_r \subset Q_{r-1}$, $r \leq k$, and Q_r is a $(1/r)$-neighborhood of A. In Q_k, pick an arbitrary $(1/(k+1))$-neighborhood U of A.

Since A is cellularly imbedded, U contains the closure of a cellular neighborhood Q_{k+1} of A:

$$A \subset Q_{k+1} \subset \overline{Q}_{k+1} \subset U \subset Q_k.$$

We construct in this fashion an infinite sequence of cellular neighborhoods of A:

$$U_1 \supset \overline{Q}_1 \supset Q_1 \supset U_2 \supset \overline{Q}_2 \supset \dots \supset U_k \supset \overline{Q}_k \supset Q_k \supset \dots$$

The intersection $\bigcap_{k=1}^{\infty} \overline{Q}_k$ is a continuum containing A; since each Q_k is a $(1/k)$-neighborhood of A, the set $B = \bigcap_{k=1}^{\infty} \overline{Q}_k$ consists of points of adherence of the set A. Suppose there exists a point $x \in B \setminus A$. Then there exists a neighborhood V of A not containing x. Since A is cellularly imbedded, V contains the closure of a cellular neighborhood W of A: $A \subset W \subset \overline{W} \subset V$; since the point x is not in $\overline{W}$, it cannot be a limit point of A, so that we have arrived at a contradiction.

Thus $x \in A$; i.e. $B = A$.

Sufficiency. Suppose $A = \bigcap_{k=1}^{\infty} \overline{Q}_k$, $\overline{Q}_k \subset Q_{k-1}$, where each $\overline{Q}_k$ is an n-cell, and let U be any neighborhood of A. Then there exists a k such that $\overline{Q}_k \subset U$, since otherwise $\bigcap_{k=1}^{\infty} \overline{Q}_k$ would contain a point not in U and therefore not in A. Since Q_k is a neighborhood of A in M^n, we have shown that A is cellularly imbedded in M^n.

Definition 3.2. A set A in a space X is said to be *cellularly separated* if in every neighborhood V of A there is an open covering $\{u_i\}$, where each $\overline{u}_i$ is an n-cell and u_i is homeomorphic to E^n, such that

$$\begin{gathered} \bar{u}_i \subset V;\ \bar{u}_i \cap \bar{u}_j = \Lambda,\ i \neq j; \\ A \subset \bigcup u_i \subset V. \end{gathered} \tag{13}$$

In contrast to cellular imbedding, a cellularly separated set in E^n need not be a compactum. But if A is a compactum, then the covering $\{u_i\}$ consists of finitely many elements.

Theorem 3.3. *A compactum A in an n-manifold M^n is cellularly separated if and only if it can be written in the form* (11), *with each $\overline{u}_i^k$ an n-cell.*

Necessity. Let V_1 be any neighborhood of A of M^n. Since A is cellularly separated, V_1 contains a neighborhood $U_1 = \bigcup_{i=1}^{N} u_i^1$, the closures of whose components are n-cells:

$$A \subset \bigcup_{i=1}^{N} u_i^1 \subset \bigcup_{i=1}^{N} \bar{u}_i^1 \subset V_1;\ \bar{u}_i^1 \cap \bar{u}_j^1 = \Lambda,\ i \neq j.$$

Just as in the proof of Theorem 3.2, we may construct a descending sequence of $(1/k)$-neighborhoods $U_k = \bigcup u_i^k$ of the compactum A which satisfy the conditions of Lemma 3.1, such that the closure $\overline{u}_i^k$ of each component of U_k is an n-cell and $A = \bigcap_{k=1}^{\infty} U_k$.

Sufficiency. Suppose A is given by (11), with each $\overline{u}_i^k$ an n-cell. For any neighborhood V of the compactum A, there exists a k such that

$$A \subset \bigcup_{i=1}^{N_k} u_i^k \subset \bigcup_{i=1}^{N_k} \bar{u}_i^k \subset V.$$

Since $\bigcup_{i=1}^{N_k} u_i^k$ is a neighborhood of A in the manifold M^n, we have shown that A is cellularly separated in M^n.

Corollary. *Every component ξ of a cellularly separated set in E^n is a continuum cellularly imbedded in E^n.*

This follows from the fact that $\xi = \bigcap_{k=1}^{\infty} u_{i_k}^k$ and from Theorems 3.2 and 3.3. The converse, in general, is false; there exist compacta in E^3, all of whose components are cellularly imbedded, but which are not cellularly separated. In particular, there exist zero-dimensional compacta which are not cellularly separated (see Chapter IV).

3. Cellular separation of countable sets.

Theorem 3.4. *Every countable set D in a manifold M^n is cellularly separated.*

Suppose first that M^n is Euclidean space E^n. We prove that for a countable set the covering $\{u_i\}$ may be so chosen that each u_i is a ball with center at a point of D. Number the points of D:

$$x_1, x_2, \ldots, x_r, \ldots. \tag{14}$$

Let V be an arbitrary neighborhood of D. Since D is countable, the set $D \backslash x_1$ is distributed over a countable set of spheres with center x_1. Since all the balls which have center at x_1 and are contained in V constitute an uncountable set, there exists a ball $\delta_1 \subset V$ with center x_1 such that

$$D \cap \operatorname{Fr} \delta_1 = \Lambda.$$

Set $u_1 = \operatorname{int} \overline{\delta}_1 = \delta_1$. Suppose we have constructed k open balls $\delta_1, \delta_2, \ldots, \delta_k$ with centers at points of D, such that

$$\overline{\delta}_i \cap \overline{\delta}_j = \Lambda, \quad i < j \leqslant k, \quad \operatorname{Fr} \delta_i \cap D = \Lambda, \quad i \leqslant k; \qquad \{x_1, \ldots, x_k\} \subset \bigcup_{i=1}^{k} \delta_i \subset \bigcup_{i=1}^{k} \overline{\delta}_i \subset V. \tag{15}$$

To construct δ_{k+1}, select from (14) the first point $x_{s_{k+1}} \subset D \backslash \bigcup_{i=1}^{k} \overline{\delta}_i$. Take a ball δ_{k+1} in V with center $x_{s_{k+1}}$ such that

$$\overline{\delta}_{k+1} \subset V; \quad \overline{\delta}_{k+1} \cap \left(\bigcup_{i=1}^{k} \overline{\delta}_i\right) = \Lambda; \quad D \cap \operatorname{Fr} \delta_{k+1} = \Lambda.$$

Then condition (15) is satisfied with k replaced by $k+1$. In particular, the condition $\{x_1, \ldots, x_k, x_{k+1}\} \subset \bigcup_{i=1}^{k+1} \delta_i$ is satisfied because of (15) and the fact that $x_{s_{k+1}}$ was the first point in (14) contained in $D \backslash \bigcup_{i=1}^{k} \overline{\delta}_i$.

Continuing this process, we construct a covering of D by countably many open balls δ_k with centers at points of D, $D \subset \bigcup_{k=1}^{\infty} \delta_k$, satisfying the conditions

$$\overline{\delta}_k \cap \overline{\delta}_r = \Lambda, \quad k \neq r;$$

$$D \subset \bigcup_{k=1}^{\infty} \delta_k \subset \bigcup_{k=1}^{\infty} \overline{\delta}_k \subset V,$$

i.e. the conditions of the theorem.

Observe that for any $\epsilon > 0$, the balls δ_k may be chosen all of diameter less than ϵ. If D is compact, the number of balls δ_k will be finite.

For a general manifold M^n, the n-cells $\overline{u}_k^n$ may be chosen in the same way as the balls δ_k, starting from the fact that each point x_k has a neighborhood homeomorphic to E^n.

§ 4. Isotopy of some zero-dimensional subsets of E^n and S^n

1. Finite sets.

Example 4.1. Any two sets in E^n (or S^n) consisting of the same finite number of points are isotopic.

Let $A = \{x_1, x_2, \cdots, x_k\}$, $B = \{y_1, y_2, \cdots, y_k\}$. We shall construct an isotopy of E^n taking A into B such that each point x_i goes into the point y_i.

First consider the case that the points x_i do not coincide with the points y_j for $i \neq j$. The points of the set $A \cup B$ may be covered by disjoint balls δ_i, δ_j' with centers at the points x_i and y_j; if $x_i = y_i$; let $\delta_i = \delta_i'$. Two different balls δ_i and δ_i' can then be enclosed in a domain U_i such that

$$\delta_i \cup \delta_i' \subset U_i; \quad \overline{U}_i \cap (\bigcup_{j\neq i} \delta_j \cup \bigcup_{j\neq i} \delta_j') = \Lambda. \tag{16}$$

By Theorem 2.2, there exists an isotopy F_t^i of E^n onto itself which is stationary outside U_i and takes the ball δ_i into the ball δ_i'; furthermore, by construction of the isotopy, the point x_i goes into the point y_i. By (16), F_t^i is stationary on the balls δ_j, $j \neq i$. Using the isotopies F_t^i, therefore, we may construct an isotopy Φ_t: $E^n \to E^n$ such that $\Phi_1(A) = B$.

Indeed, Φ_t is constructed by means of Theorem 2.1 such that Φ_0 is the identity mapping and, for $r \leq k$, the homeomorphism $\Phi_{r/k}$ takes the set A into the set $\{y_1, \cdots, y_r, x_{r+1}, \cdots, x_k\}$.

Now suppose some points y_i coincide with points x_j, for $i \neq j$.

Construct a set C from B by replacing each point y_i by a point y_i' not in the set $A \cup B$ and such that different points y_i are replaced by different y_i'.

Then there exists an isotopy Φ_t: $E^n \to E^n$ such that $\Phi_1(A) = C$, and an isotopy Ψ_t: $E^n \to E^n$ such that $\Psi_1(C) = B$; hence by Theorem 2.1 there exists an isotopy F_t: $E^n \to E^n$ such that $F_1(A) = B$.

Observe that if the set $A \cup B$ is contained in some domain V, the isotopy F_t taking A into B may be constructed so as to be stationary outside V:

$$F_t(x) = x, \quad x \in E^n \setminus V, \quad 0 \leqslant t \leqslant 1.$$

2. Countable sets. Two homeomorphic countable sets in E^n need not be isotopic. Thus, for example, for any $n \geq 1$, the following two sets on the line $E^1 \subset E^n$ are nonisotopic: the set C consisting of the points with coordinates

$\{x_i = 1/i\}$, and the set D consisting of the points $\{x'_i = i\}$.

What is more, there is no homeomorphism $H: E^n \to E^n$ taking C into D. Indeed, if there were such a homeomorphism, it would take the limit point $x = 0$ of the set C into a limit point of the set D; but D has no limit points.

By an ϵ*-isotopy* is meant an isotopy which is an ϵ-homotopy (see Definition 1.4).

Theorem 4.1. *Let X and Y be two countable everywhere dense subsets of E^n (resp. S^n). Then for any $\epsilon > 0$ there exists an ϵ-isotopy of E^n (resp. S^n) onto itself taking X into Y.*

Let $X = \{x_k\}$, $Y = \{y_k\}$, $k = 1, 2, \cdots$. Choose a coordinate system in E^n, and let $\{\pi_\nu\}$ be the set of all $(n-1)$-hyperplanes parallel to the hyperplanes of the system.

Since the set $X \cup Y$ is countable, it is distributed over a countable set of hyperplanes π_ν. We may therefore use those hyperplanes π_ν that do not intersect $X \cup Y$ to subdivide the space E^n into a countable set of parallelepipeds Δ_r^1 of diameter less than ϵ:

$$E^n = \bigcup_{r=1}^{\infty} \bar{\Delta}_r^1, \quad \operatorname{diam} \Delta_r^1 < \varepsilon,$$

$$(\bigcup \operatorname{Fr} \Delta_r^1) \cap (X \cup Y) = \Lambda.$$

In each Δ_r^1, take the first point (i.e. the one with smallest index) $x_r^1 \in X \cap \Delta_r^1$ and the first point $y_r^1 \in Y \cap \Delta_r^1$. Such points exist because X and Y are everywhere dense. By Theorem 2.2 there exists for each Δ_r^1 an isotopy F_t^{1r} from the identity mapping, taking x_r^1 into y_r^1 and stationary outside Δ_r^1. Since $\operatorname{diam} \Delta_r^1 < \epsilon$, F_t^{1r} is an ϵ-isotopy for each Δ_r^1.

Since $F_t^{1r}(x) = x$ for $x \in \operatorname{Fr} \Delta_r^1$, $0 \le t \le 1$, we can then define an isotopy F_t^1 of all of E^n onto itself by setting

$$F_t^1(x) = F_t^{1r}(x) \quad \text{for} \quad x \in \Delta_r^1.$$

F_t^1 is obviously an ϵ-isotopy, and

$$F_1^1(x_r^1) = y_r^1, \quad r = 1, 2, \ldots$$

Set

$$f_0 = 1, \quad f_1 = F_1^1.$$

Now suppose we have constructed k homeomorphisms of E^n onto itself:

$$f_0, f_1, f_2, \ldots, f_k$$

such that f_i is joined to f_{i+1} by an $(\epsilon/2^i)$-isotopy F_t^{i+1}, and each f_i corresponds to a subdivision $E^n = \bigcup_{r=1}^{\infty} \Delta_r^i$ of E^n into parallelepipeds whose boundaries lie on the hyperplanes π_ν. Suppose further that each parallelepiped Δ_r^i is contained in some parallelepiped Δ_ρ^{i-1}, and that

$$(\bigcup_{r=1}^{\infty} \operatorname{Fr} \Delta_r^i) \cap [f_{i-1}^{-1}(Y) \cup X] = \Lambda;$$

$$\operatorname{diam} \Delta_r^i < \frac{\varepsilon}{2^{i-1}}, \qquad \operatorname{diam} f_{i-1}(\Delta_r^i) < \frac{\varepsilon}{2^{i-1}}. \tag{17}$$

$$f_i(x_r^j) = f_j(x_r^j) = y_r^j, \quad \text{for} \quad j < i. \tag{18}$$

To construct f_{k+1} we now subdivide each parallelepiped Δ_r^k, by means of the hyperplanes π_ν, into finitely many parallelepipeds Δ_ρ^{k+1} so as to satisfy condition (17) for $i = k+1$; this is possible because the set $f_k^{-1}(Y) \cup X$ is countable and the function f_k is uniformly continuous on Δ_r^k.

Next, just as before for Δ_r^1, take, for each ρ, the first point $x_\rho^{k+1} \in \Delta_\rho^{k+1} \cap X$ which does not coincide with any previously chosen x_r^j, $j \leq k$, and the first point $y_\rho^{k+1} \in Y \cap f_k(\Delta_\rho^{k+1})$ which does not coincide with any previously chosen y_r^j, $j \leq k$.

By Theorem 2.2 there exists for each $f_k(\Delta_\rho^{k+1})$ an isotopy $\Phi_t^{k+1,\rho}$ from the identity mapping of the cell $f_k(\Delta_\rho^{k+1})$ onto itself, taking the point $f_k(x_\rho^{k+1})$ into the point y_ρ^{k+1} and stationary on $\operatorname{Fr} f_k(\Delta_\rho^{k+1})$ as well as on all the points y_r^j, $j \leq k$, contained in $f_k(\Delta_\rho^{k+1})$. If now we set

$$\Phi_t^{k+1}(x) = \Phi_t^{k+1,\rho}(x) \qquad \text{for} \qquad x \in f_k(\Delta_\rho^{k+1}),$$

we obtain an $\epsilon/2^k$-isotopy of all of E^n onto itself, such that

$$\Phi_t^{k+1} f_k(\Delta_\rho^{k+1}) = f_k(\Delta_\rho^{k+1}), \qquad \rho = 1, 2, \ldots;$$

$$\Phi_t^{k+1}(x) = x \quad \text{for} \quad x \in \bigcup_{\rho=1}^{\infty} \operatorname{Fr} f_k(\Delta_\rho^{k+1}); \tag{19}$$

$$\Phi_t^{k+1}(y_r^j) = y_r^j, \quad j \leqslant k, \quad 0 \leqslant t \leqslant 1;$$

$$\Phi_1^{k+1} f_k(x_\rho^{k+1}) = y_\rho^{k+1}, \ \Phi_0^{k+1} = 1. \tag{20}$$

Finally, setting $f_{k+1} = \Phi_1^{k+1} f_k$, we obtain a homeomorphism $f_{k+1}\colon E^n \to E^n$, which is joined to f_k by an $\epsilon/2^k$-isotopy $\Phi_t^{k+1} f_k$, and such that, by (18) and (20),

$$f_{k+1}(x_r^j) = y_r^j, \quad j \leqslant k+1, \quad r = 1, 2, \ldots. \tag{21}$$

Continuing this process, we construct an infinite sequence of homeomorphisms $f_0, f_1, \ldots, f_k, \ldots$ of E^n onto itself, with $\rho(f_k, f_{k+1}) < \epsilon/2^k$. Since $\bigcup_{r=1}^{\infty} \operatorname{Fr} \Delta_r^1 \subset \bigcup_{\rho=1}^{\infty} \operatorname{Fr} \Delta_\rho^k$ for any k, and since (19) holds, we have, for any k,

$$\Phi_t^{k+1} f_k(x) = x \quad \text{for} \quad x \in \bigcup_{r=1}^{\infty} \operatorname{Fr} \Delta_r^1, \tag{22}$$

$$\Phi_t^{k+1} f_k(\Delta_r^1) = \Delta_r^1, \quad r = 1, 2, \ldots, \quad 0 \leqslant t \leqslant 1.$$

Since $\overline{\Delta}_r^1$ is a compactum, we are in the situation of Theorem 1.1; thus the sequence $f_1, \cdots, f_k, \cdots$ converges on $\overline{\Delta}_r^1$ to a continuous mapping ϕ^r of the parallelepiped $\overline{\Delta}_r^1$ onto itself, with ϕ^r homotopic to the identity mapping f_0 and

$$\varphi^r(x) = x \quad \text{for} \quad x \in \mathrm{Fr}\, \Delta_r^1, \quad r = 1, 2, \ldots .$$

Hence, setting

$$f(x) = \varphi^r(x), \; x \in \overline{\Delta}_r^1,$$

we obtain a continuous mapping f of E^n onto itself, homotopic to the identity mapping.

The homotopy F_t taking the identity mapping f_0 into f is an ϵ- isotopy, since, by (22),

$$F_t(\overline{\Delta}_r^1) = \overline{\Delta}_r^1, \quad 0 \leqslant t \leqslant 1, \; r = 1, 2, \ldots .$$

The mapping f is a homeomorphism.

Indeed, let x_1 and x_2 be two distinct points of E^n. There exists a k such that x_1 and x_2 appear in different parallelepipeds: $x_1 \in \Delta_i^k$, $x_2 \in \Delta_j^k$, $\Delta_i^k \cap \Delta_j^k = \Lambda$. Since f_k is a homeomorphism, we have $f_k(\Delta_i^k) \cap f_k(\Delta_j^k) = \Lambda$. From (19) it follows that for any $p > 0$,

$$f_{k+p}(\Delta_r^k) = f_k(\Delta_r^k), \quad r = i, j.$$

Hence

$$f(\Delta_r^k) = f_k(\Delta_r^k), \quad r = i, j,$$

so that $f(x_1) \neq f(x_2)$ and the mapping f is one-to-one. Since it is a homeomorphism on each Δ_r^1, and since $f(\Delta_r^1) = \Delta_r^1$, it is a homeomorphism on all of E^n. Since F_t is obtained on the intervals $[k/(k+1), (k+1)/(k+2)]$ from the isotopy $\Phi_t^{k+1} f_k$ by a linear transformation of the variable t (see Theorem 1.1), it follows that F_t is a homeomorphism for every $t < 1$. And since f is also a homeomorphism, it follows that *F_t is an isotopy.*

Finally, we assert that $f(X) = Y$.

By construction, each point $x_p \in X$ is, for some $k \leq p$, the first unused point of the set X contained in some parallelepiped Δ_r^k. But then $f_k(x_p) = y_q$, where y_q is the first unused point of the set $Y \cap f_{k-1}(\Delta_r^k)$. By (18), we have for any $s > 0$ that $f_{k+s}(x_p) = y_q$, so that

$$f(x_p) = y_q.$$

Hence the homeomorphism f takes every point of X into a point of Y.

One shows in the same way that every point y of Y is the image of some point $x \in X$. Consequently the isotopy F_t takes the set X into the set Y, and this completes the proof of the theorem.

3. Construction of isotopies. We now prove a lemma containing the technical apparatus for constructing an isotopy of E^n (or S^n) onto itself taking a compactum A into a compactum B.[1)]

Lemma 4.1. *Let A and B be homeomorphic compacta in E^n, $\phi: A \to B$ a homeomorphism, and*

$$A = \bigcap_{k=1}^{\infty} \bigcup_{i=1}^{n_k} a_i^k = \bigcap_{k=1}^{\infty} \bigcup_{i=1}^{n_k} u_i^k = \bigcap_{k=1}^{\infty} \bigcup_{i=1}^{n_k} O_i^k, \tag{23}$$
$$a_i^k \subset u_i^k \subset O_i^k,$$

where the $a_i^k \subset A$ are compacta, while the second two representations of A are of the form (10); *and let*

$$f_0, f_1, \ldots, f_k, \ldots$$

be a sequence of homeomorphisms of E^n onto itself, where f_0 is the identity mapping, each f_{k-1} is joined to f_k by an isotopy F_t^k, $0 \le t \le 1$, and the following conditions are satisfied:

1) *for every $a_j^k \subset u_i^k$, $p > 0$, $0 \le t \le 1$,*

$$\varphi(a_j^k) \subset F_t^{k+p}(u_i^k) \subset f_k(O_i^k);$$

2) *the diameters of O_i^k and $f_k(O_i^k)$ approach zero with $1/k$, and each O_j^{k+1} is contained in some O_i^k;*

3) $F_t^{k+p}(x) = f_k(x)$ *for* $x \in E^n \setminus \bigcup_{j=1}^{n_k} O_j^k$.

Then there exists an isotopy Φ_τ of E^n onto itself from the identity mapping, taking A into B, with

$$\Phi_1|_A \equiv \varphi.$$

It follows from condition 3) that for any $k \ge 0$ and $0 \le t \le 1$,

$$F_t^k(x) = x \quad \text{for} \quad x \in F^n \setminus V,\ V = \bigcup_{i=1}^{n_1} O_i^1, \tag{24}$$

and since F_t^k is a homeomorphism for each t, we have

$$F_t^k(\overline{V}) = \overline{V}. \tag{25}$$

Together with conditions 1) and 2) this implies that the sequence $\{f_k\}$ and the isotopies F_t^k satisfy the conditions of the corollary to Theorem 1.1. Indeed, since the $\{O_i^k\}$ and $\{u_i^k\}$ are fine sequences of coverings of A, there exists, for each k, an N such that every O_i^N is contained in some u_j^k:

$$a_i^N \subset O_i^N \subset u_j^k \subset O_j^k;$$

---

1) *Translator's note.* For the author's purpose, it seems best to let k start from zero in (23) (with $n_0 = 1$, say) and adjust the subsequent argument accordingly.

then by 1)

$$F_t^{N+p}(O_i^N) \subset F_t^{N+p}(u_j^k) \subset f_k(O_j^k), \quad p \geqslant 1, \ 0 \leqslant t \leqslant 1;$$

while by 3)

$$F_t^{N+p}(x) = f_N(x) \quad \text{for} \quad x \in E^n \setminus \bigcup_s O_s^N.$$

Hence

$$\rho\,[f_N, F_t^{N+p}] \leqslant \sup_j \operatorname{diam} f_k(O_j^k), \qquad p > 0, \ 0 \leqslant t \leqslant 1.$$

Consequently, there exists an isotopy Φ_τ such that

$$\Phi_0 = f_0; \quad \Phi_1 = \lim_{k \to \infty} f_k.$$

Next, we show that

$$\Phi_1 |_A \equiv \varphi.$$

If $x \in A$, then $x \in \bigcap_{k=1}^\infty a_i^k = \bigcap_{k=1}^\infty O_i^k$, and by condition 1), for any k,

$$\varphi(x) \in f_k(O_i^k), \qquad f_k(x) \in f_k(O_i^k).$$

Since the diameters of the $f_k(O_i^k)$ approach zero with increasing k, we have

$$\varphi(x) = \bigcap_{k=1}^{\infty} f_k(O_i^k) = \lim_{k \to \infty} f_k(x) = \Phi_1(x), \ x \in A.$$

Since for each $\tau < 1$ Φ_τ is obtained from some F_t^k by a linear transformation of the variable t, it follows that Φ_τ is a homeomorphism for every $\tau < 1$. By (24), (25) and condition 3),

$$\Phi_1(x) = x, \ x \in E^n \setminus V; \quad \Phi_1(\overline{V}) = \overline{V}.$$

Consequently, to prove that Φ_1 is a homeomorphism it suffices to show that the images of distinct points x_1 and x_2 are distinct. There are three cases.

1) $x_1 \in A$, $x_2 \in A$, $x_1 \neq x_2$; then $\Phi_1(x_1) \neq \Phi_1(x_2)$ because $\Phi_1|_A = \phi$ and ϕ is a homeomorphism.

2) $x_1 \in A$, $x_2 \in E^n \setminus A$. Then there exists a k such that $x_2 \in E^n \setminus \bigcup O_i^k$, and it follows from condition 3) that

$$\Phi_1(x_2) = f_k(x_2) \in f_k(E^n \setminus \bigcup O_i^k).$$

By condition 1), $\Phi_1(x_1) \in f_k(O_j^k)$ for some j, and so $\Phi_1(x_1) \neq \Phi_1(x_2)$.

3) $x_1 \in E^n \setminus A$, $x_2 \in E^n \setminus A$. Then there exists a k such that

$$x_1 \in E^n \setminus \bigcup_{i=1}^{n_k} O_i^k, \quad x_2 \in E^n \setminus \bigcup_{i=1}^{n_k} O_i^k.$$

Hence $\Phi_1(x_1) = f_k(x_1)$, $\Phi_1(x_2) = f_k(x_2)$, and since f_k is a homeomorphism it follows that $\Phi_1(x_1) \neq \Phi_1(x_2)$.

Thus Φ_τ is a homeomorphism for every τ, i.e. Φ_τ is an isotopy.

4. Cellularly separated zero-dimensional compacta in E^n

Definition 4.1. A zero-dimensional compactum $A \subset E^n$ is said to be *tame* in E^n if there exists a homeomorphism H of E^n onto itself such that $H(A)$ lies on a line E^1.

If there is no such homeomorphism, A is said to be *wild* in E^n. An example is given in Chapter IV of a wild zero-dimensional compactum. Every linear compactum is cellularly separated: any neighborhood contains a neighborhood which is a union of disjoint balls.[1] Consequently, if a zero-dimensional compactum $A \subset E^n$ is to be tame, it must be cellularly separated. We show now that this condition is also sufficient for A to be tamely imbedded.

Lemma 4.2. *Given an n-cell $Q^n \subset E^n$ (or S^n), and a ball δ^n and compactum K both contained in the interior of Q^n, there exists a cell $\widetilde{Q}^n$, interior to Q^n and containing $\delta^n \cup K$:*

$$\delta^n \cup K \subset \operatorname{int} \widetilde{Q}^n \subset \widetilde{Q}^n \subset \operatorname{int} Q^n \tag{26}$$

and an isotopy F_t: $Q^n \to Q^n$ from the identity mapping, taking $\widetilde{Q}^n$ into δ^n, with $F_t(x) = x$ for $x \in \operatorname{Fr} Q^n$.

Let ϕ be a homeomorphism of a ball B^n of radius 1 onto Q^n, and let $O' = \phi(O)$, where O is the center of B^n. Suppose first that the center of the ball δ^n is at O'. Then $\phi^{-1}(\delta^n)$ is an n-cell interior to the ball B^n, and $O \in \phi^{-1}(\delta^n)$.

Choose in the interior of B^n two balls B' and B'' concentric with B^n, such that

$$B' \subset \operatorname{int} \varphi^{-1}(\delta^n) \subset \varphi^{-1}(K \cup \delta^n) \subset \operatorname{int} B''. \tag{27}$$

Using the formulas of Example 2.6, construct an isotopy h_t of the ball B^n onto itself such that

$$h_0 = 1; \quad h_1(B'') = B'; \quad h_t(x) = x \quad \text{for} \quad x \in \operatorname{Fr} B^n \cup \{O\}, \quad 0 \leqslant t \leqslant 1.$$

Set

$$\theta = h_1^{-1}\varphi^{-1}(\delta^n), \quad \widetilde{Q}^n = \varphi(\theta).$$

Since δ^n is a ball and $h_1^{-1}\phi^{-1}$ a homeomorphism, θ is an n-cell contained in B^n, and $\widetilde{Q}^n$ is a cell contained in Q^n. By (27),

$$B'' = h_1^{-1}(B') \subset \operatorname{int} \theta \subset B^n, \quad K \cup \delta^n \subset \widetilde{Q}^n. \tag{28}$$

Set $F_t = \phi h_t \phi^{-1}$. Since ϕ is a homeomorphism and h_t an isotopy, F_t is an isotopy of the cell Q^n onto itself. Since $h_t(x) = x$, $x \in \operatorname{Fr} B^n$, and $h_t(B^n) = B^n$, we

---

1) *Translator's note.* Either "balls" should read "cells" or else "zero-dimensional" should be inserted before "linear".

have that $F_t(Q^n) = Q^n$, $0 \le t \le 1$, $F_t(y) = y$ for $y \in \operatorname{Fr} Q^n$, and F_0 is the identity mapping. Finally,

$$F_1(\tilde{Q}^n) = \varphi h_1 \varphi^{-1} (\varphi h_1^{-1} \varphi^{-1} (\delta^n)) = \delta^n.$$

By (28) the cell $\tilde{Q}^n$ and the isotopy F_t satisfy the conditions of the lemma.

If the center of δ^n does not coincide with $\phi(O)$, first construct in Q^n an appropriate n-cell $\tilde{Q}^n$ and an isotopy $F_t^1: Q^n \to Q^n$ taking $\tilde{Q}^n$ into a ball $\tilde{\delta}^n$ with center at $\phi(O)$, and then an isotopy F_t^2 taking $\tilde{\delta}^n$ onto δ^n and stationary on $\operatorname{Fr} Q^n$.

By Theorem 2.1, there then exists an isotopy F_t which is stationary on $\operatorname{Fr} Q^n$ and takes $\tilde{Q}^n$ into δ^n.

Corollary. *Let Q_1^n and Q_2^n be any two n-cells in E^n (or S^n); K_1 and K_2 compacta, with $K_1 \subset \operatorname{int} Q_1^n$, $K_2 \subset \operatorname{int} Q_2^n$; and let U be a domain containing $Q_1^n \cup Q_2^n$. Then there exist two n-cells $\tilde{Q}_1^n$ and $\tilde{Q}_2^n$ such that*

$$K_i \subset \operatorname{int} \tilde{Q}_i^n \subset \tilde{Q}_i^n \subset \operatorname{int} Q_i^n, \qquad i = 1, 2, \tag{29}$$

and an isotopy F_t of E^n onto itself taking the n-cell $\tilde{Q}_1^n$ into the n-cell $\tilde{Q}_2^n$ and stationary on $E^n \setminus U$; moreover, if Q_i^n is a ball, we may take $\tilde{Q}_i^n = Q_i^n$.

Let δ_1^n and δ_2^n be balls contained in the interiors of Q_1^n and Q_2^n. By Lemma 4.2, for $i = 1, 2$ there exist n-cells $\tilde{Q}_i^n$ satisfying (29) and isotopies $F_t^i: Q_i^n \to Q_i^n$ taking $\tilde{Q}_i^n$ into δ_i^n and stationary on $\operatorname{Fr} Q_i^n$. By Theorem 2.2 there exists an isotopy F_t^3 stationary on $E^n \setminus U$ and taking the ball δ_1^n into the ball δ_2^n. Regard F_t^1 and F_t^2 as the identity outside the corresponding Q_i^n. The required isotopy F_t is then constructed, by Theorem 2.1, by means of the isotopies F_t^1, F_t^3, $(F_t^2)^{-1}$, via the schema

$$F_t^i : \tilde{Q}_1^n \to \delta_1^n \to \delta_2^n \to \tilde{Q}_2^n.$$

F_t is stationary on $E^n \setminus U$ since this is so for each F_t^i, $i = 1, 2, 3$.

Lemma 4.3. *Let U be a domain in E^n (or S^n) containing a finite number of disjoint n-cells $\overline{O}_i$, disjoint balls δ_i, and compacta A_i in the interiors of the O_i: $A_i \subset O_i$, $i = 1, 2, \cdots, N$. Then there exists an n-cell $\overline{u}_i$, for each $i \le N$, such that*

$$A_i \subset u_i \subset \overline{u}_i \subset O_i,$$

and an isotopy $F_t: \overline{U} \to \overline{U}$ from the identity mapping, stationary on $\operatorname{Fr} U$ and such that

$$F_1(\overline{u}_i) = \delta_i, \quad i = 1, 2, \ldots, N.$$

By the corollary to Lemma 4.2, there exists an n-cell $\overline{u}_1$ such that

$$A_1 \subset u_1 \subset \overline{u}_1 \subset O_1,$$

and an isotopy $\phi_t^1\colon \overline{U} \to \overline{U}$ from the identity mapping, stationary on $\operatorname{Fr} U$ and such that

$$\varphi_1^1(\overline{u}_1) = \delta_1.$$

Suppose we have constructed, for $i \le r \le k < N$, n-cells $\overline{u}_i$ and isotopies $\phi_t^r\colon \overline{U} \to \overline{U}$, stationary on $\operatorname{Fr} U$, such that

$$\varphi_1^r(\overline{u}_i) = \delta_i \quad \text{for} \quad i \leqslant r, \tag{30}$$
$$A_i \subset u_i \subset \overline{u}_i \subset O_i.$$

Since $\overline{O}_i \cap \overline{O}_j = \Lambda$, $i \neq j$, we have

$$\delta_i \cap \varphi_1^k(O_j) = \Lambda, \quad \text{for} \quad i \leqslant k, \ j > k.$$

The union of the balls δ_i, $i \le k$, does not disconnect U, so that $U_k = U \setminus \bigcup_{i=1}^k \delta_i$ is still a domain. By the corollary to Lemma 4.2, there exists in $\phi_1^k(O_{k+1})$ an n-cell Δ^{k+1} such that

$$\varphi_1^k(A_{k+1}) \subset \operatorname{int} \Delta^{k+1} \subset \Delta^{k+1} \subset \varphi_1^k(O_{k+1}),$$

and an isotopy $\psi_t\colon \overline{U} \to \overline{U}$ from the identity mapping, stationary outside U_k and such that

$$\psi_1(\Delta^{k+1}) = \delta_{k+1}.$$

Set $\overline{u}_{k+1} = (\phi_1^k)^{-1}(\Delta^{k+1})$, and $\phi_t^{k+1}(x) = \psi_t \phi_1^k(x)$ for $x \in \overline{U}_k$, $\phi_t^{k+1}(x) = \phi_1^k(x)$ for $x \in \delta_i$, $i \le k$. This defines an isotopy ϕ_t^{k+1} on $\overline{U}$, and it is easily verified that (30) holds for $r = k+1$, $i \le k+1$, since $\psi_t(x) = x$ for $x \in \delta_i$, $i \le k$. Construct such isotopies ϕ_t^k for all $k \le N$. Since ψ_0 was the identity mapping, we have

$$\varphi_1^k = \varphi_0^{k+1}.$$

Hence by Theorem 2.1 we may construct, using the ϕ_t^k, $k = 1, 2, \cdots, N$, an isotopy $F_\tau\colon \overline{U} \to \overline{U}$, stationary on $\operatorname{Fr} U$ like all the ϕ_t^k, and taking each n-cell $\overline{u}_i$ into the ball δ_i.

Theorem 4.2 [7]. *Given a zero-dimensional cellularly separated compactum A in E^n, there exists an isotopy of E^n onto itself from the identity mapping, taking A into a compactum lying on a line.*

Since $\dim A = 0$, it follows from Theorem 3.1 that there exists a representation (10)

$$A = \bigcap_{k=1}^{\infty} \bigcup_{i=1}^{n_k} O_i^k$$

such that $a_i^k = A \cap O_i^k$ is open-closed in A, $\overline{O}_i^k \cap \overline{O}_j^k = \Lambda$ for $i \neq j$, and $\overline{O}_i^k \subset O_j^{k-1}$ whenever $a_i^k \subset a_j^{k-1}$. Thus, for a zero-dimensional compactum, a representation (10) is the same as a representation (11). Since A is cellularly separated, the O_i^k may be chosen so that the $\overline{O}_i^k$ are all n-cells, whose diameters approach zero with $1/k$.

Corresponding to each O_i^k, take a ball δ_i^k with center on the line E^1, such that

$$\delta_i^k \cap \delta_j^k = \Lambda,\ i \neq j;\ \delta_i^k \subset \operatorname{int} \delta_j^{k-1} \quad \text{whenever} \quad \overline{O}_i^k \subset O_j^{k-1},$$

and the diameters of the δ_i^k approach zero with increasing k, and set

$$B = \bigcap_{k=1}^{\infty} \bigcup_{i=1}^{n_k} \delta_i^k;$$

then B is a compactum on the line E^1.

Let U be a bounded domain containing $\bigcup_{i=1}^{n_1} O_i^1 \cup \bigcup_{i=1}^{n_1} \delta_i^1$. By Lemma 4.3, we may select in each O_i^1 an n-cell $\overline{u}_i^1$ such that

$$O_i^1 \cap (\bigcup_p \overline{O}_p^2) \subset u_i^1 \subset \overline{u}_i^1 \subset O_i^1,$$

and construct an isotopy $F_t^1 \colon E^n \to E^n$ from the identity mapping, stationary on $E^n \backslash U$, such that

$$F_1^1(\overline{u}_i^1) = \delta_i^1, \quad i = 1, 2, \ldots, n_1.$$

Letting $f_1 = F_1^1$, we have $f_1(\overline{u}_i^1) = \delta_i^1$, $i = 1, 2, \cdots, n_1$.

Suppose we have constructed, for all $r \leq k$, n-cells $\overline{u}_i^r$, homeomorphisms f_r of E^n onto itself, and isotopies F_t^r joining f_{r-1} to f_r, such that

1) $O_j^r \cap (\bigcup_i \overline{O}_i^{r+1}) \subset u_j^r \subset \overline{u}_j^r \subset O_j^r$,
2) $F_t^{r+p}(\overline{u}_i^r) = f_r(\overline{u}_i^r) = \delta_i^r$,
3) $F_t^{r+p}(x) = f_r(x),\quad 0 \leqslant t \leqslant 1,\ x \in E^n \setminus \bigcup_{i=1}^{n_r} u_i^r,\ p \geqslant 1,\ r + p \leqslant k.$

We now construct an isotopy F_t^{k+1}. By 1),

$$f_k[O_i^k \cap (\bigcup_s \overline{O}_s^{k+1})] \subset f_k(\overline{u}_i^k) \subset f_k(O_i^k),$$

and by 2),

$$f_k(\overline{u}_i^k) = \delta_i^k.$$

Using Lemma 4.3, construct, for each ball δ_i^k, an isotopy $F_t^i \colon \delta_i^k \to \delta_i^k$ from the identity mapping, stationary on $\operatorname{Fr} \delta_i^k$, which for each $f_k(O_j^{k+1}) \subset \delta_i^k$ takes into δ_j^{k+1} a cell Δ_j^{k+1} such that

$$f_k[O_j^{k+1} \cap (\bigcup_s \overline{O}_s^{k+2})] \subset \Delta_j^{k+1} \subset f_k(O_j^{k+1}).$$

Now set

$$\overline{u}_j^{k+1} = f_k^{-1}(\Delta_j^{k+1}),$$

$$F_t^{k+1}(x) = F_t^i f_k(x) \quad \text{for} \quad x \in \overline{u}_i^k,$$

$$F_t^{k+1}(x) = f_k(x) \quad \text{for} \quad x \in E^n \setminus \bigcup_{i=1}^{n_k} \overline{u}_i^k.$$

It is easily verified that conditions 1)–3) now hold for $r \leq k+1$ and $p \geq 1$.

For any decreasing sequence $O_{i_1}^1 \supset O_{i_2}^2 \supset \cdots \supset O_{i_k}^k \supset \cdots$, set

$$\varphi\left(\bigcap_{k=1}^{\infty} O_{i_k}^k\right) = \bigcap_{k=1}^{\infty} \delta_{i_k}^k;$$

then ϕ is a homeomorphism of the compactum A onto the compactum B.

By condition 1) we have $a_i^k = A \cap O_i^k = A \cap u_i^k$. Hence the sequence $\{f_k\}$ and the isotopies F_t^k satisfy the conditions of Lemma 4.1. Specifically, condition 2) of the lemma is satisfied because the diameters of the O_i^k and δ_i^k approach zero; condition 1) of the lemma follows from 1) and 2), since $\phi(a_i^k) \subset \delta_i^k$; and condition 3) of the lemma follows from 3). Consequently there exists an isotopy Φ_τ from the identity mapping, taking A into $B = \phi(A) = \Phi_1(A)$.

Corollary 1. *A zero-dimensional compactum in E^n is tame if and only if it is cellularly separated in E^n.*

Necessity is obvious. Sufficiency follows from Theorem 4.2, since Φ_1 is a homeomorphism of E^n onto itself and $\Phi_1(A) \subset E^1$.

Corollary 2. *Every countable compactum in E^n is tame.*

For, by Theorem 3.4, every such compactum is cellularly separated in E^n.

Corollary 3. *For any tame zero-dimensional compactum A in E^n, there exists an isotopy Φ_t: $E^n \to E^n$ taking A into a linear compactum.*

Indeed, if the zero-dimensional compactum A is tame, it is cellularly separated in E^n, so that the required isotopy exists by Theorem 4.2.

Corollary 4. *Let A and C be two zero-dimensional cellularly separated compacta in E^n, and ϕ a homeomorphism of A onto C. Then ϕ extends to a homeomorphism Φ of E^n onto itself such that $\Phi|_A = \phi$.*

Let

$$A = \bigcap_{k=1}^{\infty} \bigcup_{i=1}^{n_k} \overline{u}_i^k; \quad C = \bigcap_{k=1}^{\infty} \bigcup_{j=1}^{m_k} \overline{v}_j^k$$

be representations of the form (10)–(11), where $\overline{u}_i^k$ and $\overline{v}_j^k$ are n-cells. By Theorem 4.2, there exists a compactum B lying on a line and homeomorphic to A and C. Let

$$A = \psi(B), \quad C = \varphi\psi(B).$$

Then

$$B = \bigcap_{k=1}^{\infty} \bigcup_{i=1}^{n_k} \psi^{-1}(A \cap u_i^k) = \bigcap_{k=1}^{\infty} \bigcap_{i=1}^{m_k} \psi^{-1}\varphi^{-1}(C \cap v_j^k),$$

where $\psi^{-1}(A \cap u_i^k)$ and $\psi^{-1}\phi^{-1}(C \cap v_j^k)$ are open-closed subsets of B. The homeomorphism ψ can be chosen so that for each k the open-closed sets $\psi^{-1}(A \cap u_i^k)$ may be covered by disjoint balls δ_i^k with centers on E^1. Moreover, the balls may be chosen so that $\delta_j^{k+1} \subset \delta_i^k$ whenever $\overline{u}_j^{k+1} \subset u_i^k$, and the diameters of the δ_i^k approach zero with $1/k$. Likewise, the $\psi^{-1}\phi^{-1}(C \cap v_j^k)$ can be separated by polyhedral balls Δ_j^k. Then

$$B = \bigcap_{k=1}^{\infty} \bigcup_{i=1}^{n_k} \delta_i^k = \bigcap_{k=1}^{\infty} \bigcup_{j=1}^{m_k} \Delta_j^k.$$

B is determined for A in the same way as in Theorem 4.2. Hence there exists an isotopy Φ_t and a homeomorphism $\Phi_1\colon E^n \to E^n$ such that $\Phi_1|_A = \psi^{-1}$, and also a homeomorphism $\Psi_1\colon E^n \to E^n$ such that $\Psi_1|_C = \psi^{-1}\phi^{-1}$, i.e., $\Psi_1^{-1}|_B = \phi\psi$. Set $H = \Psi_1^{-1}\Phi_1$; then H is a homeomorphism of E^n onto itself and

$$H|_A = \Psi_1^{-1}|_B \Phi_1|_A = \varphi\psi\psi^{-1} \equiv \varphi.$$

Thus H satisfies the conditions of the corollary.

§5. Pseudo-isotopy

1. **Definition 5.1.** A *pseudo-isotopy* of a space X is a homotopy $F_t\colon X \to X$, $0 \leq t \leq 1$, such that F_t is a homeomorphism of X onto itself for each $t < 1$, and is a continuous mapping of X onto itself for $t = 1$.

Examples of pseudo-isotopies.

Example 5.1. Specify points $x \in E^3$ by polar coordinates (α, β, ρ), where ρ is the radius vector. Consider the homotopy $F_t(x) = (\alpha, \beta, \rho_t(x))$ such that the coordinates α and β of a point remain unchanged:

$$\alpha_t(x) = \alpha(x); \quad \beta_t(x) = \beta(x), \quad 0 \leqslant t \leqslant 1,$$

while the radius vector $\rho_t(x)$ is defined as follows:

$$\rho_t(x) \equiv \rho(x), \qquad \text{for } \rho(x) \geqslant 2,$$

$$\rho_t(x) = \rho(x) - t\,[2 - \rho(x)], \quad \text{for } 1 \leqslant \rho(x) \leqslant 2,$$

$$\rho_t(x) = (1 - t)\,\rho(x), \quad \text{for } \rho(x) \leqslant 1.$$

For each $t < 1$, this determines a homeomorphism of E^3 onto itself, under which the ball Q_1 of radius 1 and center at the origin O goes into the ball Q_{1-t} of radius $1 - t$ and center at O, and the segment $[1 \leq \rho \leq 2]$ on any ray issuing from O goes into the segment $[1 - t \leq \rho \leq 2]$ on the same ray.

For $t = 1$, the whole ball Q_1 collapses into the point O. F_1 is a continuous mapping of E^3 onto itself, under which O is the only point whose inverse image is not a point: $F_1^{-1}(O)$ is the ball Q_1. $F_1(x) = x$ for $\rho(x) \geq 2$, and F_1 takes the segment $[1 \leq \rho \leq 2]$ on any ray issuing from O into the segment $[0 \leq \rho \leq 2]$ on the same ray.

Example 5.2. Specify the points $x \in E^3$ as in Example 5.1; the homotopy Φ_t is again to leave unchanged the coordinates α and β. The radius vector of the point $\Phi_t(x)$ is defined as follows:

$$\rho_t(x) = \rho(x) \quad \text{for} \quad \rho(x) \geqslant 3 \quad \text{or} \quad \rho(x) \leqslant 1,$$

$$\rho_t(x) = \rho(x) - t\,[3 - \rho(x)] \quad \text{for} \quad 2 \leqslant \rho(x) \leqslant 3,$$

$$\rho_t(x) = \rho(x) - t\,[\rho(x) - 1] \quad \text{for} \quad 1 \leqslant \rho(x) \leqslant 2.$$

The homotopy Φ_t is stationary on the ball of radius 1 and center O, as well as outside the ball of radius 3. For any $t < 1$, Φ_t is a homeomorphism of E^3 onto itself, under which the segment $[1 \leq \rho \leq 2]$ on any ray goes into the segment $[1 \leq \rho \leq 2 - t]$ of the same ray, while the segment $[2 \leq \rho \leq 3]$ goes into the segment $[2 - t \leq \rho \leq 3]$. For $t = 1$, the whole segment $[1 \leq \rho \leq 2]$ on each ray collapses to the point $\rho = 1$. The decomposition elements for the continuous mapping Φ_1 are the segments $[1 \leq \rho \leq 2]$ on all rays issuing from O, together with the points of E^3 not on these segments.

2. Pseudo-isotopies are frequently used to study the imbedding of a continuous image of a space X into itself or into a space Y containing X. We shall consider continuous mappings f for which the corresponding decomposition of X into inverse images of points is continuous and the inverse images are compacta.

The decomposition of X into point-inverses under a continuous mapping f will be denoted by X_f, and the decomposition space by $f(X)$. Many of the proofs that the space $f(E^n)$ of a countinuous decomposition E^n_f of E^n is contained in E^k, $k \geq n$, (or is the same as E^n) proceed in the following way. Regarding E^n as a hyperplane in E^k, construct a pseudo-isotopy $\Phi_t\colon E^k \to E^k$ from the identity mapping such that $\Phi_1|_{E^n}$ is topologically equivalent to f; more precisely, such that the correspondence $\Phi_1(x) \leftrightarrow f(x)$, $x \in E^n$, is a homeomorphism between the images of E^n under the mappings f and $\Phi_1|_{E^n}$. Let us give an example of such a proof.

Theorem 5.1 [4–6, 48]. *The space $\Psi(E^n)$ of the decomposition of E^n into the components of a cellularly separated compactum P in E^n and the points $x \in E^n \setminus P$ is homeomorphic to E^n.*

Observe first of all that the decomposition E^n_Ψ is continuous. This is obvious for the points $x \in E^n \setminus P$. If a component ξ is isolated in P, i.e. if there exists a neighborhood $V \supset \xi$ such that $\overline{V} \cap (P \setminus \xi) = \Lambda$, then the decomposition E^n_Ψ is obviously continuous at ξ. Now let ξ be a nonisolated component of P. Then there exists a sequence of points $x_1, x_2, \cdots, x_k, \cdots$ lying on distinct components of P: $x_k \in \xi_k$, $\xi_i \neq \xi_j$, $i \neq j$, which converges to a point $x_0 \in \xi$:

$$\lim_{k\to\infty} x_k = x_0.$$

The upper topological limit $\overline{\lim}_{k\to\infty} \xi_k$ of the continua ξ_k is connected, since the lower limit $\underline{\lim}_{k\to\infty} \xi_k$ contains the point x_0. Since P is compact, $\overline{\lim}_{k\to\infty} \xi_k \subset P$, and since ξ is a component of P and $\xi \cap \overline{\lim}_{k\to\infty} \xi_k \neq \Lambda$, we have $\overline{\lim}_{k\to\infty} \xi_k \subset \xi$. This implies that the decomposition E^n_Ψ is continuous at ξ.

We now construct a pseudo-isotopy Φ_t taking the identity mapping of E^n into one which induces the decomposition E^n_Ψ. Since P is cellularly separated, there exists a representation of the form (11):

$$P = \bigcap_{k=1}^{\infty} \bigcup_{i=1}^{N_k} \overline{u}_i^k; \quad \overline{u}_i^k \cap \overline{u}_j^k = \Lambda, \quad i \neq j,$$

where the $\overline{u}_i^k$ are all n-cells and each $\overline{u}_i^k$ is contained in some u_j^{k-1}. In each u_i^1 pick a ball Δ_i^1 of radius < 1. By Lemma 4.2 we may construct for each u_i^1 an n-cell q_i^1 such that

$$u_i^1 \cap \left(\bigcup_{j=1}^{N_2} \overline{u}_j^2 \right) \subset \operatorname{int} q_i^1 \subset q_i^1 \subset u_i^1, \tag{31}$$

and an isotopy ϕ_t^{1i}: $u_i^1 \to u_i^1$, stationary on $\operatorname{Fr} u_i^1$, such that

$$\varphi_1^{1i}(q_i^1) = \Delta_i^1.$$

Set

$$F_t^1(x) = \varphi_t^{1i}(x), \quad \text{for} \quad x \in \overline{u}_i^1;$$

$$F_t^1(x) = x, \quad \text{for} \quad x \in E^n \setminus \bigcup_i u_i^1; \quad F_1^1 = f_1.$$

Then $f_1(q_i^1) = \Delta_i^1$ for each i so that by (31)

$$\operatorname{diam} f_1 \left[u_i^1 \cap \left(\bigcup_{j=1}^{N_2} \overline{u}_j^2 \right) \right] < 1, \quad 1 \leqslant i \leqslant N_1.$$

Suppose we have constructed, for all $r \leq k$, homeomorphisms f_r: $E^n \to E^n$ and isotopies F_t^r joining f_{r-1} to f_r, with

$$\operatorname{diam} f_r \left[u_i^r \cap \left(\bigcup_j \overline{u}_j^{r+1} \right) \right] < \frac{1}{r},$$

$$F_t^{r+p}(\overline{u}_i^r) = f_{r-1}(\overline{u}_i^r), \quad r + p \leqslant k, \quad p \geqslant 0, \tag{32}$$

$$F_t^{r+p}(x) = f_{r-1}(x) \quad \text{for} \quad x \in E^n \setminus \bigcup_i u_i^r.$$

These conditions imply that

$$\operatorname{diam} F_t^{r+p}(\overline{u}_i^r) < \frac{1}{r-1}, \quad p \geqslant 0, \ r + p \leqslant k.$$

To construct F_t^{k+1} and $f_{k+1} = F_1^{k+1}$, choose in each $f_k(u_i^{k+1})$ a ball Δ_i^{k+1} of radius less than $1/(k+1)$ and, just as for $r=1$, find an n-cell q_i^{k+1} such that

$$f_k\left[u_i^{k+1}\cap\left(\bigcup_{j=1}^{N_{k+2}}\overline{u}_j^{k+2}\right)\right]\subset \operatorname{int} q_i^{k+1}\subset q_i^{k+1}\subset f_k(u_i^{k+1}), \tag{33}$$

and an isotopy $\phi_t^{k+1,i}\colon f_k(\overline{u}_i^{k+1})\to f_k(\overline{u}_i^{k+1})$ from the identity mapping, stationary on $\operatorname{Fr} f_k(\overline{u}_i^{k+1})$, such that

$$\varphi_1^{k+1,i}(q_i^{k+1})=\Delta_i^{k+1},\quad i=1,2,\dots,N_{k+1}.$$

Define the isotopy F_t^{k+1} by setting

$$F_t^{k+1}(x)=\varphi_t^{k+1,i}f_k(x)\quad\text{for}\quad x\in u_i^{k+1},$$

$$F_t^{k+1}(x)=f_k(x)\quad\text{for}\quad x\in E^n\setminus\bigcup_{j=1}^{N_{k+1}}u_j^{k+1}.$$

By (33), conditions (32) remain valid for $r+p=k+1$ and $p\geq 0$.

It is clear from (32) that the sequence $\{f_k\}$ and isotopies F_t^k satisfy the conditions of the corollary to Theorem 1.1, since for every k we have $F_t^k(x)=x$ for $x\in E^n\setminus\bigcup u_i^1$. Hence the sequence $\{f_k\}$ converges to a continuous mapping Φ_1 of E^n onto itself, and there exists a homotopy Φ_t taking the identity mapping f_0 into Φ_1. On each interval $(k-1)/k\leq t\leq k/(k+1)$, Φ_t is obtained from F_t^k by a linear transformation of the variable t. Hence for each $t<1$, Φ_t is a homeomorphism of E^n onto itself. Consequently, Φ_t *is a pseudo-isotopy*. We now show that the continuous decomposition $E^n_{\Phi_1}$ is the same as E^n_{Ψ}.

To do so, it suffices to show that the elements of the decomposition $E^n_{\Phi_1}$ are the components of P and the points of $E^n\setminus P$.

Suppose $x\in E^n\setminus P$. Then $x\in E^n\setminus\bigcup_i\overline{u}_i^k$ for some k, so that, by (32), for every $p\geq 0$ we have

$$F_t^{k+p}(x)=f_{k-1}(x).$$

Consequently,

$$\Phi_1(x)=f_{k-1}(x).$$

Since f_{k-1} is a homeomorphism, Φ_1 is a homeomorphism on $E^n\setminus\bigcup_i\overline{u}_i^k$; and since $\Phi_1(\overline{u}_i^k)=f_{k-1}(\overline{u}_i^k)$, x is outside $\Phi_1(\bigcup_i\overline{u}_i^k)$; thus x is an element of the decomposition $E^n_{\Phi_1}$.

Now let ξ be a component of P. Then there exists a sequence of integers i_k such that

$$\xi=\bigcap_{k=1}^{\infty}u_{i_k}^k,\quad \overline{u}_{i_{k+1}}^{k+1}\subset u_{i_k}^k.$$

The sequence $f_1(\xi), f_2(\xi),\cdots, f_k(\xi),\cdots$ converges topologically to the continuum $\Phi_1(\xi)$. But $\operatorname{diam} f_k(\xi)<1/k$ by (32) so that $\Phi_1(\xi)$ is a point. By (32)

$\Phi_1(\xi) \neq \Phi_1(x)$ for $x \in E^n \backslash P$, since $x \in E^n \backslash \bigcup_i u_i^k$ for sufficiently large k. Also, $\Phi_1(\xi_1) \neq \Phi_1(\xi_2)$ if ξ_1 and ξ_2 are different components of P, since for sufficiently large k we have $\xi_1 \subset u_i^k$, $\xi_2 \subset u_j^k$, $i \neq j$. Hence the elements of the decomposition $E^n_{\Phi_1}$ are the components of P and the points $x \in E^n \backslash P$; consequently $E^n_{\Phi_1}$ coincides with E^n_Ψ. Since Φ_1 is a continuous mapping of E^n onto itself, $\Psi(E^n)$ is homeomorphic with E^n.

Corollary 1. *Theorem* 5.1 *remains valid if in its hypothesis E^n is replaced by an arbitrary space X.*

Since P is cellularly separated in X, there exists a neighborhood U of the compactum P whose closure is a finite union of disjoint n-cells $\overline{U}^i$, $i = 1, 2, \cdots, p$; each U^i is a component of U homeomorphic to E^n. Then $P_i = P \cap U^i$ is a compactum cellularly separated in U^i. Applying Theorem 5.1, there exists for each U^i a pseudo-isotopy Φ_t^i: $U^i \to U^i$ such that the continuous decomposition $U^i_{\Phi_1^i}$ is equivalent to the decomposition U^i_Ψ, so that $\Psi(U^i) = U^i$. By construction, each pseudo-isotopy Φ_t^i is stationary on $U^i \backslash \bigcup_{j=1}^{N_1} \overline{u}_j^1$, where $\{u_j^1\}$ is a covering of P contained in U. Hence, setting

$$\Phi_t(x) = \Phi_t^i(x) \quad \text{for} \quad x \in U^i,$$

$$\Phi_t(x) = x \quad \text{for} \quad x \in X \backslash U,$$

we obtain a pseudo-isotopy Φ_t: $X \to X$ such that the decomposition X_{Φ_1} is the same as X_Ψ; hence $\Psi(X) = \Phi_1(X) = X$.

Corollary 2. *The space $\phi(X)$ of the decomposition of a space X into points and finitely many disjoint cellularly imbedded continua is homeomorphic to X.*

Let $\xi_1, \xi_2, \cdots, \xi_k$ be the continua which are elements of the decomposition X_φ, and let V_i be any neighborhood of ξ_i.

For each ξ_i take a cellular neighborhood $W_i \subset \overline{W}_i \subset V_i$. Since $\bigcup_i W_i$ is a manifold, the ξ_i may be separated in $\bigcup_i W_i$ by disjoint neighborhoods

$$\xi_i \subset U_i \subset \overline{U}_i \subset W_i; \quad \overline{U}_i \cap \overline{U}_j = \Lambda, \quad i \neq j.$$

By hypothesis the U_i may be chosen so that every $\overline{U}_i$ is an n-cell. Consequently the compactum $\bigcup_{i=1}^k \xi_i$ is cellularly separated in X and Corollary 1 applies; thus $\phi(X) = X$.

An arbitrary compactum P in E^n, all of whose components are cellularly imbedded, *need not be cellularly separated*, and Theorem 5.1 need not hold for it. An example has been constructed by Bing [25] of a compactum $P \subset E^3$ all of whose components are cellularly imbedded simple arcs (as shown by Fort [39], the construction can be made in such a way that the arcs are three-segment polygonal lines), but such that P is not cellularly separated, and the continuous decomposition of E^3 into the components of P and the points of $E^3 \setminus P$ is not the same as E^3.

If a compactum $P \subset E^n$ has only countably many components, all of which are cellularly imbedded in E^n, then the corresponding decomposition space is the same as E^n: for $n = 3$ this follows from a theorem of Bing [24], and for $n > 3$ from a theorem of Kwun [48].

It can be shown that, in this case also, P is cellularly separated.

3. There are grounds for supposing that pseudo-isotopy has application to the study of the question of imbedding a manifold in E^n. Many examples of topological spheres S^2 and disks D^2 in E^3 can be arrived at by a pseudo-isotopy $F_t\colon E^3 \to E^3$ from the identity mapping such that

$$S^2 = F_1(\Sigma^2), \quad D^2 = F_1(I^2),$$

where Σ^2 is the boundary of a cube I^3, or where I^2 is a square. It is unknown, however, whether an arbitrary imbedding of S^{n-1} into E^n, or of an $(n-1)$-cell D^{n-1} into E^n, is so obtainable by means of a pseudo-isotopy.[1)] We show here that a pseudo-isotopy of E^n onto itself allows one to obtain an arbitrary zero-dimensional compactum, starting from a zero-dimensional compactum on a line.

Lemma 5.1. *Let A and $\phi(A)$ be two homeomorphic compacta in E^n, and let*

$$A = \bigcap_{k=1}^{\infty} \bigcup_{i=1}^{n_k} a_i^k = \bigcap_{k=1}^{\infty} \bigcup_{i=1}^{n_k} u_i^k, \quad a_i^k \subset u_i^k, \quad \operatorname{diam} u_i^k < \frac{1}{k},$$

$$\varphi(A) = \bigcap_{k=1}^{\infty} \bigcup_{i=1}^{n_k} \varphi(a_i^k) = \bigcap_{k=1}^{\infty} \bigcup_{i=1}^{n_k} v_i^k, \quad \varphi(a_i^k) \subset v_i^k, \quad \operatorname{diam} v_i^k < \frac{1}{k}$$

be representations of the form (10) *for A and $\phi(A)$. Let U be a bounded domain containing $\bigcup_{i=1}^{n_1}(u_i^1 \cup v_i^1)$, and $f_0, f_1, \cdots, f_k, \cdots$ a sequence of homeomorphisms of E^n onto itself, where f_0 is the identity mapping and each f_{k-1} is joined to f_k by an isotopy F_t^k, with*

$$\begin{gathered} F_t^{k+p}(x) \in v_i^k, \ p \geqslant 1, \ \text{if} \ f_k(x) \in v_i^k, \\ F_t^{k+p}(x) = f_k(x), \ p \geqslant 1, \ \text{if} \ f_k(x) \in E^n \setminus \bigcup_{i=1}^{n_k} v_i^k, \\ F_t^k(x) = x, \ \text{if} \ x \in E^n \setminus U, \\ f_k(\bar{u}_i^k) \subset v_i^k, \ 0 \leqslant t \leqslant 1. \end{gathered} \tag{34}$$

Then there exists a pseudo-isotopy Φ_t from the identity mapping, with Φ_1 a homeomorphism on A and on $E^n \setminus \Phi_1^{-1}\Phi_1(A)$, and $\Phi_1|_A \equiv \phi$.

---

1) The author [79] has recently proved the existence of such a pseudo-isotopy for $n = 3$.

It follows from (34) and the fact that the diameters of the v_i^k approach zero with increasing k, that the sequence $\{f_k\}$ and the homotopies F_t^k satisfy the conditions of the corollary to Theorem 1.1. Hence there exists a homotopy Φ_t from the identity mapping, with $\Phi_1 = \lim_{k\to\infty} f_k$. For each $t < 1$, Φ_t is obtained by a linear transformation of the variable t from some F_t^k and is consequently a homeomorphism; thus Φ_t is a pseudo-isotopy. $\Phi_0 = f_0$ is the identity mapping.

We assert that $\Phi_1(x) = \phi(x)$ if $x \in A$. For, if $x \in A$, there is a sequence of integers i_k such that

$$x = \bigcap_{k=1}^{\infty} a_{i_k}^k = \bigcap_{k=1}^{\infty} u_{i_k}^k, \quad \overline{u}_{i_{k+1}}^{k+1} \subset u_{i_k}^k,$$

$$\Phi_1(x) = \lim_{k\to\infty} f_k(x) = \lim_{k\to\infty} f_k\left(\bigcap_{k=1}^{\infty} u_{i_k}^k\right).$$

Since $f_k(\overline{u}_{i_k}^k) \subset v_{i_k}^k$ we get

$$\Phi_1(x) = \lim_{k\to\infty} f_k(u_{i_k}^k) = \bigcap_{k=1}^{\infty} v_{i_k}^k = \bigcap_{k=1}^{\infty} \varphi(a_{i_k}^k) = \varphi(x).$$

Consequently Φ_1 is a homeomorphism on A. Now suppose $x \in \Phi_1^{-1}[E^n \backslash \Phi_1(A)]$. Then $\Phi_1(x) \in E^n \backslash \Phi_1(A)$, so that there exists a k such that $\Phi_1(x) \in E^n \backslash \bigcup_{i=1}^{n_k} \overline{v}_i^k$; but by (34), $\Phi_1(x') = f_k(x')$ for all x' such that $f_k(x') \in E^n \backslash \bigcup_{i=1}^{n_k} \overline{v}_i^k$; hence Φ_1 is a homeomorphism on $\Phi_1^{-1}(E^n \backslash \Phi_1(A))$.

But there may be points $x \in E^n \backslash A$ such that, for any k, $f_k(x) \in v_{i_k}^k$ for some i_k. It follows from conditions (34) that then $\overline{v}_{i_{k+1}}^{k+1} \subset v_{i_k}^k$, and hence $\lim_{k\to\infty} f_k(x) = \Phi_1(x) \in A$.

Theorem 5.2 [7]. *For any zero-dimensional compactum A in E^n, there exists a compactum B on a line $E^1 \subset E^n$, and a pseudo-isotopy Φ_t: $E^n \to E^n$ from the identity mapping, taking B into A, stationary outside an arbitrary bounded domain U containing $A \cup B$, and such that Φ_1 is a homeomorphism on B and on $E^n \backslash \Phi_1^{-1}(A)$.*

Let $\dim A = 0$, let $A = \bigcap_{k=1}^{\infty} \bigcup_{i=1}^{n_k} v_i^k$ be a representation of the form (10), (11), and suppose $\bigcup_{i=1}^{n_1} v_i^1 \subset U$. Since A is an arbitrary zero-dimensional compactum, the $\overline{v}_i^k$ need not be n-cells.

For each v_i^k pick a ball δ_i^k in U with center on a line E^1, $E^1 \cap U \neq \Lambda$, such that

$$\delta_i^k \cap \delta_j^k = \Lambda, \quad i \neq j, \quad \operatorname{diam} \delta_i^k < \frac{1}{k}, \quad \delta_i^k \subset \operatorname{int} \delta_j^{k-1} \quad \text{whenever } \overline{v}_i^k \subset v_j^{k-1},$$

and set

$$B = \bigcap_{k=1}^{\infty} \bigcup_{i=1}^{n_k} \delta_i^k, \tag{35}$$

$$\varphi\left(\bigcap_{1=1}^{\infty} \delta_{i_k}^k\right) = \bigcap_{i=1}^{\infty} v_{i_k}^k, \quad \overline{v}_{i_k}^k \subset v_{i_{k-1}}^{k-1}.$$

Since the diameters of the v_i^k and δ_i^k approach zero with $1/k$, ϕ is a homeomorphism of the linear compactum B onto the compactum A. Since $\bigcup_{i=1}^{n_1} (\delta_i^1 \cup \overline{v}_i^{-1}) \subset U$, we have $B \subset U$. Pick in each $v_i^1 \setminus \bigcup_{j=1}^{n_2} \overline{v}_j^2$ a ball π_i^1.

By Lemma 4.3, there exists an isotopy F_t^1 from the identity mapping, which is stationary on $E^n \setminus U$ and takes each ball δ_i^1 into the ball π_i^1:

$$F_t^1(x) = x, \ x \in E^n \setminus U; \ F_1^1(\delta_i^1) = f_1(\delta_i^1) = \pi_i^1 \subset v_i^1 \setminus \bigcup_j \overline{v}_j^2.$$

This condition agrees with (34) for $k = 1$, $\overline{u}^{-1} = \delta_i^1$, $i = 1, 2, \cdots, n_1$. Suppose we have constructed, for all $r \leq k$, balls π_i^r, n-cells $\overline{u}_i^r \subset \delta_i^r$ and isotopies F_t^r such that conditions (34) hold for $r + p \leq k$ and

$$f_r(\overline{u}_i^r) = \pi_i^r \subset v_i^r \setminus \bigcup_{j=1}^{n_{r+1}} \overline{v}_j^{r+1}, \ \delta_i^r \cap (\bigcup_j \delta_j^{r+1}) \subset u_i^r. \tag{36}$$

We now construct F_t^{k+1} and f_{k+1}. To do so, pick in each $v_j^{k+1} \setminus \bigcup_s \overline{v}_s^{k+2}$ a ball π_j^{k+1}. By (36), if $\delta_j^{k+1} \subset \delta_i^k$, then

$$f_k(\delta_j^{k+1}) \subset f_k(\overline{u}_i^k) = \pi_i^k \subset v_i^k \setminus \bigcup_j \overline{v}_j^{k+1}.$$

By Lemma 4.3, we may pick in the interior of each $f_k(\delta_j^{k+1})$ an n-cell q_j^{k+1} such that

$$f_k[\delta_j^{k+1} \cap (\bigcup_s \delta_s^{k+2})] \subset \operatorname{int} q_j^{k+1} \subset q_j^{k+1} \subset f_k(\delta_j^{k+1}), \tag{37}$$

and construct for each $\overline{v}_i^k$ an isotopy $\psi_t^i \colon \overline{v}_i^k \to \overline{v}_i^k$ from the identity mapping which is stationary on $\operatorname{Fr} \overline{v}_i^k$ and takes each n-cell $q_j^{k+1} \subset v_i^k$ into π_j^{k+1}:

$$\psi_1^i(q_j^{k+1}) = \pi_j^{k+1}, \ q_j^{k+1} \subset v_i^k.$$

Set

$$F_t^{k+1}(x) = \psi_t^i f_k(x), \quad \text{if} \quad f_k(x) \in v_i^k,$$

$$F_t^{k+1}(x) = f_k(x), \quad \text{if} \quad f_k(x) \in E^n \setminus \bigcup_i v_i^k,$$

$$\overline{u}_j^{k+1} = f_k^{-1}(q_j^{k+1}).$$

By (37) we get

$$\delta_j^{k+1} \cap (\bigcup_s \delta_s^{k+2}) \subset u_j^{k+1} \subset \overline{u}_j^{k+1} \subset \delta_j^{k+1}.$$

It is easily verified that the u_i^r, f_r, F_t^r satisfy (34) for $r + p \leq k + 1$. Continuing the construction ad infinitum, we obtain u_i^k, f_k and F_t^k, $k = 1, 2, \cdots$, satisfying the conditions of Lemma 5.1. Consequently there exists a pseudo-isotopy Φ_t from the identity mapping such that $\Phi_1|_B = \phi$ (so that $\Phi_1(B) = A$) and Φ_1 is a homeomorphism on B and on $E^n \setminus \Phi_1^{-1}(A)$.

A pseudo-isotopy which is an ϵ-homotopy is called an *ϵ-pseudo-isotopy*.

Corollary. *For any zero-dimensional compactum A in E^n and any $\epsilon > 0$, there exists a tame compactum $B \subset E^n$ and an ϵ-pseudo-isotopy F_t of E^n onto* *itself, such that* $F_1(B) = A$, F_1 *is a homeomorphism on* B *and on* $E^n \setminus F_1^{-1}(A)$, *and* $F_t(x) = x$, $0 \leq t \leq 1$, *outside an* ϵ*-neighborhood of* A.

Pick for A an open covering $A \subset \bigcup_{i=1}^{k} u_i$ such that $\bar{u}_i \cap \bar{u}_j = \Lambda$, $i \neq j$, and diam $\bar{u}_i < \epsilon$.[1)]

Each $a_i = A \cap u_i$ is open-closed in A and is consequently a compactum. For each i, pick a compactum $b_i \subset u_i$ lying on a line segment and homeomorphic to a_i. By Theorem 5.2 there exists for each i a pseudo-isotopy $F_t^i: \bar{u}_i \to \bar{u}_i$, stationary on $\mathrm{Fr}\, u_i$ and such that

$$F_1^i(b_i) = a_i,$$

satisfying the conditions of the theorem.

Setting

$$F_t(x) = F_t^i(x) \quad \text{for} \quad x \in \bar{u}_i,$$

$$F_t(x) = x \quad \text{for} \quad x \in E^n \setminus \bigcup_i \bar{u}_i,$$

we obtain an ϵ-pseudo-isotopy taking the compactum $B = \bigcup_{i=1}^{k} b_i$ into A and satisfying the conditions of the corollary. The compactum B is tame, since it lies on finitely many line segments and is therefore cellularly separated.

The inverse image $F_1^{-1}(x)$ of any point has diameter no greater than ϵ, since each $F_1^{-1}(x)$ which is not a point is contained is some u_i. F_t is stationary on $E^n \setminus \bigcup_i u_i$, and $\bigcup_i u_i$ is an ϵ-neighborhood of A. We have $\rho[x, F_t(x)] < \epsilon$, $0 \leq t \leq 1$, since either the point x is stationary or else x and $F_t(x)$ are contained in the same u_i.

1) *Translator's note*. Each u_i is to be a domain.

CHAPTER II

PLANAR ISOTOPIES

§1. Planar polyhedra

1. To study topological imbeddings in the plane, we shall examine the polyhedral structure of the plane E^2, planar polyhedra, and semilinear mappings of the latter. We shall see in this chapter that topological imbeddings in the plane have simple properties that no longer hold for imbeddings in E^n when $n > 2$.

In §3 we show that every continuum that does not separate E^2 is cellularly imbedded in E^2; and in §4, that for every simple closed curve C (or simple arc L), there exists an isotopy of the plane onto itself taking C onto the boundary of a 2-simplex (or L onto a line segment). It follows that every simple closed curve in E^2 bounds a domain whose closure is a disk. We shall see in Chapter IV that a topological 2-sphere in E^3 may bound a domain whose closure is not a topological ball.

We examine first the polyhedral simple arcs (polygonal lines) and polyhedral simple closed curves in E^2, and then prove the general theorem by passage to the limit. We shall start with the proof of some elementary propositions concerning the polyhedral structure of the plane. The analogues of some of these remain true for E^n when n is arbitrary, but we shall consider here only the planar case.

2. **Lemma 1.1.** *A finite union of lines in E^2 separates E^2 into finitely many convex domains.*

We prove the lemma by induction on the number of lines $l_1, l_2, \cdots, l_k$. It is obviously true if $k = 1$. Assuming it holds for $i < k$, we show that it holds for k. Let $u_1, u_2, \cdots, u_N$ be the convex domains which make up $E^2 \setminus \bigcup_{i=1}^{k-1} l_i$. Suppose $u_j \cap l_k \neq \Lambda$; since the domain u_j is convex, $\overline{u}_j \cap l_k$ is a segment on the line l_k, finite or infinite. [1] We assert that $u_j \setminus l_k$ is the union of two convex components. For, let V_1 and V_2 be the components of $E^2 \setminus l_k$. It is obvious that $V_i \cap u_j \neq \Lambda$, $i = 1, 2$. V_i and u_j are convex domains, and the intersection of convex domains is again a convex domain. $E^2 \setminus \bigcup_{i=1}^{k} l_i$ is the union of all the

1) *Translator's note*: Although the distinction is not strictly maintained either in the Russian text or in this translation, it should be pointed out that, in general, "segment" and "interval," when referring to subsets of a line, mean different things: the first is closed, the second open. See also the note to Theorem 1.1 below.

$V_i \cap u_j$, $i = 1, 2$, $j = 1, 2, \cdots, N$, i.e. the union of finitely many convex domains.[1)]

We shall consider finite planar polyhedra and, in what follows, will not bother to specify the finiteness condition every time.

Recall that a finite polyhedron P of dimension n is a point set which is the union of finitely many disjoint simplices Δ^i of dimension $0, 1, \cdots, n$, i.e. of points (the vertices of P), line intervals and open simplices of dimension $\leq n$, such that the closures of any two simplices in P may intersect only in the closure of a common face (every simplex Δ^i is a face of itself).

The representation of P as a union of simplices Δ^i is called a triangulation T of the polyhedron P, and the simplices Δ^i are called the elements of T. A polyhedron P can have various triangulations. If every element of a triangulation T_2 of a polyhedron P is contained in some element of a triangulation T_1, then T_2 is called a subdivision of T_1.

Lemma 1.2. *Let P and Q be finite planar polyhedra, and T_P and T_Q corresponding triangulations. Then $P \cup Q$ is a polyhedron with a triangulation T which on P is a subdivision of T_P and on Q is a subdivision of T_Q.*

The set of all 1-simplices of P and Q lies on a finite number of lines l_1, $l_2, \cdots, l_k$. All points of intersection x_r of the lines l_i which are contained in $P \cup Q$, as well as all vertices Δ_i^0 of the triangulations T_1 and T_2, are to be vertices of T; any 1-simplex Δ_i^1 belonging to T_P or T_Q is divided by the vertices x_r and Δ_i^0 into finitely many intervals λ_s. Each such interval is to be a simplex of T. Also to be 1-simplices of T are those segments λ_s of the lines l_i between points of intersection that lie in 2-simplices of T_P and T_Q.

By Lemma 1.1 the union of the lines l_i, $i = 1, 2, \cdots, k$, divides E^2 into finitely many convex domains u_j. From the definition of the lines l_i, if $u_j \cap (P \cup Q) \neq \Lambda$ it follows that u_j is contained either in a 2-simplex of T_P, or in a 2-simplex of T_Q, or in the intersection of a 2-simplex of T_P with one of T_Q. The boundary of each domain u_j contained in $P \cup Q$ consists of simplices Δ_i^0, x_r, λ_s of the triangulation T. If $\operatorname{Fr} u_j$ contains only three vertices of T, then u_i is to be a simplex of σ_j^2 of T. If $\operatorname{Fr} u_j$ contains more than three vertices of T, pick in u_j a point y_j and join it by line segments λ'_ρ with all the vertices of T on $\operatorname{Fr} u_j$. The segments λ'_ρ divide u_j into finitely many 2-simplices σ_ρ^2. All the points Δ_i^0, x_r, y_j, the 1-simplices λ_s, λ'_ρ, and the 2-simplices σ_j^2, σ_ρ^2 are now to be the elements of the triangulation T. By construction, $P \cup Q$ is the union of these simplices; and T is a triangulation of $P \cup Q$, since the closures of two simplices either are disjoint or intersect in a common face. On the polyhedron P, T is a

1) This argument, as well as others below, can be carried over to the n-dimensional case.

subdivision of T_P, since every simplex of T that meets P is contained in some simplex of T_P. This is obvious for 0-simplices. A 1-simplex $\Delta' \in T$ meeting P lies either on a 1-simplex of T_P, or on a 1-simplex of T_Q, or in a domain u_j. In the latter two cases, Δ^1 is contained in a 2-simplex of T_P. Finally, any 2-simplex of T that meets P either coincides with a simplex of T_P, or else is contained in some $u_j \subset P$ and so in a 2-simplex of T_P. A similar proof shows that, on Q, T is a subdivision of T_Q.

Corollary 1. *The union and intersection of finitely many planar polyhedra is a polyhedron.*

Corollary 2. *Let T_1 and T_2 be two triangulations of a planar polyhedron P. Then there exists a triangulation T of P which is a subdivision of both T_1 and T_2.*

By Lemma 1.2, on the intersection $P \cap Q$ the triangulation T is a subdivision of T_P and T_Q. For $Q = P$ this yields Corollary 2.

3. A simple closed curve is called polyhedral if it is a polyhedron. A polyhedral simple closed curve C is the union of finitely many segments and their vertices, each vertex adjoining exactly two segments. Any pair of points separates a polyhedral simple closed curve into two nonclosed polygonal lines, and the union of two such polygonal lines that meet only at their endpoints is a polyhedral simple closed curve.

Remark 1.1. A subset of a polyhedral simple arc $L \subset E^2$ cannot separate E^2, since by Theorem 2.3 of Chapter I there exists an isotopy $E_t\colon E^2 \to E^2$ taking L into a line segment.

Theorem 1.1. *A planar polyhedral simple closed curve C separates the plane into two domains, interior and exterior, and is their common boundary.*

We prove the theorem by induction on the number n of segments [1] in C. If $n = 3$, the theorem is true, since C is the boundary of a 2-simplex and the complement of its closure is connected. Suppose $n > 3$ and the theorem has been proved for all integers k such that $3 \leq k < n$. We prove first that if $n > 3$, then there exist on C two vertices p and q which can be joined by a line interval μ not intersecting C. Let E^2 be provided with its usual Euclidean metric, let l be a line not intersecting C, and let K be the set of points of C nearest l. The set K is a union of finitely many segments and vertices of C, lying on a line l' parallel to l. If K consists of more than one component, we may take for μ the interval on l' joining two components of K. If K consists of a single component, it is either a point (a vertex of C) or a line segment on l'. Let $p = p_1$ be a

1) *Translator's note*: We have run the risk of some slight confusion by using the same word, "segment," where the Russian uses two different words to distinguish between a (closed) line segment in general and a "link" in a given polygonal line.

vertex of C on K which adjoins at least one segment of C/K, let p_2 be the other vertex of this segment, and let p_3 be the second vertex of the other segment adjoining p_1 (Figure 2.1).

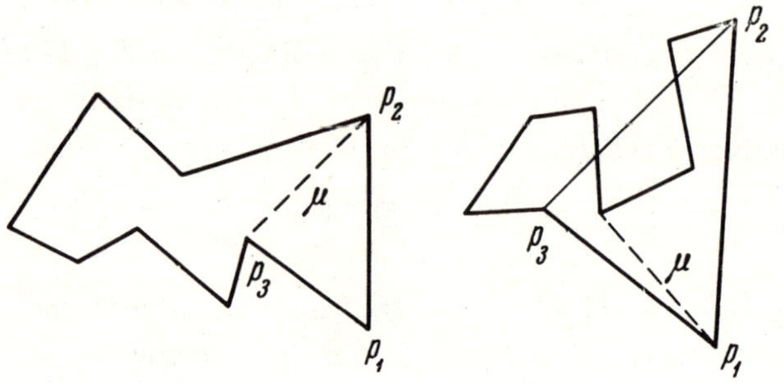

Figure 2.1a Figure 2.1b

If C has no points interior to the triangle $p_1p_2p_3$, there exists on the segment $\overline{p_2p_3}$ an interval p_ip_j joining two vertices of C and not lying on C. Set $\mu = p_ip_j$. (Possibly $p_ip_j = p_2p_3$.)

In the opposite case there exist vertices of C interior to the triangle $p_1p_2p_3$. Select from these a vertex q at maximum distance from the line containing the segment $\overline{p_2p_3}$. Then the interval qp_1 cannot meet C, since the interior of the triangle cut off from $p_1p_2p_3$ by the line through q parallel to $\overline{p_2p_3}$ contains no points of C. Hence we may set $\mu = qp_1$.

The segment $\bar{\mu}$ intersects C in the two vertices p and q, and consequently divides C into two polygonal lines, each of which consists of at least two segments: $C = C_0 \cup C_1$. Then

$$C \cap \bar{\mu} = C_0 \cap \bar{\mu} = C_1 \cap \bar{\mu} = \{p\} \cup \{q\}.$$

Hence $C_0 \cup \bar{\mu}$ and $C_1 \cup \bar{\mu}$ are polyhedral simple closed curves, each consisting of less than n segments.

By the induction assumption, each $C_i \cup \mu$, $i = 0, 1$, separates E^2 into two domains, the interior U_i and exterior V_i, and is their common boundary.

The simple arc C_i lies, except for its endpoints, either in U_{1-i} or in V_{1-i}. There are two possible cases.

1) $U_0 \cap U_1 = \Lambda$; then $\bar{U}_0 \cap \bar{U}_1 = \bar{\mu}$. For any point $x \in \mu$, there exists an open disk δ_x with center at x such that $\delta_x \subset U_0 \cup \mu \cup U_1$. Hence $U_0 \cup \mu \cup U_1 = W$ is a domain. Its boundary is contained in $C_0 \cup C_1 = C$, and since every proper subset of C is contained in a polyhedral simple arc and (by Remark 1.1) cannot separate E^2, we have $\operatorname{Fr} W = C$.

The set $E^2 \setminus \bar{W}$ is connected. Indeed, each segment of C, being a line segment, belongs to the boundary of only two components of $E^2 \setminus C$, of which one is W. If $E^2 \setminus \overline{W}$ where not connected, it would have a component bounded by only part of C, which is impossible, since such a part is contained in a polygonal line. Consequently $E^2 \setminus \overline{W}$ is connected, and C is the common boundary of W and $E^2 \setminus \overline{W}$.

2) $U_0 \cap U_1 \neq \Lambda$. Then $C_i \subset U_{1-i}$ for some i, and $C_i \cup \mu \subset \overline{U}_{1-i}$, so that $U_i \subset U_{1-i}$ (Figure 2.2). In this case C bounds the domain $W = U_{1-i} \setminus \overline{U}_i$. The exterior domain is $E^2 \setminus \overline{W} = E^2 \setminus \overline{U}_{1-i} \cup \mu \cup U_i$; that it is connected is proved in the same way as for W in case 1), and that W is connected in the same way as for $E^2 \setminus \overline{W}$ in case 1).

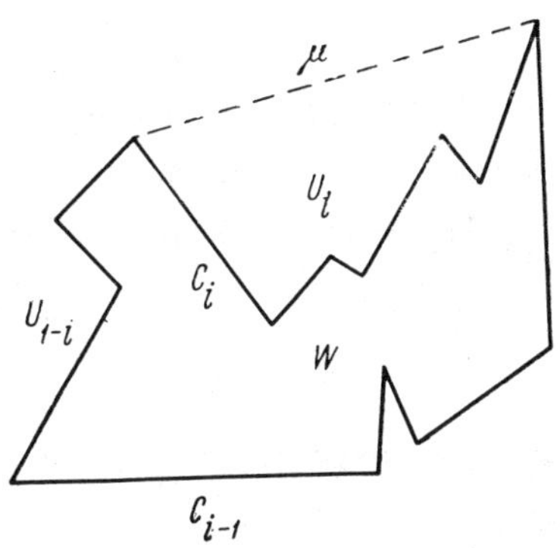

Figure 2.2

Corollary 1. *A planar polyhedral closed curve bounds a polyhedral domain.*

The union of all the lines l_i containing the various segments of C divides the interior domain U of C into finitely many convex domains (see Lemma 1.1). The triangulation of U and of $\overline{U}$ is constructed in the same way as in the proof of Lemma 1.2.

Corollary 2. *Let U be a planar domain bounded by a polyhedral simple closed curve C, and L a polygonal line contained in U except for its endpoints, which lie on C. Then L divides U into two domains, each of which is bounded by a polyhedral closed curve.*

Let p and q be the endpoints of L. The point pair $\{p\} \cup \{q\}$ lies on C and separates C into two polygonal lines:

$$C = C_1 \cup C_2, \quad C_1 \cap C_2 = C_1 \cap L = C_2 \cap L = \{p\} \cup \{q\}.$$

Hence $C_1 \cup L$ and $C_2 \cup L$ are polyhedral simple closed curves. Each bounds an interior domain U_i, $i = 1, 2$. Since $L \subset U$ and $C_i \cup L \subset \overline{U}$, we have $U_i \subset U$.

Every point $x \in L$ belongs to the boundary of exactly two components of $U \setminus L$, and at the same time, $x \in \operatorname{Fr} U_1 \cap \operatorname{Fr} U_2$. Since the boundary of each component of $U \setminus L$ intersects L, we have $U \setminus L = U_1 \cup U_2$.

§2. Semilinear isotopies of the plane

1. Let P and Q be rectilinear n-dimensional polyhedra, and T_P and T_Q corresponding triangulations, and suppose given a simplicial mapping of P into Q (i.e. one such that each simplex Δ^i of the triangulation T_P maps onto a simplex of T_Q). This mapping induces a mapping $f\colon P \to Q$ which is linear on every

individual simplex of T_P.

Definition 2.1. A continuous mapping f of a polyhedron P into a polyhedron Q is called a *semilinear mapping* of P into Q if there exist triangulations T_P and T_Q of P and Q such that f is linear on each simplex of T_P and takes the simplex into a simplex of the triangulation T_Q of Q.

Let P be a planar polyhedron, the closure of a finite union of 2-simplices, and let T_1 and T_2 be two triangulations of P which coincide as regards the simplices on the boundary of P (taken as a planar set) and for which there exists a semilinear homeomorphism $f\colon P \to P$ of P onto itself which is the identity on the points of $\operatorname{Fr} P$ and takes the triangulation T_1 into the triangulation T_2.

The homeomorphism f extends to a homeomorphism of all of E^2 onto itself, if we set

$$f(x) = x, \quad x \in E^2 \setminus P.$$

Such a homeomorphism will be called a semilinear homeomorphism of the plane E^2 onto itself.

Remark 2.1. Under a semilinear homeomorphism of E^2 onto itself, a line segment goes into a polygonal line of finitely many segments.

Indeed, the intersection of the segment l with a closed 2-simplex is either void or again a line segment. Since a linear homeomorphism takes a line segment into a line segment, the part of l that lies in any one 2-simplex of P goes into a line segment under the homeomorphism f, so that l goes into a polygonal line. In what follows we shall be making constant use of this.

Theorem 2.1. *The composite of finitely many semilinear homeomorphisms of E^2 onto itself is again a semilinear homeomorphism of E^2 onto itself.*

It suffices to prove this for two semilinear homeomorphisms, since it then follows easily by induction on the number of homeomorphisms.

Let f_1 and f_2 be two semilinear homeomorphisms of the plane onto itself, with f_1 taking a triangulation T_P of a polyhedron P into a second such triangulation T'_P, and f_2 taking a triangulation T_Q of a polyhedron Q into a triangulation T'_Q. By Lemma 1.2 there exists a triangulation τ_1 of the polyhedron $P \cup Q$ which on P is a subdivision of T'_P and on Q is a subdivision of T_Q.

Since f_1 is a linear homeomorphism on the simplices of T_P, f_1^{-1} is a linear homeomorphism on the simplices of T'_P, and so also on the simplices of τ_1 which are contained in P. On the simplices of τ_1 contained in $Q \setminus P$, the mapping f_1 is the identity. Hence for each simplex $\sigma_r \in \tau_1$ the inverse image $f_1^{-1}(\sigma_r)$ is a simplex contained either in P or in $Q \setminus P$. If $\sigma_r \subset P$, then $f_1^{-1}(\sigma_r)$ is contained in a simplex of the triangulation T_P, while if $\sigma_r \subset Q \setminus P$, then $f_1^{-1}(\sigma_r) = \sigma_r$ and σ_r

is contained in a simplex of the triangulation T_Q.

Let τ_2 be the set of simplices $\sigma_r' = f_1^{-1}(\sigma_r)$, $\sigma_r \in \tau_1$; then τ_2 is a triangulation of the polyhedron $P \cup Q$ which on P is a subdivision of T_P, on $Q \setminus P$ is a subdivision of T_Q, and on $P \cap \overline{(Q \setminus P)}$ is a subdivision of both T_P and T_Q (since here f_1^{-1} is stationary).

Thus f_1 is a semilinear mapping of E^2 which takes the traingulation τ_2 of the polyhedron $P \cup Q$ into the triangulation τ_1. On the simplices $\sigma_r \subset Q$ of the triangulation τ_1 the homeomorphism f_2 is linear, since τ_1 is a subdivision of T_Q. If, on the other hand, $\sigma_r \subset P \setminus Q$, , then f_2 is the identity on σ_r. The homeomorphism f_1 is linear on each $\sigma_r' = f_1^{-1}(\sigma_r)$, and the homeomorphism f_2 is linear on each σ_r. Hence $f_2 f_1$ is a linear homeomorphism on each simplex σ_r' of the triangulation τ_2 of the polyhedron $P \cup Q$, while on $E^2 \setminus (P \cup Q)$ the mapping $f_2 f_1$ is the identity. Consequently, $f_2 f_1$ is a semilinear homeomorphism of E^2 onto itself.

2. Definition 2.2. An isotopy F_t of a polyhedron P onto a polyhedron Q (finite or infinite) is called *semilinear* if the mapping F_t: $P \to Q$ is a semilinear homeomorphism for each t, $0 \leq t \leq 1$.

In the definition of a semilinear isotopy taking a semilinear homeomorphism f: $P \to Q$ into a semilinear homeomorphism Φ: $P \to Q$, the requirement is often added that the homeomorphism Φ: $P \times I^1 \to Q \times I^1$ given by $\Phi(t, x) = (t, F_t(x))$, $x \in P$, also be semilinear. But this requirement would very much complicate the presentation in our case, and is for our purposes unnecessary.

We therefore omit the additional condition in defining a semilinear isotopy for planar polyhedra. Our definition of semilinear isotopy occurs in the literature; in particular, it is used by Sanderson [66]. The question is studied in detail in [78].

We remark that in the cases we consider the continuous mapping $F(x, t)$ will be "bi-semilinear", i.e. semilinear in each variable separately.

The product of finitely many semilinear isotopies is obviously again a semilinear isotopy.

Corollary. *Let $A_1, A_2, \cdots, A_k$ be a finite number of planar sets, such that for each $i < k$ there exists a semilinear isotopy F_t^i: $E^2 \to E^2$ from the identity mapping, taking A_i into A_{i+1}. Then there exists a semilinear isotopy F_t: $E^2 \to E^2$ from the identity mapping which takes A_1 into A_k.*

By Theorem 2.1 of Chapter I, there exists an isotopy F_t: $E^2 \to E^2$ taking A_1 into A_k, which is the product of the isotopies $F_t^i F_1^{i-1} \cdots F_1^1$, $i = 1, 2, \cdots \cdots, k-1$. Since each homeomorphism F_t^i is semilinear for any t, it follows by

Theorem 2.1 that the isotopy F_t is semilinear.

We note that if each isotopy F_t^i is stationary outside a polyhedron P_i, then F is stationary outside $\bigcup_{i=1}^k P_i$.

Examples of semilinear isotopies of E^2

Example 2.1. There exists a semilinear isotopy of the plane taking a point A into a point B and stationary outside an arbitrary domain U containing A and B.

Join A and B by a polygonal line $L \subset U$.

By the Corollary, it suffices to prove that an arbitrary vertex a_i of L can be taken into a_{i+1} by a semilinear isotopy F_t^i from the identity mapping, stationary outside U. On the line containing the segment l_i with vertices a_i and a_{i+1}, pick a point x to the left of a_i and a point y to the right of a_{i+1} such that $\overline{xy} \subset U$, and construct three triangles in U whose common base γ contains the point x and whose vertices are a_i, a_{i+1} and y respectively. A line parallel to l_i and intersecting γ intersects the sides of these triangles at points $a \in \gamma$, b, c, d. The isotopy defined by equations (8) of Chapter I is semilinear, is stationary outside U, and takes the point a_i into the point a_{i+1}.

Example 2.2. For any two polyhedral simple arcs L_1 and L_2 in E^2 and any domain U containing $L_1 \bigcup L_2$, there exists a semilinear isotopy $F_t\colon E^2 \to E^2$ taking L_1 into L_2 and stationary outside U. If L_1 and L_2 have a common endpoint x_0, the isotopy F_t can be constructed so as to be stationary on a given polygonal line l, with endpoint x_0, intersecting $L_1 \bigcup L_2$ in the single point x_0.

The first half of this assertion follows from the fact that the isotopy in each step of the proof of Theorem 2.3 of Chapter I is semilinear. Now let l be a polygonal line, with $l \bigcap (L_1 \bigcup L_2) = \{x_0\}$. Take a disk δ with center x_0 such that δ intersects each of the polygonal lines l, L_1, L_2 only at points of a single segment λ_i, $i = 0, 1, 2$. For each $L_i \setminus \{x_0\}$, take a neighborhood $U_i \subset U$ such that $U_i \bigcap l = \Lambda$. By Remark 2.1 of Chapter I we can take L_i into λ_i via a semilinear isotopy F_t^i, stationary outside U_i.

If the angle ϕ between λ_1 and λ_2 whose interior does *not* contain λ_0 is less than π, then by Lemma 2.1 of Chapter I we may construct a semilinear isotopy F_t^0 taking λ_1 into λ_2, which is stationary outside a small triangle Δ containing $\lambda_1 \bigcup \lambda_2$ with vertex x_0 and such that $\lambda_0 \bigcap \Delta = \{x_0\}$. If $\phi \geq \pi$ let μ be a segment with vertex x_0 dividing ϕ into two angles less than π, and construct the isotopy as a product of two semilinear isotopies, the first taking λ_1 into μ and the second taking μ into λ_2. The isotopy $F_t^1 \circ F_t^0 \circ (F_t^2)^{-1}$ now takes L_1 into L_2 and is stationary on l.

3. Semilinear isotopies of E^2 deforming a polyhedral closed curve.

Definition 2.3. By a *disk* is meant a homeomorphic image of the square I^2; by an *open disk*, a homeomorphic image of $\operatorname{Int} I^2$. (An open disk is homeomorphic to E^2.) 1)

We show now that the closure of the planar domain bounded by a polyhedral simple closed curve is a polyhedral disk and, indeed, is semilinearly isotopic to a 2-simplex.

Lemma 2.1. *Let P be a closed planar polyhedron bounded by a simple closed curve and containing at least two 2-simplices. Then P contains at least two 2-simplices Δ_i^2 satisfying one of the following two conditions*:

1) *one face of Δ_i^2 belongs to the boundary of P, while the other two and the opposite vertex are interior to P*;

2) *Two faces of Δ_i^2 belong to the boundary of P, while the third is interior to P.*

Every 1-simplex (p_0p_1) on $\operatorname{Fr} P$ is incident to a single 2-simplex $\Delta^2 \subset P$, where Δ^2 satisfies either 1) or 2) or

3) all the vertices of Δ^2 belong to $\operatorname{Fr} P$, while the faces (p_0p_2) and (p_2p_1) are interior to P.

We prove the lemma by induction on the number n of 2-simplices contained in P.

If $n = 2$, the lemma is obviously true, since each simplex satisfies 2). Suppose $n > 2$, and that the lemma is proved for any polyhedron containing $k < n$ 2-simplices.

If every simplex with a face on $\operatorname{Fr} P$ satisfies 1) or 2), the lemma is true. Suppose, then, that $(p_0p_1p_2)$ is a 2-simplex of P which satisfies 3) (Figure 2.3):

$$\overline{(p_0p_1)} \subset \operatorname{Fr} P, \ \{p_2\} \in \operatorname{Fr} P, \quad (p_0p_2) \cup (p_2p_1) \subset \operatorname{Int} P .$$

((pq) means an open 1-simplex, (pqr) a 2-simplex, $\overline{pq}$ and $\overline{pqr}$ their closures). We show that the closed simplex $\overline{p_0p_1p_2}$ disconnects the polyhedron P. It disconnects $\operatorname{Fr} P$ into two polygonal lines p_0xp_2 and p_2yp_1, whose closures intersect in the point p_2.

By Corollary 2 to Theorem 1.1, the segment $\overline{p_0p_2}$ divides P into two domains: U_0, bounded by the polyhedral closed curve $\overline{p_0xp_2} \cup \overline{p_2p_0}$, and U_1', containing the simplex $(p_0p_1p_2)$; and the segment $\overline{p_1p_2}$ in turn divides U_1' into the simplex $(p_0p_1p_2)$ and a domain U_1 bounded by $\overline{p_1p_2} \cup \overline{p_2yp_1}$. The closures P_i of the domains U_i, $i = 0, 1$, are

1) *Translator's note*: Up to now, "disk" has meant "circular disk." The Russian uses different words to distinguish between a circular disk and the homeomorphic image of one. We shall make no linguistic distinction when the meaning is clear.

polyhedra, each containing $k_i < n$ 2-simplices. If $k_i = 1$, then U_i is a 2-simplex in P which satisfies 2), since $p_i p_2 \subset \operatorname{Int} P$ and the remaining faces of U_i lie on $\operatorname{Fr} P$. If $k_i > 1$ for some i, then by the induction assumption P_i contains two 2-simplices each of which has a face on $\operatorname{Fr} P_i$ and satisfies either 1) or 2). One of them may have the face $(p_i p_2)$, but for the other, say Δ_i^2, the intersection $\overline{\Delta}_i^2 \cap \operatorname{Fr} P_i$ belongs to either $p_0 x p_2$ or $p_2 y p_1$, i.e. lies on $\operatorname{Fr} P$; thus Δ_i^2, $i = 0, 1$, satisfies either condition 1) or 2) with respect to P. The pair of simplices Δ_0^2 and Δ_1^2 then satisfies the conditions of the lemma.

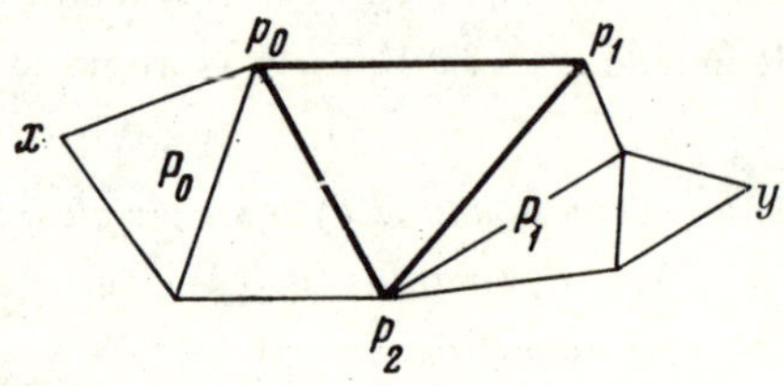

Figure 2.3

Theorem 2.2. *Let P be a polyhedral planar domain bounded by a closed polyhedral curve C, and let U be an arbitrary neighborhood of P. Then there exists a semilinear isotopy F_t: $E^2 \to E^2$ from the identity mapping, taking P into a 2-simplex and C into its boundary, which is stationary outside U.*

We prove the theorem by induction on the number n of simplices. For $n = 1$ the theorem is trivial, with $F_t = 1$ for all t.

Suppose $n > 1$, and the theorem proved for all $k < n$. By Lemma 2.1 there is a 2-simplex $(p_0 p_1 p_2)$ with $\overline{p_0 p_1} \subset \operatorname{Fr} P$, satisfying either condition 1) or condition 2).

We shall construct a semilinear isotopy F_t^1: $E^2 \to E^2$ which is stationary outside a small neighborhood of $\overline{p_0 p_1 p_2}$ contained in U and takes P into the polyhedron $P' = P \setminus \overline{p_0 p_1 p_2}$.

The polyhedron P' contains $n - 1$ 2-simplices (all those of P except for $(p_0 p_1 p_2)$) and is bounded by the polyhedral simple closed curve $F_1^1(C)$. By the induction assumption there exists a semilinear isotopy F_t^2: $E^2 \to E^2$ from the identity mapping which is stationary on $E^2 \setminus U$ and takes P' into a 2-simplex Δ^2 and $F_1^1(C)$ into its boundary. By the corollary to Theorem 2.1 the isotopy $F_t = F_t^1 \circ F_t^2$ is semilinear, is stationary on $E^2 \setminus U$, and takes P into Δ^2 and C into $\operatorname{Fr} \Delta^2$.

It remains to construct the isotopy F_t^1. Set $\overline{P} = \bigcup_{i=1}^n \overline{\Delta}_i^2$, where the Δ_i^2 are the 2-simplices of P.

Consider the case that $\overline{p_0 p_1 p_2}$ satisfies condition 1). Then $(p_0 p_2) \cup \{p_2\} \cup (p_2 p_1) \subset P \subset \operatorname{Int} \overline{P}$. Hence we may pick a point q in the star [1)] of the point p_2 in the polyhedron P such that the line $q p_2$ intersects $p_0 p_1$ in a point O, $\overline{p_0 q p_1} \supset$

1) The star $\operatorname{St}_P \Delta^i$ of a simplex Δ^i in a polyhedron P is the union of all simplices of P incident to Δ^i, i.e. all those of which Δ^i is a face.

$\overline{p_0p_2p_1}$ and $\overline{p_0qp_1}\setminus\overline{p_0p_1} \subset P$. In $U\setminus\overline{P}$, pick a point q' on the line qp_2 such that

$$\overline{p_0q'p_1}\setminus\overline{p_0p_1}\subset U\setminus\overline{P}.$$

Then $Q = \overline{p_0qp_1} \cup p_0q'p_1$ is a quadrilateral, and $\overline{p_0p_1p_2} \subset Q \subset U$ (Figure 2.4). Every line parallel to qp_2 either fails to meet Q at all, or meets Q in a single point (p_0 or p_1), or intersects $\operatorname{Fr} Q \cup \operatorname{Fr}(p_0p_1p_2)$ in four points:

$$a\in p_iq',\ b\in p_ip_{1-i},\ c\in p_ip_2,$$

$$d\in p_iq,\quad i=0,1.$$

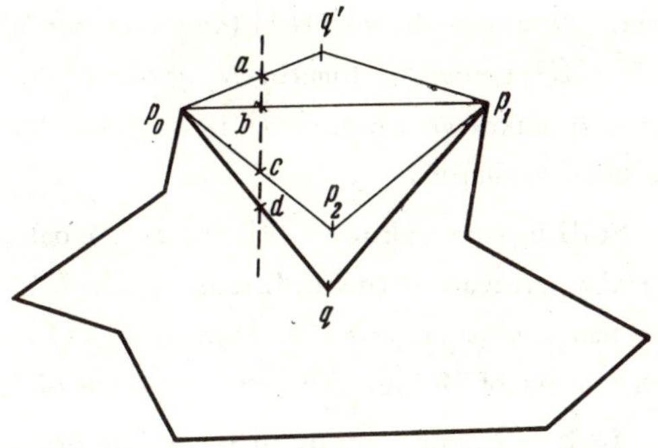

Figure 2.4

Define the isotopy F_t^1 on Q by equations (8) of Chapter I. Then $F_t^1(x) = x$ for $x \in \operatorname{Fr} Q$; hence the isotopy F_t^1 can be defined on the whole plane by putting

$$F_t^1(x) = x,\quad x\in E^2\setminus Q,\ 0\leqslant t\leqslant 1.$$

The isotopy F_t^1 is semilinear and is stationary on $E^2\setminus Q$, and therefore on $E^2\setminus U$, and

$$F_1^1(\overline{p_0p_1}) = \overline{p_0p_2}\cup\overline{p_2p_1},\ F_1^1(P) = P\setminus\overline{p_0p_1p_2},$$

$$F_1^1(C) = C\setminus\overline{p_0p_1}\cup\overline{p_0p_2}\cup\overline{p_2p_1}.$$

The isotopy F_t^1 is constructed in a similar fashion in case 2). We thus construct, finally, an isotopy $F_t = F_t^1\circ F_t^2\colon E^2 \to E^2$ such that

$$F_0\equiv 1,\quad F_1(\overline{P}) = \overline{\Delta}^2,\quad F_1(\operatorname{Fr} P) = \operatorname{Fr}\Delta^2.$$

Corollary 1. *The closure of the domain bounded by a polyhedral closed curve is a polyhedral disk, since F_1 is a homeomorphism and $\overline{\Delta}^2$ is homeomorphic to I^2.*

Corollary 2. *The union of two polyhedral disks D_1 and D_2 which meet in a polygonal line on the intersection of their boundaries is a polyhedral disk.*

We have $D_1 \cap D_2 = \operatorname{Fr} D_1 \cap \operatorname{Fr} D_2 = L$. Hence $D_1 \cup D_2$ is the closed domain bounded by the polyhedral closed curve $\overline{\operatorname{Fr} D_1 \cup \operatorname{Fr} D_2 \setminus L}$, and is consequently itself a disk.

Up to the present time it is unknown whether, for $n > 3$, a polyhedral $(n-1)$-sphere Σ^{n-1} in E^n bounds a topological ball. For $n = 3$ this was proved by Alexander in 1924 [17]. Sanderson [66, 67] has proved the existence of a semilinear isotopy $F_t\colon E^3 \to E^3$ taking a polyhedral 2-sphere Σ^2 into the boundary of a 3-simplex. For $n > 3$ the question in its general form is unanswered, i.e. on the assumption that Σ^{n-1} is a polyhedron homeomorphic to the $(n-1)$-sphere. In 1960 M. H. A. Newman studied the case that Σ^{n-1} is a combinatorial sphere, i.e. that Σ^{n-1} is a semilinear image of some triangulation of the boundary of an n-simplex; it is unknown, however, whether every polyhedral sphere is combinatorial. Newman showed [61] (see Chapter V) that a combinatorial sphere $\Sigma^{n-1} \subset E^n$ bounds a domain D whose closure $\overline{D}$ is a topological ball (an n-cell). But it is unknown whether $\overline{D}$ is a *combinatorial ball*, i.e. the semilinear image of a closed n-simplex.

Stallings has shown [72], for $n \geq 5$ only, that the interior domain D is combinatorially equivalent to Euclidean space E^n, i.e. that D is a semilinear homeomorphic image of some triangulation of E^n. For $n = 3$ the corresponding fact follows from results of Moise. Thus the question of its validity remains open only for $n = 4$.

4. Semilinear isotopy of the plane deforming a polygonal line and stationary at the endpoints.

Theorem 2.3. *Let $L \subset E^2$ be a polygonal line, and l the line segment joining the endpoints a and b of L. Then there exists a semilinear isotopy F_t: $E^2 \to E^2$ taking L into l and stationary at the points a and b and outside a given disk D containing $L \cup l \setminus \{a\} \setminus \{b\}$ in its interior.*

The intersection $L \cap l$ consists of finitely many points and segments of l. Now, it is easy to produce an isotopy F_t: $E^2 \to E^2$ which is stationary at a and b and outside an arbitrary given neighborhood of $L \cap l$ and takes L into a polygonal line L' such that $L' \cap l$ consists of finitely many points. To do so we need only transform every maximal segment $\overline{\alpha_i \beta_i} \subset L \cap l$ into a two-segment polygonal line via an isotopy ϕ_t^i which is stationary outside a small quadrilateral $q_i \supset \alpha_i \beta_i$, $q_i \subset D$; ϕ_t^i is constructed in the same way as F_t^1 in Theorem 2.2, and the quadrilaterals q_i are chosen so that $q_i \cap q_j = \Lambda$, $i \neq j$, and $q_i \cap L = \overline{\alpha_i \beta_i}$.

We may therefore suppose that $L \cap l$ consists of finitely many points only, and prove the theorem by induction on the number of points.

1) $L \cap l = \{a\} \cup \{b\}$. Then $L \cup l$ is a simple closed curve bounding a polyhedral domain P.

We prove the theorem in case 1) by induction on the number n of segments of L. If $n = 2$, then $L \cup l$ is the boundary of a simplex, and the isotopy may be constructed in the same way as F_t^1 in Theorem 2.2.

Suppose L contains $n > 2$ segments. We show first that there exists a line segment μ containing two vertices of L and lying, except for its endpoints, in P.

Let c be a vertex of L at maximum distance from the line containing l, such that at least one of the adjoining segments of L is not parallel to l, and let $\overline{dc}$ and $\overline{ce}$ be these segments of L. Join the points d and e by an interval;

if $(de) \cap L = \Lambda$, set $\overline{de} = \mu$. In the opposite case, triangle (dce) contains a vertex of L. Pick such a vertex c' at maximum distance from the line containing $\overline{de}$. Then $(\overline{cc'}) \cap L = \Lambda$, and we may set $\mu = cc'$ (Figure 2.5).

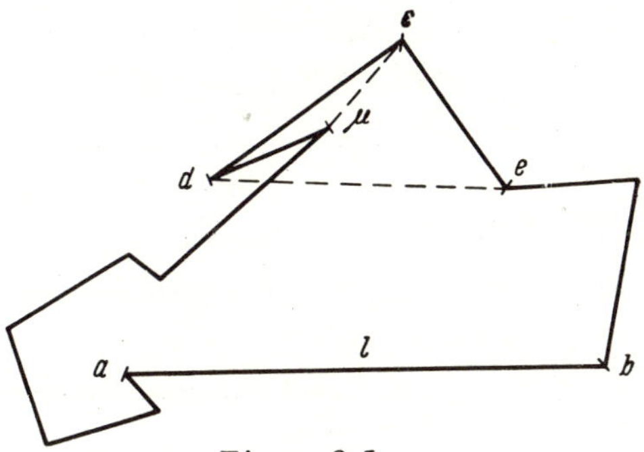

Figure 2.5

We assert that $\mu \subset P$. Indeed, the line through c parallel to l cannot meet P; hence the part of the triangle (dce) adjoining the vertex c must be contained in P; and hence $\mu \subset P$.

The interval μ divides P into two domains:

$$P \setminus \mu = P_1 \cup P_2,$$

where $\operatorname{Fr} P_2 = L_{\alpha\beta} \cup \mu$, $L_{\alpha\beta}$ being the portion of L between the endpoints α and β of μ (one of which can also be an endpoint of l), and $\operatorname{Fr} P_1 = L' \cup l$, where $L' = (L \setminus L_{\alpha\beta}) \cup \bar{\mu}$. We can now find disks D_i such that

$$\bar{P}_1 \setminus \{a\} \setminus \{b\} \subset \operatorname{Int} D_1 \subset D,$$

$$\bar{P}_2 \setminus \{\alpha\} \setminus \{\beta\} \subset \operatorname{Int} D_2 \subset D \text{ and } D_2 \cap ((L \setminus L_{\alpha\beta}) \cup l) = \{\alpha\} \cup \{\beta\}.$$

Since the polygonal lines $L_{\alpha\beta}$ and L' contains less than n segments, there exist two semilinear isotopies $F_t^i\colon E^2 \to E^2$ from the identity mapping, such that F_t^i is stationary on $E^2 \setminus D_i$ and $F_1^1(L') = l$; $F_1^2(L_{\alpha\beta}) = \mu$. The isotopy $\Phi_t = F_t^2 \circ F_t^1$ is then stationary on $E^2 \setminus D$, and

$$\Phi_1(L) = l.$$

2) Suppose the theorem proved when $L \cap l$ contains $k < n$ points. We now prove it holds for n. To begin with, if there is a point $c \in L \cap l$ such that the part of L contained in a small circular disk δ with center at c lies on one side of l, pick points c' and c'' on $L \cap \delta$ on opposite sides of c. As in Theorem 2.2, we may then construct an isotopy ϕ_t, stationary outside an arbitrary quadrilateral q containing the triangle $c'cc''$ and such that $q \cap L = c'c \cup cc''$, and taking L into the polygonal line $L' = L \setminus (c'c \cup cc'') \cup c'c''$. Then L' intersects l in $n-1$ points.

Now suppose L crosses l at every point of $L \cap l$, i.e. that in the small disk δ it passes from one side of l to the other, and let c be the first point, travelling away from a on l, of $L \cap l$.

Then $L_{ac} \cap \overline{ac} = \{a\} \cup \{c\}$, and $L_{ac} \cup \overline{ca}$ is a polyhedral closed curve. Since the portion L_{cb} of L intersects $\overline{ac}$ only in the point c, we have that $L_{cb} \setminus \{c\}$ is contained in one of the two domains in which $L_{ac} \cup \overline{ca}$ divides E^2.

There are two possible cases.

a) $L_{cb}\setminus\{c\}\subset U$, where U is the exterior domain of $L_{ac}\cup\overline{ca}$. Take a point c' on the segment in L_{cb} adjoining the point c such that $\overline{ac'}\cap L=\{a\}\cup\{c'\}$ (Figure 2.6). This is possible because $\overline{ac}\cap L=\{a\}\cup\{c\}$. Then for $\overline{ac'}$ and $L_{ac}\cup\overline{cc'}$, 1) is satisfied, so that there exists a semilinear isotopy $F_t^1\colon E^2\to E^2$ taking $L_{ac}\cup cc'$ and stationary at the points a and c' and outside a neighborhood $O\subset D$ of

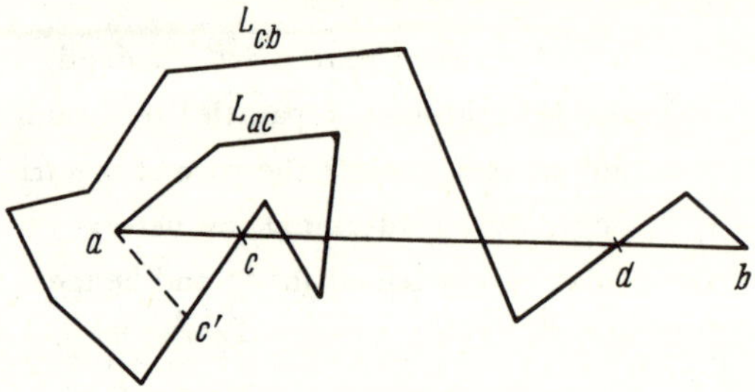

Figure 2.6

$\overline{V}'\setminus(\{a\}\cup\{c'\})$, where V' is the interior domain of $L_{ac'}\cup\overline{c'a}$. The neighborhood O can be so chosen that $O\cap L_{c'b}=\Lambda$; then

$$F_1^1(L)=\overline{ac'}\cup L_{c'b}.$$

The polygonal line $\overline{ac'}\cup L_{c'b}$ now intersects l in at most $(n-1)$ points, since $\overline{ac'}\cap l=\{a\}$.

b) $L_{cb}\setminus\{c\}\subset V$, where V is the interior domain of $L_{ac}\cup\overline{ca}$ (Figure 2.7).

Let d be the first point, travelling away from b on l, of the intersection $L\cap l$ (d may coincide with c). Let V_b be the interior domain of $L_{bd}\cup\overline{db}$. We assert that $V_b\subset V$.

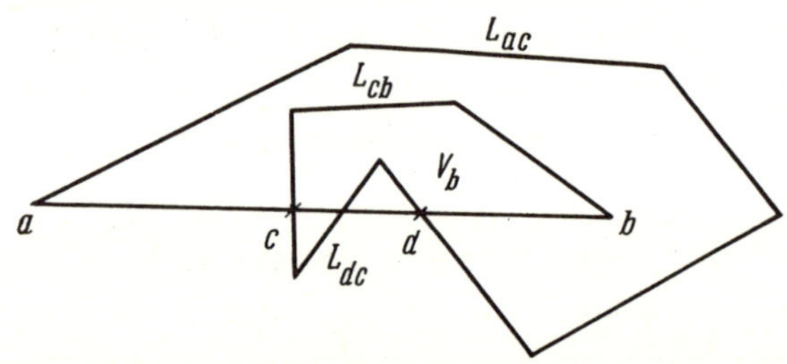

Figure 2.7

To see this, observe that $b\in V$ and $db\cap L=\{b\}\cup\{d\}$, so that

$$\overline{db}\setminus\{d\}\subset V.$$

If $L_{db}\subset L_{cb}$, then by hypothesis $\operatorname{Fr}V_b=\overline{bd}\cup L_{db}\subset V$, and therefore $V_b\subset V$. Suppose, on the other hand, that

$$L_{db}=L_{dc}\cup L_{cb}\text{ and } L_{dc}\neq\Lambda,\quad L_{dc}\not\subset L_{cb}.$$

Then $L_{dc}\subset L_{ac}\subset\operatorname{Fr}V$. Hence

$$\operatorname{Fr}V_b=\overline{bd}\cup L_{dc}\cup L_{cb},$$

with $\overline{bd}\cup L_{cb}\subset V$ and $L_{dc}\subset\operatorname{Fr}V$. Therefore we again get $V_b\subset V$.

Since $a \in \operatorname{Fr} V$ and $L_{ad} \cap \operatorname{Fr} V_b = \{d\}$, the fact that $V_b \subset V$ implies that $L_{ad} \setminus \{d\} \subset U_b = E^2 \setminus \overline{V}_b$, i.e. that with respect to $L_{bd} \cup \overline{db}$ we are in the same position as in case a). This means [1] that there exists a semilinear isotopy F_t^1: $E^2 \to E^2$, stationary outside $O \subset D$, where O is a neighborhood of $\overline{V}_b \setminus \{b\} \setminus \{d\}$, as well as at the point b and on L_{ad}, and taking $L_{bd'}$ into $\overline{bd'}$, where $d' \subset L_{ad}$ and the point d' is chosen in the same way as the point c in case a), such that

$$F_1^1(L) = L_{ad'} \cup \overline{d'b}$$

and $l \cap F_1^1(L)$ contains at most $n-1$ points.

Thus in both the cases a) and b) $F_1^1(L) \cap l$ contains at most $n-1$ points. By the induction assumption there exists a semilinear isotopy F_t^2: $E^2 \to E^2$ taking $F_1^1(L)$ into l and satisfying the conditions of the theorem. Then $F_t^1 \circ F_t^2$ is an isotopy taking L into l and likewise satisfying the conditions of the theorem.

Corollary. *Let L_1 and L_2 be two polygonal lines with common endpoints a and b. Then there exists a semilinear isotopy F_t: $E^2 \to E^2$ taking L_1 into L_2 and stationary at the points a and b.*

The isotopy F_t is the product of the isotopy F_t^1 taking L_1 into the segment $\overline{ab}$ and the isotopy $(F_t^2)^{-1}$ taking $\overline{ab}$ into L_2.

§3. Imbedding of compacta in the plane

1. We now study polyhedral neighborhoods of compacta in E^2. In contrast to the spaces E^n of dimension $n > 2$, in the plane every compactum has a polyhederal neighborhood of rather simple structure. Observer that every compactum $K \subset E^n$, $n = 1, 2, \cdots, \infty$, has an arbitrarily close polyhedral neighborhood. Indeed, let U be any neighborhood of K, and I^n an n-dimensional cube such that $K \subset I^n$. Take a triangulation T of I^n so fine that the closure of the star of every simplex $\sigma \in T$ intersecting K is contained in U:

$$\overline{\operatorname{St}}_T\, \sigma \subset U \quad \text{whenever } \sigma \cap K \neq \Lambda.$$

Then $P = \bigcup_{\sigma \cap K \neq \Lambda} \operatorname{St}_T \sigma$ is a neighborhood of K, and $P \subset U$.

Definition 3.1. A point p of a connected set M is called a *separating point* if $M \setminus \{p\}$ is disconnected. The point $p \in M$ is called a *locally separating point* if it is a separating point for every sufficiently small neighborhood of p in M.

Lemma 3.1. *The boundary of a compact planar polyhedron P which has no isolated or locally separating points consists of finitely many disjoint polyhedral*

---

1) *Translator's note.* There seems to be a small misstatement here: V_b is not the analogue of V'.

closed curves.

Let I^2 be a square containing the polyhedron P; let T_1 be a triangulation of I^2 and T_2 a triangulation of P. By Lemma 1.2 there exists a triangulation T of I^2 which is a subdivision of T_1 and on P is a subdivision of both T_1 and T_2. $\operatorname{Fr} P$ is a one-dimensional complex:

$$\operatorname{Fr} P = \bigcup_{i=1}^{k} \Delta_i^1 \cup \bigcup_{j=1}^{r} \Delta_j^0.$$

It follows from the hypothesis of the lemma that P contains no 0- or 1-simplices not incident to 2-simplices, since P would otherwise contain locally separating or isolated points.

We assert that every vertex $\Delta^0 \subset \operatorname{Fr} P$ is incident to exactly two 1-simplices of $\operatorname{Fr} P$:

$$\Delta^0 = \overline{\Delta}_i^1 \cap \overline{\Delta}_j^1, \quad \Delta_i^1 \cup \Delta_j^1 \subset \operatorname{Fr} P.$$

In the first place, Δ^0 cannot be incident to only one simplex Δ', since then $\overline{\Delta}'$ would not separate a sufficiently small disk with center at Δ^0, contradicting the assumption that $\overline{\Delta}' \subset \operatorname{Fr} P$.

Suppose Δ^0 is the vertex of k 1-simplices of the boundary of P, where $k \geq 3$:

$$\Delta_1^1, \ \Delta_2^1, \ \ldots, \ \Delta_k^1,$$

the simplices Δ_i^1 being numbered in circular order about the point Δ^0. Since $\Delta_i^1 \subset \operatorname{Fr} P$, each Δ_i^1 is incident to a 2-simplex $\Delta_p^2 \subset P$ and a 2-simplex $\Delta_q^2 \subset I^2 \setminus P$. Consider a disk δ with center Δ^0 and of sufficiently small radius such that $\Delta_i^1 \cap \operatorname{Fr} \delta \neq \Lambda$ for each i, $1 \leq i \leq k$. The simplices Δ_i^1 separate δ into k parts, each of which belongs either entirely to P or entirely to $\delta \setminus P$, and of any two neighbors only one belongs to P.

Let K_1 be the part of δ between Δ_1^1 and Δ_2^1, and suppose $K_1 \subset P$. Then in $I^2 \setminus P$ there is a simplex Δ_1^2 incident to Δ_1^1, and a simplex Δ_2^2 incident to Δ_2^1. But there is also in $\operatorname{Fr} P$ the simplex Δ_3^1, incident to a 2-simplex $\Delta_3^2 \subset P$, so that $\delta \cap P \setminus K_1 \neq \Lambda$. It is easily seen that

$$\overline{K}_1 \cap (\overline{\delta \cap P \setminus K_1}) = \Delta^0,$$

since K_1 and $K_2 = \delta \cap P \setminus K_1$ are separated from each other by the simplices Δ_1^2 and Δ_2^2 in $I^2 \setminus P$. Hence $K_1 \cup K_2 \setminus \Delta^0$ is disconnected and the vertex Δ^0 separates $P \cap \delta$, i.e. Δ^0 is a locally separating point in P, contradicting the hypothesis. It follows that Δ^0 must be the vertex of only two 1-simplices of the boundary $\operatorname{Fr} P$.

If we now run through the vertices of $\operatorname{Fr} P$, starting from one of them, in the other in which they are joined by the 1-simplices of $\operatorname{Fr} P$, we come back to the original vertex and so obtain a simple closed curve, which, however, need not contain all the vertices of $\operatorname{Fr} P$. Since P contains only finitely many vertices, by iterating the process we obtain finitely many simple closed curves constituting $\operatorname{Fr} P$.

Corollary. *A compact connected planar polyhedron without locally separating points is a disk from which have been removed finitely many open disks with disjoint closures.*

Since $\operatorname{Fr} P$ is the union of finitely many polyhedral closed curves $c_0, c_1, \cdots$ $\cdots, c_k$, and P is a finite connected polyhedron, P is contained in one of the disks d_i bounded by these curves, say in d_0, and $d_i \subset \operatorname{Int} d_0$, $i > 0$. Hence $d_0 \setminus \bigcup_{i=1}^k \operatorname{Int} d_i \subset P$.

No disk d_i can be contained in another d_j, $j \neq 0$, since P would otherwise not be connected. Hence

$$d_i \cap d_j = \Lambda, \quad i \neq 0 \neq j; \quad P = d_0 \setminus \bigcup_{i=1}^k \operatorname{Int} d_i.$$

Lemma 3.2. *Every compactum $K \subset P^2$ has an arbitrarily close polyhedral neighborhood whose closure has no locally separating points.*

Let I^2 be a square containing K in its interior, U an arbitrary neighborhood of K, and T a triangulation of I^2 so fine that

$$\overline{\operatorname{St}_T \sigma} \subset U \quad \text{whenever } \sigma \in T, \ \sigma \cap K \neq \Lambda.$$

Then $V = \bigcup_{\sigma \cap K \neq \Lambda} \operatorname{St}_T \sigma$ is a polyhedral neighborhood of K, with $K \subset V \subset \overline{V} \subset U$. But $\overline{V}$ may have a finite number of locally separating vertices $\Delta_1^0, \Delta_2^0, \cdots, \Delta_r^0$. Take a subdivision T' of T so fine that

$$\begin{gathered} \overline{\operatorname{St}_{T'} \Delta_i^0} \subset U, \quad (\overline{\operatorname{St}_{T'} \Delta_i^0}) \cap (\overline{\operatorname{St}_{T'} \Delta_j^0}) = \Lambda, \quad i \neq j, \\ \overline{V} \cap \overline{\operatorname{St}_{T'} \Delta_i^0} \subset \operatorname{St}_T^{\overline{V}} \Delta_i^0, \end{gathered} \tag{1}$$

where $\operatorname{St}_T^{\overline{V}} \Delta_i^0$ is the star of vertex Δ_i^0 in the polyhedron $\overline{V}$ with respect to the triangulation T. If we now set

$$W = V \cup \bigcup_{i=1}^r \operatorname{St}_{T'} \Delta_i^0,$$

we obtain a new polyhedral neighborhood of K, with $\overline{W} \subset U$, Because of (1), $\overline{W}$ has no locally separating points. Indeed, any such points can occur only on the boundary of some $\operatorname{St}_{T'} \Delta_i^0$. But by (1), $(\overline{W} \setminus \operatorname{St}_{T'} \Delta_i^0) \cap \overline{\operatorname{St}_{T'} \Delta_i^0}$ contains, with any vertex Δ^0, a 1-simplex of the traingulation T' contained in a 2-simplex of T belonging to $\operatorname{St}_T^{\overline{V}} \Delta_i^0$; that is to say, Δ^0 is not a locally separating point for $\overline{W}$.

Corollary 1. *Every planar continuum has an arbitrarily close neighborhood which is an open polyhedral disk* Int d_0 *from which have been removed finitely many closed disks* $d_1, d_2, \cdots, d_k$, *where*

$$d_i \subset \operatorname{Int} d_0, \quad i > 0; \quad d_i \cap d_j = \Lambda, \quad 0 < i \neq j > 0.$$

Let U be any neighborhood of the continuum K. By Lemma 3.2 there exists a polyhedral neighborhood P of K such that $K \subset P \subset \overline{P} \subset U$ and $\overline{P}$ has no locally separating points. Since K is a continuum, P can be supposed connected. Then, by the corollary to Lemma 3.1, P satisfies the required conditions.

From this we obtain, in turn,

Corollary 2. *Every planar compactum has an arbitrarily close neighborhood consisting of finitely many components with disjoint closures which satisfy the conditions of Corollary* 1.

2. **Imbedding of compacta which do not separate the plane.**

Lemma 3.3. *Let P be a connected polyhedral neighborhood of a compactum K, where K does not separate E^2, and suppose $\overline{P}$ has no locally separating points. Then there exists a polyhedral disk D such that $K \subset \operatorname{Int} D \subset P$.*

We show first that K does not separate P. Suppose this is false. Then

$$P \setminus K = U \cup V, \quad U \cap V = \Lambda.$$

By the corollary to Lemma 3.1,

$$P = u_0 \setminus \bigcup_{i=1}^{k} \bar{u}_i, \quad \bar{u}_i \subset u_0, \quad \bar{u}_i \cap \bar{u}_j = \Lambda, \quad i \neq j, \ \operatorname{Fr} u_i = c_i,$$

where the $\bar{u}_i$, $i = 0, 1, \cdots, k$, are polyhedral disks. Since $K \cap (\bigcup_{i=0}^{k} c_i) = \Lambda$, each curve c_i belongs to the boundary of only one component of $P \setminus K$. Suppose $c_0 \subset \operatorname{Fr} V$. Then

$$\operatorname{Fr} U = K' \cup c_{i_1} \cup \cdots \cup c_{i_r},$$

where each $i_s > 0$ and $K' \subset K$. Let

$$U_1 = U \cup \bigcup_{s=1}^{r} \bar{u}_{i_s}.$$

U_1 is open, since each polyhedral closed curve c_{i_s} separates E^2 into two domains, so that adjoining $\bar{u}_{i_s}$ to U makes every point $x \in C_{i_s}$ interior to U_1. Hence

$$\operatorname{Fr} U_1 = K' \subset K, \quad U_1 \subset E^2 \setminus K;$$

consequently K separates E^2, contradicting the hypothesis.

We now prove the lemma by induction on the number k of polyhedral curves c_i constituting the boundary of P. If $k = 1$, then $\operatorname{Fr} P = c_0$ and $\overline{P}$ is itself a polyhedral disk.

Suppose the lemma holds if the number of curves c_i is at most k; we prove it holds for a P whose boundary consists of $k+1$ curves.

We shall construct, in P, a new polyhedral neighborhood P_1 of K, whose boundary consists of k polyhedral closed curves. Pick a point $x_0 \in c_0$ and a point $y_0 \in c_k$. Since $P \setminus K$ is connected, the points x_0 and y_0 may be joined by a polygonal line $\lambda \subset P \setminus K$.

Let v be the component of the intersection of a small neighborhood of λ with P which contains λ, the neighborhood being chosen so that

$$\bar{v} \cap K = \Lambda, \text{ and } \bar{v} \cap c_i = \Lambda \text{ for } 0 \neq i \neq k.$$

We may then construct in v a polygonal line λ' which does not meet λ and joins a point $x_0' \in c_0$ with a point $y_0' \in c_k$. The polyhedral closed curve $\overline{x_0 x_0'} \cup \lambda' \cup \overline{y_0' y_0} \cup \lambda$ bounds a domain $V \subset P \cap v$, and hence $\overline{V} \cap K = \Lambda$ (Figure 2.8). Set $P_1 = P \setminus \overline{V}$. It is easily seen that

$$\operatorname{Fr} P_1 = c_0' \cup \bigcup_{i=1}^{k-1} c_i,$$

where

$$c_0' = (c_0 \setminus \overline{x_0 x_0'}) \cup \lambda' \cup (c_k \setminus \overline{y_0 y_0'}) \cup \lambda,$$

i.e. $\operatorname{Fr} P_1$ consists of k disjoint polyhedral closed curves. Since $\overline{V} \cap K = \Lambda$, it follows that P_1 is a neighborhood of K that satisfies the conditions of the lemma. By the induction assumption there exists a polyhedral disk D such that

$$K \subset \operatorname{Int} D \subset P_1.$$

Since $P_1 \subset P$, we have $\operatorname{Int} D \subset P$. This completes the proof of the lemma.

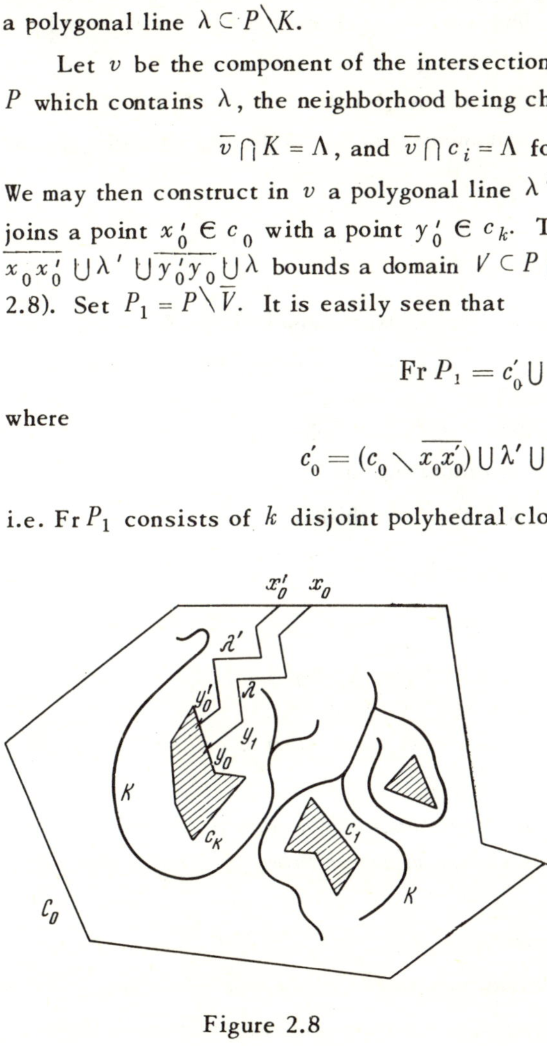

Figure 2.8

Theorem 3.1. *A planar continuum which does not separate E^2 is cellularly imbedded in E^2.*

By Definition 3.1 of Chapter I, we must show that any neighborhood U of the continuum K contains, together with its closure, a cellular neighborhood V, i.e. one such that $\overline{V}$ is a disk. By Lemma 3.2 U contains a connected polyhedral neighborhood P of K: $K \subset P \subset \overline{P} \subset U$, such that $\overline{P}$ has no locally separating points. By Lemma 3.3 $\overline{P}$ contains a disk D such that

$$K \subset \operatorname{Int} D \subset D \subset \overline{P}.$$

Then $\operatorname{Int} D$ is a cellular neighborhood of K contained in U.

Theorem 3.2. *A planar compactum which does not separate E^2 is cellularly separated in E^2.*

By Definition 3.2 of Chapter I, we must show that any neighborhood U of the compactum A contains another neighborhood V, such that $A \subset V \subset \overline{V} \subset U$ and $\overline{V}$ is the union of finitely many disjoint disks.

By Lemma 3.2 U contains a polyhedral neighborhood P of A, $A \subset P \subset \overline{P} \subset U$, where $\overline{P}$ consists of finitely many components and has no locally separating points:

$$\overline{P} = \bigcup_{i=1}^{k} \overline{P}_i, \quad P_i \cap A = A_i \neq \Lambda,$$

$$\overline{P}_i \cap \overline{P}_j = \Lambda, \quad i \neq j.$$

Since $A \cap \operatorname{Fr} \overline{P} = \Lambda$, each A_i is an open-closed subset of A. We assert that the compactum A_i does not separate E^2. If it did, we should have a domain $V_i \subset E^2 \setminus A_i$, $V_i \neq E^2 \setminus A_i$, such that $\operatorname{Fr} V_i \subset A_i$. Since $A \setminus A_i$ are disjoint compacta, it follows that $V_i \cap A$ is a compactum and $V_i \setminus A = V' \neq \Lambda$; then $V' \subset E^2 \setminus A$ and $\operatorname{Fr} V' \subset A$, $V' \neq E^2 \setminus A$, contradicting the hypothesis.

By Lemma 3.3 each $\overline{P}_i$ contains a disk D_i such that

$$A_i \subset \operatorname{Int} D_i \subset D_i \subset \overline{P}_i.$$

Consequently

$$A \subset \bigcup_{i=1}^{k} \operatorname{Int} D_i \subset U, \quad D_i \cap D_j = \Lambda, \quad i \neq j.$$

Corollary 1. *Every planar zero-dimensional compactum is cellularly separated in E^2, since it does not separate E^2.*

Corollary 2. *For any planar zero-dimensional compactum A there exists an isotopy $F_t\colon E^2 \to E^2$ from the identity mapping which takes A into a compactum $F_1(A)$ on a line.*

This follows from Corollary 1 and Theorem 4.2 of Chapter I.

Corollary 3. *Every zero-dimensional compactum in the plane is tame.*

This follows from the definition of tame zero-dimensional compactum (Definition 4.1, Chapter I) and Corollary 2.

§4. Imbedding of simple arcs and simple closed curves in the plane

1. We shall show now that an arbitrary simple arc L is imbedded in E^2 like a line segment; more precisely, that there exists a homeomorphism of E^2 onto

itself taking L into a line segment. Likewise, for any simple closed curve C in E^2 there exists a homeomorphism of E^2 onto itself taking C into the boundary Σ of a 2-simplex.

We shall see in Chapter IV that the analogous theorems no longer hold for the imbedding of a simple arc or a simple closed curve in E^n, $n > 2$. Observe that in the case $n = 3$ the imbedding of a polyhedral simple closed curve already need not be equivalent to the imbedding of the boundary of a 2-simplex, since it may be knotted in E^3. On the other hand, the imbedding of a polyhedral simple arc in E^n is, as we have seen (Chapter I, Theorem 2.3), equivalent to the imbedding of a segment in E^n, for any n.

Lemma 4.1. *A simple arc and a simple closed curve in E^2 have an arbitrarily fine open covering of multiplicity* 2, *such that the closure of the elements of the covering are polyhedral disks δ_i, each δ_i intersects only the two neighboring ones, and each intersection $\delta_i \cap \delta_{i+1}$ is likewise a polyhedral disk and does not separate δ_i or δ_{i+1}.*

The proof is the same for both a simple arc and a simple closed curve. We give it for the case of a simple arc L. Pick a number $\epsilon > 0$, and represent L in the form

$$L = \bigcup_{i=1}^{n} l_i, \ \operatorname{diam} l_i < \varepsilon/3,$$

where the l_i are segments of L and each intersection $l_i \cap l_{i+1} = a_i$ is a point. Let l_i' be a neighborhood of l_i on L with $\operatorname{diam} l_i' < \epsilon/3$. By Theorem 3.1 of Chapter I there exists an open covering of L similar to $\{l_i\}$:

$$L \subset \bigcup_{i=1}^{n} u_i, \ l_i \subset u_i, \ \bar{u}_i \cap \bar{u}_j = \Lambda \quad \text{for } |i-j| > 1,$$
$$\bar{u}_i \cap (L \setminus l_i') = \Lambda, \ \operatorname{diam} \bar{u}_i < \varepsilon/3.$$

In each u_i, take a polyhedral neighborhood p_i' of l_i such that $\bar{p}_i'$ has no locally separating points. Then $\bar{p}_i'$ is the difference between a disk d_i' and, possibly, a finite number of disks $d_{ir} \subset \operatorname{Int} d_i'$.

If we add to $\bar{p}_i'$ all the d_{ir}, i.e. if we replace the neighborhood p_i' by the cellular neighborhood $\operatorname{Int} d_t'$, we do not thereby increase its diameter. We may suppose, further, that $d_i' \cap (L \setminus l_i') = \Lambda$; otherwise, since $\bar{p}_i'(L \setminus l_i') = \Lambda$, at least one of the two components of $L \setminus l_i'$ is contained in some d_{ir}, and consequently has diameter $< \epsilon/3$, so that replacing l_i by the segment $\bigcup_{j=1}^{i} l_j$ (or $\bigcup_{j=i}^{n} l_j$) then yields a segment which is still of diameter $< \epsilon/3$.

Since $\bar{p}_i' \cap \bar{p}_j' = \Lambda$ for $|i-j| > 1$, we have $p_j' \subset E^2 \setminus d_i'$ and $d_i' \cap d_j' = \Lambda$,

$|i-j|>1$. We may suppose that each intersection $d_i' \cap d_{i+1}'$ has no locally separating points. Otherwise, replace the disks d_i' by larger ones obtained, as in the proof of Lemma 3.2, by adding small neighborhoods of any such points, chosen so that all conditions are preserved.

The intersection $d_i' \cap d_{i+1}'$ consists of finitely many components $\kappa_{i\rho}$, each of which is a disk. Indeed, $\kappa_{i\rho}$ is a planar polyhedron without locally separating points, and is therefore, by the corollary to Lemma 3.1, the difference between a disk D and certain open disks D_l interior to D. But any disks D_l removed from $D \subset d_i'$ are then also not contained in d_i', which is impossible, since d_i' is a disk. Thus $\kappa_{i\rho} = D$ is a disk.

The boundary of each $\kappa_{i\rho}$ consists of finitely many polygonal lines, each lying either on $\operatorname{Fr} d_i'$ or on $\operatorname{Fr} d_{i+1}'$. We now transform the disks d_i' in such a way that the intersections are connected. One of the components, say κ_{i0}, contains the point $a_i = l_i \cap l_{i+1}$. If the component $\kappa_{i\rho}$, $\rho > 0$, does not intersect l_i (resp. l_{i+1}), it can be removed from d_i' (resp. d_{i+1}') together with a small neighborhood $v_{i\rho}$ cut off from d_i' by a finite number of disjoint polygonal lines. By Corollary 2 to Theorem 1.1, each component of $d_i' \setminus v_{i\rho}$ is a polyhedral disk. On the other hand, suppose a component $\kappa_{i\rho}$, $\rho > 0$, intersects both l_i and l_{i+1} (Figure 2.9).

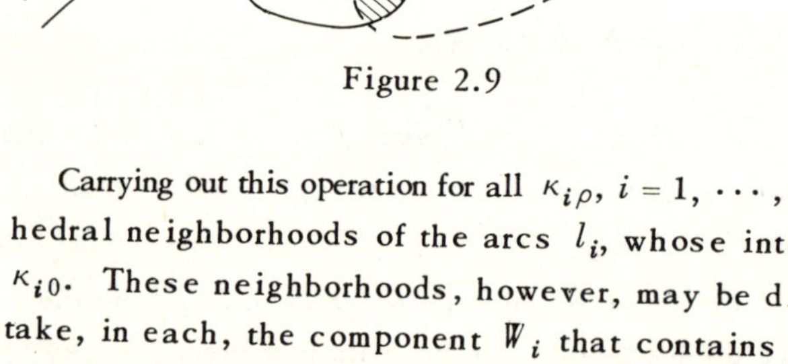

Figure 2.9

As above, remove from d_i' and d_{i+1}' a small neighborhood $v_{i\rho} \supset \kappa_{i\rho}$. Let $\lambda_i = l_i \cap \bar{v}_{i\rho}$, $\lambda_{i+1} = l_{i+1} \cap \bar{v}_{i\rho}$. Since $a_i \in \kappa_{i\rho}$, we have $\lambda_i \cap \lambda_{i+1} = \Lambda$. Take disjoint neighborhoods of λ_i and λ_{i+1} in $\bar{v}_{i\rho}$: $\lambda_i \subset q_{i\rho} \subset \bar{v}_{i\rho}$, $\lambda_{i+1} \subset q_{i\rho}' \subset \bar{v}_{i\rho}$, $\bar{q}_{i\rho} \cap \bar{q}_{i\rho}' = \Lambda$. Now add $q_{i\rho}$ to d_i' and $q_{i\rho}'$ to d_{i+1}'.

Carrying out this operation for all $\kappa_{i\rho}$, $i = 1, \cdots, n$, $\rho > 0$, we obtain new polyhedral neighborhoods of the arcs l_i, whose intersections are just the disks κ_{i0}. These neighborhoods, however, may be disconnected. In that case, take, in each, the component W_i that contains l_i. $\overline{W}_i$ may not be a disk, since the addition of $q_{i\rho}$ and $q_{i-1,\rho}'$ may have given rise to polyhedral closed curves, consisting of segments of $\operatorname{Fr} v_{i\rho}$ and $\operatorname{Fr} q_{i\rho}$, which are in $\overline{W}_i$ and bound disks whose interiors are not in $\overline{W}_i$. By construction any such disks lie in d_i' and do not meet $\kappa_{i\tau}$ or $\kappa_{i-1,\tau'}$, $\tau \neq \rho \neq \tau'$. Hence, adding them to $\overline{W}_i$, we obtain disks d_i such that

$$\overline{W}_i \subset d_i, \quad d_i \cap d_{i+1} = \overline{W}_i \cap \overline{W}_{i+1} = \varkappa_{i0}.$$

The disk $d_i \cap d_{i+1}$ may not separate d_i and d_{i+1} into finitely many components $\sigma_{i\rho}$ and $\sigma'_{i+1,\tau}$. Among the components $\sigma_{i\rho}$ there is only one, say σ_{i0}, that intersects d_{i-1}, since $d_{i-1} \cap d_i$ is connected; and among the $\sigma'_{i+1,\tau}$ there is only one, $\sigma'_{i-1,0}$, that intersects d_{i+2}. Each $\sigma_{i\rho}$ is cut off from d_i by a single arc $\lambda_{i\rho}$ lying on $\mathrm{Fr}\, d_{i+1}$. Indeed, if $\mathrm{Fr}\,\sigma_{i\rho}$ contained two arcs $\lambda_{i\rho}$ and $\lambda'_{i\rho}$ such that $\lambda_{i\rho} \cup \lambda'_{i\rho} \subset \mathrm{Fr}\, d_{i+1}$, $\lambda_{i\rho} \cap \lambda'_{i\rho} = \Lambda$ then, by Corollary 2 to Lemma 1.3, $\lambda_{i\rho} \cup \lambda'_{i\rho}$ would separate the disk d_i into three domains, one containing $\sigma_{i\rho}$ and the other intersecting d_{i+1}, which would imply that $d_i \cap d_{i+1}$ is not connected.

Thus each arc $\lambda_{i\rho}$ cuts off from d_i a polyhedral disk $\sigma_{i\rho}$. Similarly, an arc $\mu_{i+1\tau}$ cuts off from d_{i+1} a disk $\sigma'_{i+1,\tau}$. Let θ_i be the component of $d_i \setminus d_{i-1} \setminus d_{i+1}$ containing $l_i \setminus l'_{i-1} \setminus l'_{i+1}$. The boundary of θ_i is a simple closed curve consisting of four polyhedral arcs: μ_{i0}, λ_{i0}, and two arcs lying on $\mathrm{Fr}\, d_i \setminus d_{i-1} \setminus d_{i+1}$, since θ_i is cut off from d_i by the arcs μ_{i0} and λ_{i0} (Figure 2.10).

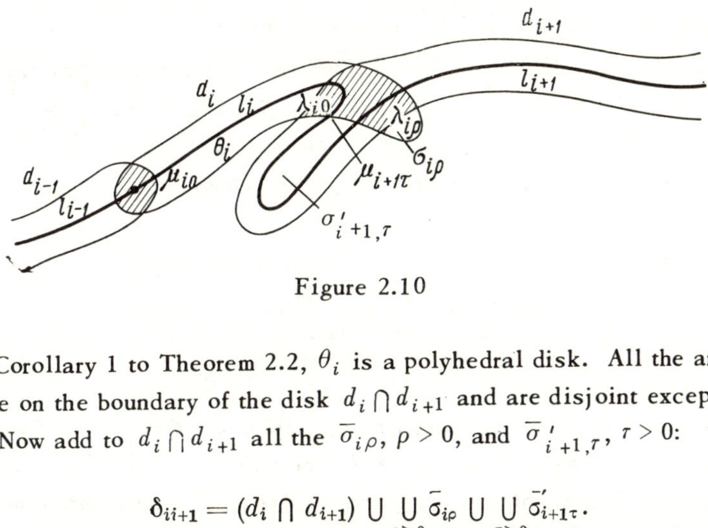

Figure 2.10

By Corollary 1 to Theorem 2.2, θ_i is a polyhedral disk. All the arcs $\lambda_{i\rho}$ and $\mu_{i+1,\tau}$ lie on the boundary of the disk $d_i \cap d_{i+1}$ and are disjoint except for endpoints. Now add to $d_i \cap d_{i+1}$ all the $\bar{\sigma}_{i\rho}$, $\rho > 0$, and $\bar{\sigma}'_{i+1,\tau}$, $\tau > 0$:

$$\delta_{ii+1} = (d_i \cap d_{i+1}) \cup \bigcup_{\rho>0} \bar{\sigma}_{i\rho} \cup \bigcup_{\tau>0} \bar{\sigma}'_{i+1\tau}.$$

Since the various $\sigma_{i\rho}$ and $\sigma'_{i+1,\tau}$ are disjoint, and their closures are disks meeting $d_i \cap d_{i+1}$ in various lines, it follows from Corollary 2 to Theorem 2.2 that $\delta_{i,i+1}$ is likewise a polyhedral disk. Furthermore,

$$\bar{\theta}_i \cap \delta_{ii+1} = \lambda_{i0}, \quad \bar{\theta}_i \cap \delta_{i-1i} = \mu_{i0}, \lambda_{i0} \cap \mu_{i0} = \Lambda.$$

Hence $\delta_i = \delta_{i-1,i} \cup \theta_i \cup \delta_{i,i+1}$ is a polyhedral disk; it contains l_i; and

$$\delta_i \cap \delta_{i+1} = \delta_{ii+1}, \quad \delta_i \cap \mathrm{Fr}\,\delta_{i-1} = \mu_{i0}, \quad \delta_i \cap \mathrm{Fr}\,\delta_{i+1} = \lambda_{i0}.$$

By construction $\delta_i \cap (L \setminus l'_i) = \Lambda$, $\delta_i \cap \delta_j = \Lambda$, $|i-j| > 1$; for, $d_i \cap (L \setminus l'_i) = \Lambda$

and $\delta_i = d_i \cup \bigcup_{\rho>0}\sigma_{i-1,\rho} \cup \bigcup_{\tau>0}\sigma'_{i+1,\tau}$, where $\sigma_{i-1,\rho}\cap L$ and $\sigma'_{i+1,\tau}\cap L$ consist of segments of L whose endpoints are on $\operatorname{Fr}(d_{i-1}\cap d_i)$ and $\operatorname{Fr}(d_i\cap d_{i+1})$, therefore intersect d_i and are consequently contained in l'_i. Since $\operatorname{diam} u_i < \epsilon/3$, we have $\operatorname{diam}\delta_i < \epsilon$.

The covering $\{\delta_i\}$ of the arc L now satisfies the conditions of the lemma, since every δ_i, $\overline{\delta_i\setminus\delta_{i-1}\setminus\delta_{i+1}} = \theta_i$ and $\delta_i\cap\delta_{i+1} = \delta_{i,i+1}$ is a disk, and $\delta_i\cap\delta_j = \Lambda$, $|i-j|>1$.

Theorem 4.1. *Every simple arc in E^2 is cellularly imbedded.*

Let L be a simple arc in E^2, and U an arbitrary neighborhood of L. In U, take an ϵ-covering of L satisfying the conditions of Lemma 4.1, say $\{\delta_i\}$, $i = 1, 2, \cdots, n$; $\bigcup_{i=1}^n \operatorname{Int}\delta_i$ is then a polyhedral domain $V\subset U$. Obviously,

$$\overline{V} = \delta_1 \cup (\overline{\delta_2\setminus\delta_1}) \cup \cdots \cup (\overline{\delta_n\setminus\delta_{n-1}}),$$

each $\sigma_i = \overline{\delta_i\setminus\delta_{i-1}}$ is a disk, and $\sigma_i\cap\sigma_{i-1}$ is the polygonal line $\operatorname{Fr}\sigma_{i-1}\cap\operatorname{Fr}\sigma_i$. Consequently, applying Corollary 2 to Theorem 2.2, we find that $\bigcup_{i=1}^k\sigma_i$ is a disk for any $k\le n$; hence $\overline{V}$ is a disk.

Corollary. *A simple arc $L\subset E^2$ does not separate E^2.*

By Theorem 4.1 we have $L = \bigcap_{k=1}^\infty \overline{V}_k$, where $\overline{V}_k$ is a polyhedral cellular $(1/k)$-neighborhood of L and $\overline{V}_k\subset V_{k-1}$; since $\overline{V}_k$ is a disk, $\operatorname{Fr} V_k$ is a polyhedral closed curve and by Theorem 1.1 $E^2\setminus\overline{V}_k$ is connected. Since $E^2\setminus L = \bigcup_{k=1}^\infty(E^2\setminus\overline{V}_k)$ and $E^2\setminus\overline{V}_k\subset E^2\setminus\overline{V}_{k+1}$, it follows that $E^2\setminus L$ is connected.

Theorem 4.2 (Jordan). *A planar simple closed curve C separates E^2 into two domains, interior and exterior, and is their common boundary.*

Let C be a simple closed curve in E^2. Then

$$C = \bigcap_{k=1}^{\infty}\bigcup_{i=1}^{n_k}\delta_{ki}, \tag{2}$$

where, for each k, $\{\delta_{ki}\}$ is a $1/k$-covering satisfying the conditions of Lemma 4.1, and for each i, $1\le i\le n_k$, there exists a j such that $\delta_{ki}\subset\operatorname{Int}\delta_{k-1,j}$. Each δ_{ki} is a neighborhood of a simple arc c_{ki}, and $c_{ki}\cap c_{k,i+1}$ is a point. Then $\overline{O}_k\subset O_{k-1}$, where $O_k = \bigcup_{i=1}^{n_k}\operatorname{Int}\delta_{ki}$.

We assert that the boundary of O_k is the union of two polyhedral closed curves A_k and B_k, which consist of polygonal lines lying on the sets $\operatorname{Fr}\delta_{ki}$. Indeed, $\operatorname{Fr}O_k$ is the union of all $\operatorname{Fr}\delta_{ki}\setminus(\operatorname{Int}\delta_{k,i-1}\cup\operatorname{Int}\delta_{k,i+1})$, $1\le i\le n_k$ (where we set $1 = n_k+1$), i.e. the union of $2n_k$ polygonal lines l^A_{ki} and l^B_{ki}, where l^A_{ki} intersects $l^A_{k,i-1}$ and $l^A_{k,i+1}$ in an endpoint or along a segment, and likewise l^B_{ki} intersects $l^B_{k,i-1}$ and $l^B_{k,i+1}$; $l^A_{kn_k}$ (resp. $l^B_{kn_k}$) intersects l^A_{k1} (l^B_{k1}); and $l^A_{ki}\cap l^B_{kj} = \Lambda$ for any i and j. Otherwise, we should have $l^A_{k1}\cap l^B_{kn_k}\ne\Lambda$, $l^B_{k1}\cap l^A_{kn_k}\ne\Lambda$, so that $\operatorname{Fr}O_k$ is a polyhedral closed curve and O_k a disk. In that case the polygonal line $\delta_{kn_k}\cap$

$\mathrm{Fr}\,\delta_{k1}$ separates O_k and so separates from each other the points $x_i = l^A_{ki} \cap l^A_{k,i+1}$ and $l^B_{ki} \cap l^B_{k,i+1} = y_i$, $1 \le i \le n_k$, which is impossible, since x_i and y_i are the endpoints of the polygonal line $\delta_{ki} \cap \mathrm{Fr}\,\delta_{ki+1}$. Thus $\bigcup_{i=1}^{n_k} l^A_{ki} = A_k$ and $\bigcup_{i=1}^{n_k} l^B_{ki} = B_k$ are polyhedral closed curves, and $A_k \cap B_k = \Lambda$, $\mathrm{Fr}\,O_k = A_k \cup B_k$.

Since $\overline{O}_k$ is a connected polyhedron and its boundary is the union of two disjoint polyhedral closed curves, O_k is the difference between two disks. Hence one of these curves, say A_k, lies in the interior domain of the other, B_k. Let V_{A_k} be the interior domain of A_k, V_{B_k} the interior domain of B_k, and $U_{B_k} = E^2 \setminus \overline{V}_{B_k}$. Then

$$\overline{V}_{A_k} \subset V_{B_k},\ O_k = V_{B_k} \setminus \overline{V}_{A_k} \tag{3}$$

and $E^2 \setminus \overline{O}_k = V_{A_k} \cup U_{B_k}$ is the union of two domains: an inner, V_{A_k}, and an outer, U_{B_k}. Since $\overline{O}_{k+1} \subset O_k$, we have

$$\overline{V}_{A_k} \subset V_{A_{k+1}},\ \overline{U}_{B_k} \subset U_{B_{k+1}}. \tag{4}$$

But $C = \bigcap_{k=1}^{\infty} O_k$, so that

$$E^2 \setminus C = \bigcup_{k=1}^{\infty} V_{A_k} \cup \bigcup_{k=1}^{\infty} U_{B_k}.$$

By (3), $\overline{V}_{A_k} \cap \overline{V}_{B_k} = \Lambda$ for any k, so that, by (4),

$$\Big(\bigcup_{k=1}^{\infty} V_{A_k}\Big) \cap \Big(\bigcup_{k=1}^{\infty} U_{B_k}\Big) = \Lambda.$$

Thus C separates E^2 into two domains: the interior $\bigcup_{k=1}^{\infty} V_{A_k} = V$ and the exterior $\bigcup_{k=1}^{\infty} U_k = U$.

Finally, *C is the common boundary of V and U.*

It suffices to show that an arbitrary neighborhood v of any point $x \in C$ contains a point of V and a point of U. There exist integers k and i such that $x \in \delta_{ki} \subset v$. But on $\mathrm{Fr}\,\delta_{ki}$ there are points of $A_k \subset V$ and points of $B_k \subset V$, so that $v \cap V \neq \Lambda$ and $v \cap U \neq \Lambda$.

Theorem 4.3 (Antoine [20]). *For any simple closed curve C in the plane E^2 there exists an isotopy F_t: $E^2 \to E^2$ from the identity mapping, that takes the boundary of a 2-simplex into C. For any simple arc L in E^2 there exists an isotopy Φ_t: $E^2 \to E^2$ from the identity mapping, that takes a line segment into L.*

The proofs of the two cases are similar. We give the proof for the case of a simple closed curve C. C separates E^2 into two domains, an interior G_A and an exterior G_B. It can be represented in the form

$$C = \bigcap_{k=1}^{\infty} \bigcup_{i=1}^{n_k} c_{ki},$$

where each c_{ki} is a simple arc of diameter $< 1/k$, $c_{ki} \cap c_{k,i+1}$ is a point, and each $c_{k+1,j}$ is contained in some c_{ki}. Next, C can be represented in the form (2), where each covering $\{\delta_{ki}\}$ satisfies the conditions of Lemma 4.1, and $c_{ki} \subset \operatorname{Int} \delta_{ki}$, $\operatorname{diam} \delta_{ki} < 1/k$. Then $\delta_{ki} \cap \delta_{k,i+1}$ is a disk, and the boundary of $\delta_{ki} \cap \delta_{k,i+1}$ is the union of two polygonal lines lying on $\operatorname{Fr} \delta_{ki}$ and $\operatorname{Fr} \delta_{k,i+1}$.

The intersection of these polygonal lines consists of two components, either of which may be a point or a polygonal line. A small modification of one of the disks, say δ_{ki}, will assure that $\operatorname{Fr} \delta_{ki} \cap \operatorname{Fr} \delta_{k,i+1}$ consists of two points. We shall suppose that all the disks δ_{ki} satisfy this condition.

$U_k = \bigcup_{i=1}^{n_k} \operatorname{Int} \delta_{ki}$ is a domain bounded by the union of two polyhedral closed curves A_k and B_k, with

$$A_k = \bigcup_{i=1}^{n_k} a_{ki}, \quad B_k = \bigcup_{i=1}^{n_k} b_{ki},$$

where a_{ki} and b_{ki} are polygonal lines lying on $\operatorname{Fr} \delta_{ki}$, and $A_k \subset G_A$, $B_k \subset G_B$. (In the case of a simple arc the domain U_k is a polyhedral disk, and

$$a_{k1} \cap b_{k1} = x_{k0} \in \operatorname{Fr} \delta_{k1}, \quad a_{kn_k} \cap b_{kn_k} \in \operatorname{Fr} \delta_{kn_k}.)$$

By Theorem 2.2 there exists a semilinear isotopy $\Psi_t\colon E^2 \to E^2$ taking into A_1 the boundary of a 2-simplex contained in the domain interior to A_1. To prove the theorem it therefore suffices to construct an isotopy $F_t\colon E^2 \to E^2$ taking A_1 into C.

Let ϕ be a homeomorphism of A_1 onto C such that

$$\varphi(a_{1i}) = c_{1i}, \quad i = 1, 2, \ldots, n_1.$$

For any k, and for $j = 1, 2, \cdots, n_k$, set

$$\gamma_{kj} = \varphi^{-1}(c_{kj}).$$

Then each γ_{kj} is a polygonal line, and the diameters of the γ_{kj} approach zero with $1/k$; furthermore, A_1 can be represented in the form (2):

$$A_1 = \bigcap_{k=1}^{\infty} \bigcup_{j=1}^{n_k} \sigma_{kj},$$

where the coverings $\{\sigma_{kj}\}$ satisfy the conditions of Lemma 4.1, and $\gamma_{kj} \subset \operatorname{Int} \sigma_{kj}$.

Let $V_k = \bigcup_{i=1}^{n_k} \operatorname{Int} \sigma_{ki}$. Then $\operatorname{Fr} V_k = \mathfrak{A}_k \cup \mathfrak{B}_k$, where $\mathfrak{A}_k$ and $\mathfrak{B}_k$ are polyhedral closed curves, $\mathfrak{A}_k$ lies in the interior domain of $\mathfrak{B}_k$ and

$$\mathfrak{A}_k = \bigcup_{i=1}^{n_k} \alpha_{ki}, \quad \mathfrak{B}_k = \bigcup_{i=1}^{n_k} \beta_{ki},$$

where α_{ki} and β_{ki} are polygonal lines lying on $\operatorname{Fr} \sigma_{ki}$.

We may suppose that every $\alpha_{ki} \cap \alpha_{k,i+1}$ and $\beta_{ki} \cap \beta_{k,i+1}$ is a point.

We shall construct an isotopy F_t: $E^2 \to E^2$, taking A_1 into C, such that

$$F_1|A_1 \equiv \phi.$$

The homeomorphism F_1 (and isotopy F_t) will be constructed as the limit of a uniformly convergent sequence of homeomorphisms f_k, where $f_k(A_1) = A_{p_k}$, $p_k \geq k+1$, and each f_{k-1} is joined to f_k by an isotopy F_t^k.

For each δ_{ki}, pick a neighborhood O_{ki} of diameter $< 1/k$, such that the covering $\{O_{ki}\}$ of the curve C is similar to the covering $\{\delta_{ki}\}$; then $C = \bigcap_{k=1}^{\infty} \bigcup_{i=1}^{n_k} O_{ki}$.

Next, pick an integer $p_1 \geq 2$ such that

$$\overline{V}_{p_1} \cap \overline{U}_{p_1} = \Lambda, \quad \bigcup_{j=j_i+1}^{j_{i+1}} \sigma_{p_1 j} \subset O_{1i}, \quad i = 1, 2, \ldots, n_1,$$

where $\bigcup_{j=j_i+1}^{j_{i+1}} \sigma_{p_1 j}$ is the smallest union of $\sigma_{p_1 j}$'s covering γ_{1i}. Since $A_1 \cap C = \Lambda$ and $\gamma_{1i} \subset \delta_{1i} \subset O_{1i}$, such an integer p_1 exists. We may suppose the domains V_k to be renumbered so that $p_1 = 2$.

We now construct, to begin with, an isotopy F_t^1: $E^2 \to E^2$ taking A_1 into A_2 and $\mathfrak{B}_2$ into B_2, such that

$$F_1^1(\gamma_{2i}) = a_{2i}, \quad F_1^1(\beta_{2i}) = b_{2i}, \quad i = 1, 2, \ldots, n_2, \tag{5}$$

and F_t^1 is stationary outside the domain W_1 with boundary $\mathfrak{A}_2 \cup \mathfrak{B}_1$.

W_1 contains the disjoint polyhedral closed curves A_1, $\mathfrak{B}_2$, A_2 and B_2, which separate W_1 into five domains $W_{1\rho}$, $1 \leq \rho \leq 5$.

Indeed, each one of these curves in succession bounds a domain containing the preceding one, so that their interior domains are contained, with their closures, each one successively in the next:

$$G_{\mathfrak{A}_2} \subset G_{A_1} \subset G_{\mathfrak{B}_2} \subset G_{A_2} \subset G_{B_2}.$$

The curve A_1 separates W_1 into two annular domains, $W_{11} = W_1 \cap G_{A_1}$ and $W_1 \setminus \overline{G}_{A_1}$; the curve $\mathfrak{B}_2$ separates the domain $W_1 \setminus \overline{G}_{A_1}$, which is bounded by A_1 and B_1, into the domains $W_{12} = (W_1 \setminus \overline{G}_{A_1}) \cap G_{\mathfrak{B}_2}$ and $W_1 \setminus \overline{G}_{\mathfrak{B}_2}$; etc. The boundary of each domain $\overline{W}_{1\rho}$ is the union of two successive polyhedral closed curves, so that $W_{1\rho}$ is a planar annulus.

Consider all the points

$$\xi_{2i} = \alpha_{2i} \cap \alpha_{2i+1} \in \mathfrak{A}_2, \quad x_{1i} = \gamma_{2i} \cap \gamma_{2i+1} \in A_1,$$

$$\eta_{2i} = \beta_{2i} \cap \beta_{2i+1} \in \mathfrak{B}_2, \quad x_{2i} = a_{2i} \cap a_{2i+1} \in A_2, \quad y_{2j} = b_{2i} \cap b_{2i+1} \in B_2.$$

In addition, on each b_{1j}, between the points $b_{1,j-1} \cap b_{1j}$ and $b_{1j} \cap b_{1,j+1}$, pick $k_j - 1$ points y_{1i}, where k_j is the number of segments γ_{2i} contained in $\gamma_{1j} = a_{1j}$; number the points y_{1i}, together with the points $b_{1j} \cap b_{1,j+1}$, in such a way that the point $b_{1j} \cap b_{1,j+1}$ receives the index $1i$ if $\delta_{2i} \cap \delta_{2,i+1} \ni c_{1j} \cap c_{1,j+1}$. The

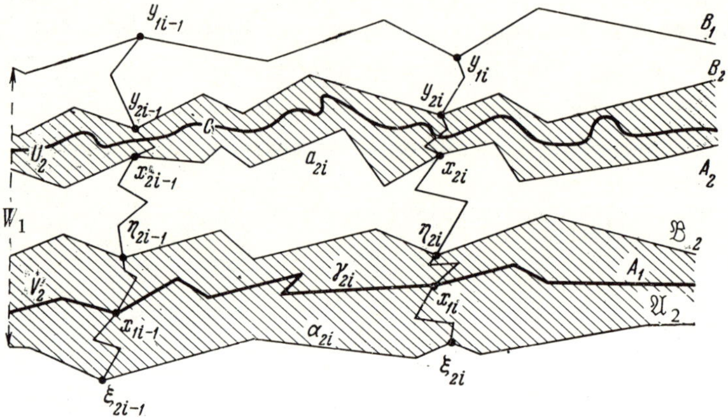

Figure 2.11

points ξ_{2i} and η_{2i} constitute $\mathrm{Fr}\,\sigma_{2i} \cap \mathrm{Fr}\,\sigma_{2,i+1}$, and $x_{1i} \in \mathrm{Int}(\sigma_{2i} \cap \sigma_{2,i+1})$. The closures of the components of $\sigma_{2i} \cap \sigma_{2,i+1} \setminus A_1$ are disks, since they are cut off from the disk $\sigma_{2i} \cap \sigma_{2,i+1}$ by polygonal lines lying on A_1. Among these disks is one, say $\kappa_i \subset \overline{W}_{11}$, that contains the point ξ_{2i}, and one, say $\kappa_i' \subset \overline{W}_{12}$, that contains the point η_{2i}. Both κ_i and κ_i' contain the point $x_{1i} \in A_1$, since the component of $A_1 \cap \sigma_{2i} \cap \sigma_{2,i+1}$ containing the point x_{1i} is unique and intersects both $\mathrm{Fr}\,\sigma_{2i}$ and $\mathrm{Fr}\,\sigma_{2,i+1}$ (since $\gamma_{2i} \subset \sigma_{2i}$, $\gamma_{2,i+1} \subset \sigma_{2,i+1}$ and $x_{1i} = \gamma_{2i} \cap \gamma_{2,i+1}$) (Figure 2.11).

Join each point ξ_{2i} to the point x_{1i} by a polygonal line $(\xi_2 x_1)_i$ interior to κ_i (i.e. contained, except for its endpoints, in the interior of κ_i), and each point x_{1i} to the point η_{2i} by a polygonal line $(x_1\ \eta_2)_i$ interior to κ_i'. Then $(\xi_2 x_1)_i \cup (x_1\ \eta_2)_i \subset \sigma_{2i} \cap \sigma_{2,i+1} \subset O_{1j}$ if $\sigma_{2i} \subset O_{1j}$. (If $x_{1i} = a_{1,j-1} \cap a_{1j}$, then $(\xi_2 x_1)_i \cup (x_1 \eta_2)_i \subset O_{1,j-1} \cap O_{1j}$.)

The points x_{2i} and y_{2i} constitute $\mathrm{Fr}\,\delta_{2i} \cap \mathrm{Fr}\,\delta_{2,i+1}$. They can therefore be joined by a polygonal line $(x_2 y_2)_i$ interior to $\delta_{2i} \cap \delta_{2,i+1}$.

The points $\eta_{2i} \in \mathfrak{B}_2$ and $x_{2i} \in A_2$ are both contained in the same O_{1j} since $\eta_{2i} \in \mathrm{Fr}\,\sigma_{2,i} \cap \mathrm{Fr}\,\sigma_{2,i+1}$ and $x_{2i} \in \mathrm{Fr}\,\delta_{2i} \cap \mathrm{Fr}\,\delta_{2,i+1}$, and they both lie on $\mathrm{Fr}\,W_{13}$. They can therefore be joined interior to W_{13} by a polygonal line $(\eta_2 x_2)_i$ contained in O_{1j}. In the same way, the points y_{2i} and y_{1i}, which lie on $\mathrm{Fr}\,W_{15}$, can be joined in W_{15} by a polygonal line $(y_2 y_1)_i$ contained in O_{1j}. The polygonal lines $(\eta_2 x_2)_i$ (as well as $(y_2 y_1)_i$) can be chosen so as to be disjoint. The other polygonal lines are already disjoint by construction, since they are contained in different $\delta_{2i} \cap \delta_{2,i+1}$ or $\sigma_{2,i+1}$. Now set

$$l_i = (\xi_2 x_1)_i \cup (x_1 \eta_2)_i \cup (\eta_2 x_2)_i \cup (x_2 y_2)_i \cup (y_2 y_1)_i.$$

By construction $l_i \cap l_{i'} = \Lambda$, $i \neq i'$. The polygonal lines l_i separate the domain W_1 into disks D_i, bounded by the polyhedral closed curves $l_{i-1} \cup \alpha_{2i} \cup l_i \cup (y_{1i-1} y_{1i})$ where $(y_{1i-1} y_{1i})$ is the portion of B_1 between the points y_{1i-1} and y_{1i}:

$$\overline{W}_1 = \bigcup_{i=1}^{n_2} D_i, \quad D_i \cap D_{i+1} = l_i, \quad D_i \cap D_j = \Lambda, \quad |i-j| > 1.$$

$$\begin{cases} D_i \cap \mathfrak{A}_2 = \alpha_{2i}, & D_i \cap A_1 = \gamma_{2i}, & D_i \cap \mathfrak{B}_2 = \beta_{2i}, \\ D_i \cap A_2 = a_{2i}, & D_i \cap B_2 = b_{2i}, & D_i \cap B_1 = y_{1i-1} y_{1i}. \end{cases} \tag{6}$$

For each disk D_i we can construct a semilinear homeomorphism h_i: $E^2 \to E^2$ taking D_i into the square I^2, such that the polygonal lines l_{i-1} and l_i go into two opposite faces of I^2. To do so, first construct a homeomorphism h_i^1 taking α_{2i} into a side K_1 of the square (Theorem 2.3 of the first chapter), then, in succession, homeomorphisms h_i^2 and h_i^3 which are stationary on K_1 and take l_{i-1} and l_i into the two sides K_2 and K_3 perpendicular to K_1 (Example 2.2); finally, by Theorem 2.3, construct a homeomorphism h_i^4 taking the polygonal line $h_i^3 h_i^2 h_i^1 (y_{1i-1} y_{1i})$ into the side K_4 and stationary on $K_1 \cup K_2 \cup K_3$. Setting $h_i = h_i^4 h_i^3 h_i^2 h_i^1$, we obtain the required semilinear homeomorphism h_i: $D_i \to I^2$. The homeomorphisms h_i and h_{i+1} can be coordinated so that the points $h_i(x_{1i})$ and $h_{i+1}(x_{1i})$ (and likewise $h_i(\eta_{2i})$ and $h_{i+1}(\eta_{2i})$, etc.) on the sides K_3 and K_2 of the square, respectively, are at the same distances from K_1.

By (6), $h_i(\gamma_{2i})$, $h_i(\beta_{2i})$, $h_i(a_{2i})$ and $h_i(b_{2i})$ are disjoint polygonal lines, which lie interior to I^2 except for their endpoints, of which one (for each of them) lies on K_2, the other on K_3. By repeated application of Theorem 2.3, we can construct a semilinear homeomorphism g_i: $I^2 \to I^2$, stationary on $\mathrm{Fr}\, I^2$ and taking each of these four polygonal lines into the line segment between its endpoints. Thus $H_i = g_i h_i$ is a homeomorphism of the disk D_i onto the square I^2, under which l_{i-1} and l_i go into the sides K_2 and K_3 of the square, and the polygonal lines γ_{2i}, β_{2i}, a_{2i} and b_{2i} go into disjoint line segments γ_i', β_i', a_i', b_i', whose endpoints lie on K_2 and K_3.

Every perpendicular to the side K_1 intersects K_1, γ_i', β_i', a_i', b_i' and K_4, respectively, in points z, x, y, x', y' and z'. Equations (7) of Chapter I then determine a semilinear isotopy T_t^i: $I^2 \to I^2$, stationary on $K_1 \cup K_4$ and such that $T_1^i(\gamma_i') = a_i'$ and $T_1^i(\beta_i') = b_i'$.

Set

$$\begin{aligned} &\text{a)} \quad F_t^1(x) = H_i^{-1}T_t^iH_i(x), \quad x \in D_i; \\ &\text{b)} \quad F_t^1(x) = x, \quad x \in E^2 \setminus W_1. \end{aligned} \tag{7}$$

We assert that F_t^1 is the required isotopy.

By (7) a), it is evident that $F_t^1(D_i) = D_i$, $0 \le t \le 1$.

Let us verify that (7) a) implies that $F_t^1(x) = x$ for $x \in \mathrm{Fr}\, W_1$. Indeed,

$$\mathrm{Fr}\, W_1 = \mathfrak{A}_2 \cup B_1 = \bigcup_{i=1}^{n_2} \alpha_{2i} \cup \bigcup_{i=1}^{n_2} (y_{1i-1}y_{1i}),$$

$$H_i(\alpha_{2i}) = K_1, \quad H_i(y_{1i-1}y_{1i}) = K_4.$$

But $T_t^i(y) = y$, $0 \le t \le 1$, for $y \in K_1 \cup K_4$, so that $F_t^1(x) = x$, $x \in \mathrm{Fr}\, W_1$. From (7) a) and the fact that

$$H_i^*(\gamma_{2i}) = \gamma_i', \quad H_i(\beta_{2i}) = \beta_i',$$
$$H_i(a_{2i}) = a_i', \quad H_i(b_{2i}) = b_i',$$

(5) follows immediately. Since the homeomorphisms H_i and H_{i+1} are compatible on l_i, (7) a) determines an isotopy F_t^1 on all of $\overline{W}_1$. Each δ_{2i} and σ_{2i} is contained in some O_{1j}. By construction, l_{1i-1} and l_{1i} are then contained in O_{1j}, so that

$$D_j \subset O_{1j},$$

and since $F_t^1(D_i) = D_i$, $0 \le t \le 1$, we have

$$\rho(x, F_t^1(x)) < \operatorname{diam} O_{1j} < 1, \quad 0 \leqslant t \leqslant 1.$$

Set $f_1 = F_1^1$.

The isotopy F_t^1 and the homeomorphism f_1 being semilinear by construction, all the $f_1(A_k)$, $f_1(B_k)$, $f_1(\mathfrak{A}_k)$ and $f_1(\mathfrak{B}_k)$ are polyhedral closed curves.

Suppose we have constructed, for all $r \le k$, semilinear homeomorphisms f_r: $E^2 \to E^2$ and isotopies F_t^r joining f_{r-1} to f_r, such that

$$f_r(A_1) = A_{p_r}, \quad f_r(\mathfrak{B}_{p_r}) = B_{p_r}$$

for some $p_r \ge r+1$, with the following conditions satisfied:

1. $F_t^{r+q}(x) \in O_{p_{r-1}i}$, $r+q \le k$, $q \ge 0$,

for some i such that $f_{r-1}(x) \in O_{p_{r-1}i}$;

2. $f_r(\gamma_{p_rj}) = a_{p_rj}$; $f_r(\beta_{p_rj}) = b_{p_rj}$, $j = 1, 2, \cdots, n_{p_r}$;
3. $F_t^{r+q}(x) = f_{r-1}(x)$ for $f_{r-1}(x) \in E^2 \setminus W_r$,

where W_r is the domain bounded by $f_{r-1}(\mathfrak{A}_{p_r}) \cup B_{p_{r-1}}$, $0 \le t \le 1$, $q \ge 0$, $r+q \le k$.

We now construct a homeomorphism f_{k+1} and an isotopy F_t^{k+1} such that conditions 1, 2, and 3 continue to be satisfied.

By 2,

$$a_{p_k i} = f_k(\gamma_{p_k i}) \subset \delta_{p_k i} \subset O_{p_k i}, \quad \gamma_{p_k i} \subset \sigma_{p_k i}.$$

Since f_k is a homeomorphism and $A_1 = \bigcap_{k=1}^{\infty} \bigcup_{i=1}^{n_k} \sigma_{ki}$, there exists an integer $p_{k+1} \geq p_k + 1 \geq k + 2$ such that, for every $i \leq n_{p_k}$,

$$a_{p_k i} \subset \bigcup_{j=j_i}^{j_{i+1}-1} f_k(\sigma_{p_{k+1} j}) \subset O_{p_k i}, \qquad \overline{U}_{p_{k+1}} \cap f_k(\overline{V}_{p_{k+1}}) = \Lambda,$$

where the j_i are so chosen that the indicated union is the smallest one covering $a_{p_k i}$; then

$$f_k(\mathfrak{A}_{p_{k+1}}) \cup f_k(\mathfrak{B}_{p_{k+1}}) \subset \bigcup_{i=1}^{n_{p_k}} O_{p_k i}.$$

Just as in the case $k = 0$, we can construct a semilinear isotopy ψ_t: $E^2 \to E^2$ from the identity mapping, taking $A_{p_k} = f_k(A_1)$ into $A_{p_{k+1}}$ and $f_k(\mathfrak{B}_{p_{k+1}})$ into $B_{p_{k+1}}$, such that

1′. $\psi_t(y) \in O_{p_k i}$ for some i such that $y \in O_{p_k i}$;

2′. $\psi_1 f_k(\gamma_{p_{k+1} j}) = a_{p_{k+1} j}$, $\psi_1 f_k(\beta_{p_{k+1} j}) = b_{p_{k+1} j}$;

3′. $\psi_t(y) = y$ for $y \in E^2 \setminus W_{k+1}$, where $\operatorname{Fr} W_{k+1} = f_k(\mathfrak{A}_{p_{k+1}}) \cup B_{p_k}$.

Setting $F_t^{k+1} = \psi_t f_k$ and $f_{k+1} = \psi_1 f_k$, we obtain a semilinear homeomorphism f_{k+1} and an isotopy F_t^{k+1} joining f_k to f_{k+1}, with conditions 1, 2 and 3 satisfied for $r = k + 1$. Conditions 2 and 3 follow from 2′ and 3′ and the fact that $W_k \subset W_r$, $r < k$, and condition 1 follows from the fact that $O_{p_k i} \subset O_{p_r j}$ for some j if $r < k$.

Continuing this procedure, we construct an infinite sequence of homeomorphisms of E^2 onto itself:

$$f_0 \equiv 1, \ f_1, \ f_2, \ \ldots, \ f_r, \ \ldots \tag{8}$$

and isotopies F_t^r joining them, such that conditions 1, 2 and 3 are satisfied for any $r > 1$ and $q \geq 0$; since $W_r \subset W_1$, $r > 1$, it follows for any r that $F_t^r(x) = x$ for $x \in E^2 \setminus W_1$. Thus the sequence (8) and isotopies F_t^r satisfy the conditions of the corollary to Theorem 1.1 of Chapter I. Thus there exists a homotopy F_t from the identity mapping, which for every $t < 1$ is obtained from some F_t^r by a transformation of the variable t. Hence F_t is a homeomorphism for every $t < 1$. We assert that F_1 is likewise a homeomorphism. Let us show first that

$$F_1 |_{A_1} \equiv \varphi.$$

Indeed, if $x \in A_1$ there exists a sequence of integers $i_1, i_2, \cdots, i_r, \cdots$ such that $x = \bigcap_{r=1}^{\infty} \gamma_{r i_r} = \bigcap_{r=1}^{\infty} \gamma_{p_r i_{p_r}}$. But $f_r(\gamma_{p_r i_{p_r}}) = a_{p_r i_{p_r}} \subset \delta_{p_r i_{p_r}}$ so that

$$F_1(x) = \lim_{r\to\infty} f_r(x) = \bigcap_{r=1}^{\infty} \delta_{p_r i_{p_r}} = \varphi(x).$$

It remains to show that on $E^2\setminus A_1$ the mapping F_1 is a homeomorphism, and that

$$F_1(E^2\setminus A_1) \cap C = \Lambda. \tag{9}$$

Now, $E^2\setminus A_1 = \bigcup_{r=1}^{\infty}(E^2\setminus \overline{V}_{p_r})$, or in other words $E^2\setminus A_1$ is the union of the interior domains of $\mathfrak{A}_{p_r}$ and the exterior domains of the $\mathfrak{B}_{p_r}$. By condition 3

$$F_1(x) = f_{r-1}(x), \quad \text{if} \quad x \in E^2\setminus W_r.^{1)}$$

It follows that F_1 is a homeomorphism on every $E^2\setminus W_r$, and therefore also on $E^2\setminus C$. But $\operatorname{Fr} W_r = f_{r-1}(\mathfrak{A}_{p_r}) \cup B_{p_{r-1}}$; and $f_{r-1}(\mathfrak{A}_{p_r})$ is contained in the interior domain of $A_{p_r} = f_{r-1}(A_1)$, while $f_{r-1}(\mathfrak{B}_{p_r}) = B_{p_r}$. Hence $F_1(E^2\setminus \overline{V}_{p_r}) \cap C = \Lambda$ for every p_r, and this implies (9).

Thus F_t is an isotopy of the plane onto itself, which is stationary on $E^2\setminus W_1$ and takes the polyhedral closed curve A_1 into C. Let Ψ_t: $E^2 \to E^2$ be an isotopy taking the boundary of a 2-simplex Δ^2 into A_1. Then $\Phi_t = \Psi_t \circ F_t$ satisfies the conditions of the theorem.

If we take $\overline{\Delta}^2$ to be in the interior domain of A_1, then Ψ_t can be supposed stationary outside G_A. And since the domain W_1 could have been chosen as an arbitrary close neighborhood of C, the isotopy F_t, and therefore also Φ_t, can be supposed stationary outside an arbitrary neighborhood of $\overline{G}_A$.

Corollary 1. *For any two simple closed curves C_1 and C_2 (arcs L_1 and L_2) in E^2, or in the sphere S^2, there exists an isotopy F_t: $E^2 \to E^2$, or $S^2 \to S^2$, from the identity mapping, taking C_1 into C_2 (L_1 into L_2).*

Let Γ be the boundary of a simplex Δ^2, Φ_t: $E^2 \to E^2$ and $\Phi_1(\Gamma) = C_1$, Ψ_t: $E^2 \to E^2$ and $\Psi_1(\Gamma) = C_2$. Then $F_t = \Phi_t^{-1} \circ \Psi_t$ is an isotopy of E^2 onto itself that takes C_1 into C_2. The theorem and the corollary remain valid if instead of E^2 we take the sphere S^2. Indeed, $S^2\setminus\{p\}$, where p is a point, is homeomorphic to E^2. The point p can be chosen in $S^2\setminus(C_1 \cup C_2)$, and the isotopy Φ_t: $S^2\setminus\{p\} \to S^2\setminus\{p\}$ taking C_1 into C_2 can be supposed stationary in a neighborhood of p. We can then extend it to S^2 by setting $\Phi_t(p) = p$, $0 \le t \le 1$.

A similar construction yields an isotopy of E^2 or of S^2 taking a simple arc L_1 into a simple arc L_2.

---

1) *Translator's note.* There is some confusion here and in the next paragraph. The intended argument is presumably as follows. If $x \in E^2\setminus V_{p_r}$, then $f_r(x) \in E^2\setminus W_{r+1}$ (as is seen by examining the images of the boundary curves of $\overline{V}_{p_r}$), and therefore, by condition 3, $F_1(x) = f_r(x)$. This shows, firstly, that F_1 is a homeomorphism on $E^2\setminus \overline{V}_{p_r}$ for every r, and therefore on $E^2\setminus A_1$; secondly that $F_1(E^2\setminus \overline{V}_{p_r}) \cap C = \Lambda$ for every r.

Corollary 2. *Every simple closed curve $C \subset E^2$ separates E^2 into domains homeomorphic to the complementary domains of a circle. In particular, the interior domain of C is homeomorphic to the interior of a circular disk, i.e. to E^2. The closure of the interior domain is homeomorphic to a circular disk, i.e. is a disk.*

Corollary 3. *The closures of both complementary domains of a simple closed curve C in S^2 are disks* (since there exists an isotopy F_t: $S^2 \to S^2$ taking C into the equator of the sphere S^2).

Corollary 4 (The Schoenflies theorem [69]). *Every homeomorphism ϕ: $C_1 \to C_2$, where C_1 and C_2 are simple closed curves in S^2 (or E^2), extends to a homeomorphism Φ: $S^2 \to S^2$ (or $E^2 \to E^2$).*

Consider first the case that one of the curves, say C_1, is a circle. The closures of the complementary domains of the C_i are disks D_i and D_i'. Let H: $D_1 \to D_2$ and H': $D_1' \to D_2'$ be homeomorphisms of the (circular) disks D_1 and D_1' onto the disks D_2 and D_2'.

We assert that the homeomorphism ϕ: $C_1 \to C_2$ can be extended to each disk D_1 and D_1'. Indeed, the map $f = H^{-1}|_{C_2}\phi$ is a homeomorphism of the circle C_1 onto itself. f can be extended to a homeomorphism F: $D_1 \to D_1$ such that $F|_{C_1} \equiv f$, $F(O) = O$, where O is the center of D_1, and F takes the radius to any point $x \in C_1$ linearly into the radius to the point $f(x) \in C_1$. Set $\Phi = HF$. Then Φ is a homeomorphism of D_1 onto D_2 and $\Phi|_{C_1}f = H|_{C_1}H^{-1}|_{C_2}\phi \equiv \phi$, i.e. F is an extension of the homeomorphism ϕ to D_1. Likewise, ϕ extends to D_1', and so to the sphere S^2.

In the case of E^2, it suffices to enlarge E^2, by adding a point, into a sphere S^2 and then extend ϕ as above.

Now let C_1 and C_2 be arbitrary simple closed curves in S^2, and let $\phi(C_1) = C_2$, where ϕ is a homeomorphism.

Let C_0 be a circle in S^2, and h: $C_0 \to C_1$ a homeomorphism of C_0 onto C_1; then ϕh is a homeomorphism of C_0 onto C_2. The homeomorphisms h and ϕh extend to homeomorphisms F_1 and F_2 of the sphere S^2 onto itself.

Setting $\Phi = F_2F_1^{-1}$, we obtain

$$\Phi|_{C_1} = \varphi h h^{-1} \equiv \varphi.$$

Corollary 5. *Every homeomorphism ϕ: $L_1 \to L_2$, where L_1 and L_2 are simple arcs in E^2 (or S^2), extends to a homeomorphism Φ: $E^2 \to E^2$ (or $S^2 \to S^2$).*

By Theorem 4.3 there exists for each L_i an isotopy F_t^i: $E^2 \to E^2$ taking L_i into a line segment; then $F_1^i(L_i)$ is contained in the boundary Γ of a 2-simplex, and consequently L_i is contained in a simple closed curve $C_i = (F_1^i)^{-1}(\Gamma)$. The homeomorphism ϕ can be extended to a homeomorphism ψ: $C_1 \to C_2$ and then, by Corollary 4, to a homeomorphism Φ: $E^2 \to E^2$.

CHAPTER III

THE FUNDAMENTAL GROUP

Examples will be given in Chapter IV of "wild" imbeddings in E^3 for simple arcs, disks, spheres and zero-dimensional compacta, i.e. imbeddings for which there exists no homeomorphism $H: E^3 \to E^3$ taking the arc into a line segment, etc. There are many examples in the literature of wild imbeddings in E^3. For $n > 3$ Ivanov [3] and Blankinship [31] have examples of wildly imbedded zero-dimensional compacta in E^n for any n; in turn, these can be used to construct wildly imbedded spheres S^k, $k < n$, and n-cells in E^n. For the majority of the known examples of wild imbeddings, the proof of wildness comes down to the proof that the fundamental group of the complement is nontrivial.

We shall therefore now take up the basic facts concerning the fundamental group of a space and indicate some methods for computing it, which we shall use in Chapter IV.

It is quite likely that for $n > 3$ a wild imbedding in E^n can manifest itself not only via the fundamental group of the complement but via the homotopy groups of higher dimensions as well. But examples where this is the case have not appeared up to now.

§1. Paths in a space

1. By a path joining two points a and b in a space $\mathfrak{S}$ we mean a continuous mapping f of the line segment $I_s = [0, 1]$ into $\mathfrak{S}$ such that $f(0) = a$, $f(1) = b$. The point a is called the initial point of the path, the point b its terminal point. As the point s, $0 \le s \le 1$, of the segment I_s goes from 0 to 1, the point $f(s)$ runs over a locally connected continuum $K_f = f(I_s)$ in $\mathfrak{S}$. The continuum K_f will be called the *image* of the path f in $\mathfrak{S}$. It is important to observe that the notion of a path f must not be confused with that of its image K_f, and all the discussion that follows concerns a path, i.e. a continuous mapping of the segment I_s into $\mathfrak{S}$. The path is called closed if

$$f(0) = f(1).$$

It is called a unit path, and denoted by e, if $e(s)$ is a constant mapping:

$$e(s) = e(0), \quad 0 \leqslant s \leqslant 1.$$

Let f and ϕ be two paths in the space $\mathfrak{S}$, with $f(1) = \phi(0)$. We may then define a path g, called the product of f and ϕ, as follows:

$$g(s) = f(2s), \quad g(s) = \varphi(2s - 1), \qquad (1)$$

$$0 \leqslant s \leqslant \frac{1}{2}, \quad \frac{1}{2} \leqslant s \leqslant 1.$$

We write $g = f \circ \phi$.

For every path f there is an inverse path f^{-1} given by

$$f^{-1}(s) = f(1 - s).$$

The image of f^{-1} obviously coincides with that of f, but is "run over" in the opposite direction:

$$f^{-1}(0) = f(1), \quad f^{-1}(1) = f(0).$$

Two paths f and ϕ in $\mathfrak{S}$ are called homotopic: $f \sim \phi$, if there exists a homotopy $F_t(s)$ which is stationary at the endpoints of I_s and takes f into ϕ:

$$F_t : f \to \varphi;$$

$$F_0(s) = f(s), \quad F_1(s) = \varphi(s),$$

$$F_t(0) = f(0) = \varphi(0), \quad F_t(1) = f(1) = \varphi(1).$$

$$0 \leqslant t \leqslant 1 \qquad\qquad 0 \leqslant t \leqslant 1$$

Thus only paths with the same endpoints may be homotopic.

The homotopy $F_t(s)$ is a continuous mapping of the rectangle $I_s \times I_t$ into the space $\mathfrak{S}$:

$$F_t(s) : I_s \times I_t \to \mathscr{E};$$

$I_s \times I_t$ is called the homotopy rectangle for the paths f and ϕ. The mapping $F_t(s)$ must be constant on the sides $s = 0$ and $s = 1$ of the rectangle. In view of this, the rectangle $I^1_s \times I^1_t$ may be deformed to a circular disk Q by identifying each of its sides $s = 0$ and $s = 1$ to a point that maps into the corresponding endpoint a or b of the paths; the homotopy so deformed is then a mapping of the homotopy disk Q into $\mathfrak{S}$.

Since $f(0) = f(1)$ for a closed path f in $\mathfrak{S}$, the path can be regarded as a mapping of a circle into $\mathfrak{S}$. For this, it suffices to map the segment I_s onto the circle by identifying endpoints.

A closed path f is said to be homotopic to zero, or null-homotopic: $f \sim 0$, if it is homotopic to a unit path $\mathfrak{S}$. The mapping $e(s) = a$ which constitutes the unit path in question can be obtained in the following way: map every point of the segment I_s into the point $s = 0$, and then map the point $s = 0$ into the point a. The homotopy disk can be mapped continuously onto itself in such a fashion that the segment of the boundary that maps into a is identified to a point O, while on the remaining part of the disk the new mapping is a homeomorphism. Then the mapping $F_t: Q \to \mathfrak{S}$ goes into a mapping $F'_t: Q \to \mathfrak{S}$, where F'_0 coincides with the path f regarded as a mapping of the boundary of Q into $\mathfrak{S}$, and F'_1 is the mapping of the point O into the point a.

Consequently, if a closed path f, regarded as a mapping of a circle into $\mathfrak{S}$, is null-homotopic, the mapping extends to a mapping of the disk into $\mathfrak{S}$.

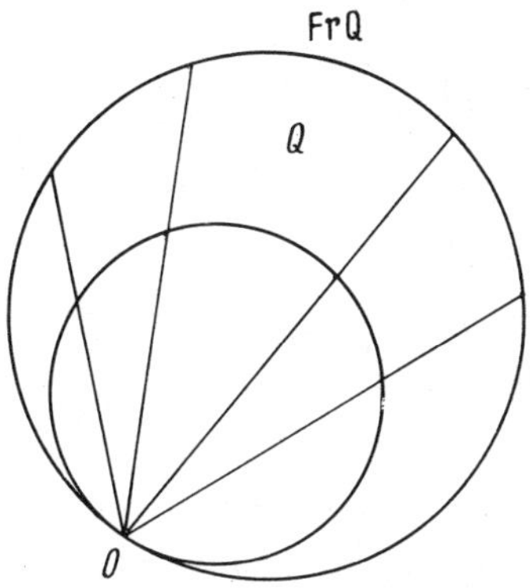

Figure 3.1

It is easily seen that the converse also holds: if a mapping f of a circle into $\mathfrak{S}$ extends to a mapping Φ of the disk into $\mathfrak{S}$, then the path f is null-homotopic. To see this, it suffices to consider the homotopy θ_t which takes the identity mapping T on $\operatorname{Fr} Q$ into the mapping of $\operatorname{Fr} Q$ into the point O, and which for each t, $0 < t < 1$, is the mapping of $\operatorname{Fr} Q$ into the circle contained in Q and tangent to $\operatorname{Fr} Q$ at O (Figure 3.1), and to set

$$\tilde{f}_t(s) = \Phi\theta_t(s),$$
$$s \in \operatorname{Fr} Q, \quad 0 \leqslant t \leqslant 1.$$

2. Elementary properties of paths.

1a. Observe first of all that homotopy of paths is an equivalence relation. Reflexivity: $F_t \equiv f$ takes f into f. Symmetry: if F_t takes f into ϕ, then F_{1-t} takes ϕ into f. We prove transitivity: if f is homotopic to ϕ, and ϕ to ψ, then f is homotopic to ψ.

Let F_t^1 be a homotopy taking f into ϕ, and F_t^2 a homotopy taking ϕ into ψ. Then the product F_t of the homotopies F_t^1 and F_t^2, defined by

$$F_t(s) = F^1(2t, s), \quad F_t(s) = F^2(2t-1, s),$$
$$0 \leqslant t \leqslant \frac{1}{2} \qquad \frac{1}{2} \leqslant t \leqslant 1$$

takes f into ψ.

1b. If a path f is homotopic to a path f', and a path ϕ homotopic to a path ϕ', and the product $f_\rho \phi$ is defined, then the product $f' \circ \phi'$ is defined and is homotopic to $f \circ \phi$.

We have

$$f'(0) = f(0), \quad f'(1) = f(1) = \varphi(0) = \varphi'(0), \quad \varphi'(1) = \varphi(1).$$

Hence the product $f' \circ \phi'$ is defined. By (1),

$$f \circ \varphi(s) = f(2s), \quad f \circ \varphi(s) = \varphi(2s-1),$$
$$f' \circ \varphi'(s) = f'(2s), \quad f' \circ \varphi'(s) = \varphi'(2s-1).$$
$$0 \leqslant s \leqslant \frac{1}{2} \qquad \frac{1}{2} \leqslant s \leqslant 1$$

Let $F_t^1(s)$ be a homotopy from f to f', and $F_t^2(s)$ a homotopy from ϕ to ϕ'. Then setting

$$F_t(s) = F_t^1(2s), \quad F_t(s) = F_t^2(2s-1), \quad 0 \leqslant t \leqslant 1,$$
$$0 \leqslant s \leqslant \frac{1}{2} \qquad \frac{1}{2} \leqslant s \leqslant 1$$

we obtain a homotopy from $f \circ \phi$ to $f' \circ \phi'$, since $f(1) = F_t^1(1) = F_t^2(0) = \phi(0)$, $0 \leq t \leq 1$.

We now derive some further elementary properties of paths, applying for this purpose the following remark of L. S. Pontrjagin [8].

Remark 1.1. Let h be a continuous mapping of the segment I_s into itself, with $h(0) = 0$, and f an arbitrary path in a space $\mathfrak{E}$. Then the composition $fh = fh(s)$ is likewise a path in $\mathfrak{E}$. If $h(1) = 1$, the path fh is homotopic to f; while if $h(1) = 0$, fh is null-homotopic.

1) $h(1) = 1$. The homotopy

$$F_t(s) = f[(1-t)h(s) + ts]$$

takes fh into f, since $F_t(0) \equiv f(0)$ and $F_t(1) \equiv f(1)$, $0 \leq t \leq 1$.

2) $h(1) \equiv 0$. The homotopy

$$F_t(s) = f[(1-t)h(s)]$$

takes fh into the unit mapping $e(s) \equiv f(0)$, since $F_t(0) = F_t(1) = f(0)$, $0 \leq t \leq 1$.

The remaining properties of paths will be derived by means of a suitable choice of the mapping h.

1c. The paths $f \circ f^{-1}$ and $f^{-1} \circ f$ are null-homotopic.

By formula (1) and the definition of the inverse of a path we have

$$f \circ f^{-1}(s) = f(2s), \quad f \circ f^{-1}(s) = f^{-1}(2s-1) = f[2(1-s)].$$
$$0 \leqslant s \leqslant \frac{1}{2} \qquad \frac{1}{2} \leqslant s \leqslant 1$$

Setting

$$h(s) = 2s, \quad h(s) = 2(1-s),$$
$$0 \leqslant s \leqslant \frac{1}{2} \qquad \frac{1}{2} \leqslant s \leqslant 1,$$

we find that $f \circ f^{-1}(s) = fh(s)$, and since $h(0) = h(1) = 0$, it follows by Remark 1.1

that $f \circ f^{-1} \sim 0$. Similarly, $f^{-1} \circ f \sim 0$ (observe that $f \circ f^{-1}$ and $f^{-1} \circ f$ are not unit paths).

1d. For any path f

$$f \circ e \sim f, \quad \text{if} \quad e(s) = f(1).$$

$$e \circ f \sim f, \quad \text{if} \quad e(s) = f(0).$$

Set

$$\underset{0 \leqslant s \leqslant \frac{1}{2}}{h(s) = 2s,} \quad \underset{\frac{1}{2} \leqslant s \leqslant 1}{h(s) = 1,}$$

then $f \circ e = fh$ by formula (1), so that, by Remark 1.1, $f \circ e \sim f$. The second relation is proved similarly.

1e. Multiplication of paths is associative:

$$(f \circ \phi) \circ \psi \sim f \circ (\phi \circ \psi).$$

Set

$$u(s) = (f \circ \varphi) \circ \psi(s), \quad v(s) = f \circ (\varphi \circ \psi)(s).$$

By formula (1),

$$u(s) = f \circ \varphi(2s), \quad u(s) = \psi(2s - 1),$$

$$v(s) = f(2s), \quad v(s) = \varphi \circ \psi(2s - 1).$$

$$0 \leqslant s \leqslant \frac{1}{2} \qquad \frac{1}{2} \leqslant s \leqslant 1$$

Applying formula (1), again we find

$$\underset{0 \leqslant s \leqslant \frac{1}{4}}{u(s) = f(4s),} \quad \underset{\frac{1}{4} \leqslant s \leqslant \frac{1}{2}}{u(s) = \varphi(4s - 1),} \quad \underset{\frac{1}{2} \leqslant s \leqslant 1}{u(s) = \psi(2s - 1),}$$

$$\underset{0 \leqslant s \leqslant \frac{1}{2}}{v(s) = f(2s),} \quad \underset{\frac{1}{2} \leqslant s \leqslant \frac{3}{4}}{v(s) = \varphi(4s - 2),} \quad \underset{\frac{3}{4} \leqslant s \leqslant 1}{v(s) = \psi(4s - 3).}$$

The mapping h, taking the segment [0, ½] into [0, ¼], [½, ¾] into [¼, ½], and [¾, 1] into [½, 1], is given by

$$\underset{0 \leqslant s \leqslant \frac{1}{2}}{h(s) = \frac{s}{2},} \quad \underset{\frac{1}{2} \leqslant s \leqslant \frac{3}{4}}{h(s) = s - \frac{1}{4},} \quad \underset{\frac{3}{4} \leqslant s \leqslant 1}{h(s) = 2s - 1.}$$

It is easily verified that $v(s) = uh(s)$, so that v is homotopic to u by Remark 1.1. It follows from 1e that the product of several paths can be written in the form $f = f_1 \circ f_2 \circ \cdots \circ f_r$.

1b′. Let $f = f_1 \circ f_2 \circ \cdots \circ f_r$, and suppose each path f_i is homotopic to a path f'_i, $i = 1, 2, \cdots, r$; then

$$f \sim f'_1 \circ f'_2 \circ \cdots \circ f'_r.$$

For $r = 2$ this is the same as 1b. For $r > 2$ we prove it by induction. Assuming 1b′ holds for $r - 1$, we get

$$f_1 \circ f_2 \circ \cdots \circ f_{r-1} \sim f'_1 \circ f'_2 \circ \cdots \circ f'_{r-1}.$$

Applying 1b, we obtain

$$(f_1 \circ f_2 \circ \cdots \circ f_{r-1}) \circ f_r \sim (f'_1 \circ f'_2 \circ \cdots \circ f'_{r-1}) \circ f'_r.$$

1f. Let $\Phi = f_1 \circ f_2 \circ \cdots \circ f_r$; then for any sequence of numbers

$$0 = a_0 < a_1 < a_2 < \ldots a_{r-1} < a_r = 1,$$

the path Φ is homotopic to the path Φ' defined as follows:

$$\Phi'(s) = f_i \theta_i(s), \quad a_{i-1} \leqslant s \leqslant a_i,$$

where $\theta_i(s)$ is the linear mapping of the segment $[a_{i-1}, a_i]$ onto $[0, 1]$.

Applying formula (1) repeatedly, as in 1e, we find

$$\Phi(s) = f_1 k_1(s), \quad \Phi(s) = f_i k_i(s), \quad i = 1, 2, \ldots r,$$

$$0 \leqslant s \leqslant \frac{1}{2^{r-1}} \qquad \frac{1}{2^{r-i+1}} \leqslant s \leqslant \frac{1}{2^{r-i}},$$

where $k_i(s)$ is the linear mapping of the segment $[1/2^{r-i+1}, 1/2^{r-i}]$ onto the segment $[0, 1]$. Let $h: I_s \to I_s$ be the mapping which is linear on each $[a_{i-1}, a_i]$ and such that

$$h[a_{i-1}, a_i] = \left[\frac{1}{2^{r-i+1}}, \frac{1}{2^{r-i}}\right].$$

By Remark 1.1 we have $\Phi h \sim \Phi$, since $h(0) = 0$, $h(1) = 1$. Obviously

$$\Phi h = f_i k_i h(s), \quad i = 1, 2, \ldots, r.$$

$$a_{i-1} \leqslant s \leqslant a_i$$

Setting $\Theta_i = k_i h$, we find that Θ_i is the linear mapping

$$\Theta_i [a_{i-1}, a_i] = [0, 1].$$

Consequently $\Phi' = \Phi h$ satisfies 1f.

1g. For any sequence of numbers

$$0 = a_0 < a_1 < \ldots < a_r = 1,$$

a path f is homotopic to the path

$$f' = f_{a_0a_1} \circ f_{a_1a_2} \circ \cdots \circ f_{a_{r-1}a_r},$$

where $f_{a_{i-1}a_i} = fk_i(s)$ and k_i is the linear mapping

$$k_i[0, 1] = [a_{i-1}, a_i].$$

This follows from 1f if we set

$$\Theta_i(s) = k_i^{-1}(s), \quad i = 1, 2, \ldots, r.$$

Each path $f_{a_{i-1}a_i}$ will be called the part of the path f on the segment $[a_{i-1}, a_i]$.

1h. The path f' obtained from a path f by replacing a part $f'_{a_1a_2}$ by a homotopic path ϕ, is homotopic to f.

By 1g, f is homotopic to the product of its parts:

$$f \sim f_{0a_1} \circ f_{a_1a_2} \circ f_{a_21}.$$

By 1b′ the path $f_{0a_1} \circ f_{a_1a_2} \circ f_{a_21}$ is homotopic to the path $f_{0a_1} \circ \phi \circ f_{a_21}$ obtained by replacing the part $f_{a_1a_2}$ by the homotopic path ϕ; and by 1a, f is homotopic to $f_{0a_1} \circ \phi \circ f_{a_21}$.

1i. Let a and ϵ be numbers such that

$$0 < a - \varepsilon < a < a + \varepsilon < 1.$$

Then any path f is homotopic to the path f' defined by

$$\begin{array}{lll} f'(s) = f\alpha_1(s), & f'_{a-\varepsilon,\, a+\varepsilon} = \varphi, & f'(s) = f\alpha_2(s), \\ 0 \leqslant s \leqslant a - \varepsilon & & a + \varepsilon \leqslant s \leqslant 1, \end{array}$$

where α_1 and α_2 are the linear mappings such that

$$\alpha_1[0, a - \varepsilon] = [0, a], \quad \alpha_2[a + \varepsilon, 1] = [a, 1],$$

and ϕ is a path homotopic to the unit path $e(s) \equiv f(a)$.

Applying Remark 1.1, we have that f is homotopic to the path fh, where $h: I_s \to I_s$ is given by

$$\begin{array}{lll} h(s) = \alpha_1(s), & h(s) = a, & h(s) = \alpha_2(s). \\ 0 \leqslant s \leqslant a - \varepsilon & a - \varepsilon \leqslant s \leqslant a + \varepsilon & a + \varepsilon \leqslant s \leqslant 1. \end{array}$$

The part $fh_{a-\epsilon, a+\epsilon}$ of this path is the unit path $e(s) = f(a)$. By 1h, if we replace this part by the homotopic path ϕ, we obtain a path homotopic to fh and consequently to f.

The operation just described of passing from f to f' is called insertion, into the path f at the point a, of a null-homotopic path. Thus 1i can be restated as follows:

The path f' obtained from f by insertion of a null-homotopic path at some point, is homotopic to f.

In particular, if w is any path with initial point $f(a)$, we can insert into the path f, at the point a, the path $w \circ w^{-1}$. If we denote by f_- the part of the path f' corresponding to $f\alpha_1$, and by f_+ the part of f' corresponding to $f\alpha_2$, we obtain

$$f \sim f_- \circ w \circ w^{-1} \circ f_+.$$

It follows also from 1i that if $f = f_1 \circ \phi \circ f_2$ and $\phi \sim 0$, then $f \sim f_1 \circ f_2$, i.e. a null-homotopic factor ϕ can be dropped from a product. In particular, a path which is the product of several paths goes into a homotopic path if we drop from the product all factors of the form $w \circ w^{-1}$ and $w^{-1} \circ w$.

§2. The fundamental group

1. The fundamental group may be defined for any spaces satisfying certain conditions. To be specific, we shall require that the space $\mathfrak{E}$ be arcwise connected. This means that any two points a and b in $\mathfrak{E}$ can be joined by a path. An arcwise connected space is obviously connected.

Observe that this is equivalent to the condition that any two points a and b in $\mathfrak{E}$ can be joined by a simple arc. Indeed, the image of a path is a locally connected continuum in $\mathfrak{E}$, and any two points of a locally connected continuum can be joined by a simple arc.

The set of all paths in $\mathfrak{E}$ does not constitute a group, nor does the set of all closed paths in $\mathfrak{E}$, since the product is not defined for all pairs of paths, and since the path $f \circ f^{-1}$ is not the unit path but is only homotopic to it.

Consider the set of all closed paths in the space with a given initial point, or origin, O, and separate them into classes: any one class c is an equivalence class of homotopic paths (see 1a). We now show that the set of all such classes of homotopic closed paths with origin O constitutes a group.

1) The classes c can be multiplied. For any two closed paths with origin O, the product is defined. The product $c_1 c_2$ of two classes c_1 and c_2 is the class consisting of all paths homotopic to the product $f_1 \circ f_2$, where $f_1 \in c_1$, $f_2 \in c_2$. By 1b the products of different paths in the classes c_1 and c_2 are homotopic.

The class c consisting of those closed paths with origin O which are null-homotopic will be called the unit class $\bar{e}$.

2) For every class c there is an inverse, i.e. a class c^{-1} such that

$$cc^{-1} = \bar{e}, \quad c^{-1}c = \bar{e}.$$

The class c^{-1} consists of all paths f^{-1} inverse to paths $f \in c$. By 1b and 1c, the product $f \circ \phi$ of a path $f \in c$ and a path $\phi \in c^{-1}$ is null-homotopic, since $f^{-1} \in c^{-1}$ and hence f^{-1} is homotopic to ϕ, so that $f \circ \phi$ is homotopic to $f \circ f^{-1} \in \bar{e}$.

We have also, for any class c,

$$c\bar{e} = c \text{ and } \bar{e}c = c,$$

since by 1b and 1d the product of a path $f \in c$ by a null-homotopic path ϕ ($\phi \in \bar{e}$) is homotopic to $f \circ e \sim f$, so that $f \circ \phi \in c$ and $\phi \circ f \in c$.

3) By 1e, multiplication of the classes c is associative.

Thus the set of all classes of homotopic closed paths with origin O is a group.

Definition 2.1. The group of all classes of homotopic closed paths with origin O in an arcwise connected space $\mathfrak{E}$ is called the *fundamental group* of $\mathfrak{E}$ and denoted by $\pi_1(\mathfrak{E}, O)$. The fundamental group $\pi_1(\mathfrak{E}, O)$ is determined by a set of generating classes $\{c_\lambda\}$ such that every class $c \in \pi_1(\mathfrak{E}, O)$ is a product of classes c_λ, together with a set of defining relations of the form

$$c_{\lambda_1} c_{\lambda_2} \ldots c_{\lambda_k} = 1$$

(where 1, for convenience, means the class $\bar{e}$).

Theorem 2.1. *The fundamental group of an arcwise connected space is independent of the choice of origin, i.e. the groups $\pi_1(\mathfrak{E}, O)$ and $\pi_1(\mathfrak{E}, O')$ for different points O and O' are isomorphic.*

Since $\mathfrak{E}$ is connected, there is a path w in $\mathfrak{E}$ joining O and O'. To each closed path f with origin O associate the path $\gamma(f) = w^{-1} \circ f \circ w$, and to each closed path f' with origin O' associate the path $\gamma'(f') = w \circ f' \circ w^{-1}$.

By 1b, homotopic paths f_1 and f_2 determine homotopic paths $\gamma(f_1)$ and $\gamma(f_2)$. Hence an element $c \in \pi_1(\mathfrak{E}, O)$ determines an element

$$\gamma(c) = \bar{w}^{-1} \circ c \circ \bar{w} \in \pi_1(\mathfrak{E}, O'),$$

and an element $c' \in \pi_1(\mathfrak{E}, O')$ determines an element

$$\gamma'(c') = \bar{w} \circ c' \circ \bar{w}^{-1} \in \pi_1(\mathfrak{E}, O),$$

where $\bar{w}$ is the class of paths in $\mathfrak{E}$ homotopic to w.

We now show that the correspondence γ (resp. γ') is an isomorphism of the group $\pi_1(\mathfrak{E}, O)$ onto the group $\pi_1(\mathfrak{E}, O')$ ($\pi_1(\mathfrak{E}, O')$ onto $\pi_1(\mathfrak{E}, O)$).

1) $\gamma(c_1 c_2) = \gamma(c_1)\gamma(c_2)$.

If $f_1 \in c_1$ and $f_2 \in c_2$, then

$$\gamma\,(f_1 \circ f_2) = w^{-1} \circ f_1 \circ f_2 \circ w.$$

By 1i, we may insert between f_1 and f_2 in this product the path $w \circ w^{-1}$. Hence

$$\gamma\,(f_1 \circ f_2) \sim (w^{-1} \circ f_1 \circ w) \circ (w^{-1} \circ f_2 \circ w) = \gamma(f_1) \circ \gamma\,(f_2).$$

Conversely, the path $\gamma(f_1 \circ f_2)$ is obtained from $\gamma(f_1) \circ \gamma(f_2)$ by dropping the factor $w \circ w^{-1}$.

2) γ is an isomorphism: distinct elements c_1 and c_2 of $\pi_1(\mathfrak{S}, O)$ determine distinct elements $\gamma(c_1)$ and $\gamma(c_2)$ of $\pi_1(\mathfrak{S}, O')$.

Otherwise, we should have two classes c_1 and c_2 such that, for $f_1 \in c_1$ and $f_2 \in c_2$, $c_1 \neq c_2$, the paths $\gamma(f_1)$ and $\gamma(f_2)$ are homotopic. But then, by 1b, the paths $\gamma'\gamma(f_1)$ and $\gamma'\gamma(f_2)$ are homotopic. Since

$$\gamma'\gamma f_1 = w \circ w^{-1} \circ f_1 \circ w \circ w^{-1} \sim f_1, \quad \gamma'\gamma f_2 = w \circ w^{-1} \circ f_2 \circ w \circ w^{-1} \sim f_2,$$

it would follow that $f_1 \sim f_2$, a contradiction.

Finally, γ is an isomorphism of $\pi_1(\mathfrak{S}, O)$ onto $\pi_1(\mathfrak{S}, O')$. Indeed, any element $c' \in \pi_1(\mathfrak{S}, O')$ determines $\gamma'(c') \in \pi_1(\mathfrak{S}, O)$, and

$$\gamma\gamma'\,(c') = w^{-1} \circ w \circ c' \circ w^{-1} \circ w = c'.$$

Observe that the isomorphism γ depends on the choice of the path w. Nonhomotopic paths w joining O and O' determine isomorphisms of $\pi_1(\mathfrak{S}, O)$ with $\pi_1(\mathfrak{S}, O')$ which differ by an inner automorphism of $\pi_1(\mathfrak{S}, O')$.

In view of Theorem 2.1, the origin need not be specified in defining the fundamental group, and we may denote the latter by $\pi_1(\mathfrak{S})$, or simply π_1 when a specific space is under consideration.

A space with trivial fundamental group, i.e. in which every closed path is null-homotopic, is called *simply connected.*

2. Theorem 2.2. *A continuous mapping* Φ *of an arcwise connected space* X *into an arcwise connected space* Y *induces a homomorphism of the fundamental group of* X *into the fundamental group of* Y.

Let O be the origin for paths in X; take as the origin for paths in Y the point $\Phi(O)$. Then every path $f \in c \in \pi_1(X)$ determines a path Φf. Homotopic paths f and ϕ in X determine homotopic paths Φf and $\Phi\phi$ in Y; indeed, if F_t is a homotopy from f to ϕ, then ΦF_t is a homotopy from Φf to $\Phi\phi$. Consequently the class $c \in \pi_1(X)$ determines the class $H(c) \in \pi_1(Y)$ containing any path Φf, $f \in c$. In particular, the unit class of $\pi_1(X)$ determines the unit class of $\pi_1(Y)$.

The product $f_1 \circ f_2$ of paths in X goes into the product $\Phi f_1 \circ \Phi f_2$. Indeed, by formula (1) we have

$$f_1 \circ f_2(s) = f_1(2s), \quad f_1 \circ f_2(s) = f_2(2s - 1).$$
$$0 \leqslant s \leqslant \frac{1}{2} \qquad\qquad \frac{1}{2} \leqslant s \leqslant 1$$

Hence

$$\Phi(f_1 \circ f_2(s)) = \Phi f_1(2s), \quad \Phi(f_1 \circ f_2(s)) = \Phi f_2(2s - 1).$$
$$0 \leqslant s \leqslant \frac{1}{2} \qquad\qquad \frac{1}{2} \leqslant s \leqslant 1.$$

Therefore

$$\Phi(f_1 \circ f_2) = (\Phi f_1) \circ (\Phi f_2).$$

Consequently the image of the product of any two elements c_1 and c_2 in $\pi_1(X)$ is the product of their images in $\pi_1(Y)$: $H(c_1 \cdot c_2) = H(c_1) \cdot H(c_2)$, i.e. the mapping H is a homomorphism of $\pi_1(X)$ into $\pi_1(Y)$ which is induced by the mapping Φ.

We shall denote by h_f the homomorphism $h_f\colon \pi_1(X) \to \pi_1(Y)$ induced by the continuous mapping $f\colon X \to Y$.

Corollary. *Let $\Phi\colon X \to Y$ and $\Psi\colon Y \to Z$ be continuous mappings. Then the homomorphisms h_Φ, h_Ψ and $h_{\Psi\Phi}$ satisfy the relation $h_{\Psi\Phi} = h_\Psi h_\Phi$.*

This follows at once from the fact that a path ϕ in X determines the path $\Phi(\phi)$ in Y and the path $\Psi[\Phi(\phi)]$ in Z.

Theorem 2.3. *Let X and Y be arcwise connected spaces, and Φ_0 and Φ_1 two continuous mappings of X into Y which are homotopic to each other, with $\Phi_0(O) = \Phi_1(O)$, $O \in X$. Then if the homotopy Φ_t connecting Φ_0 with Φ_1 is constant at O: $\Phi_t(O) = O_0$, $0 \leq t \leq 1$, the homomorphisms h_{Φ_0} and h_{Φ_1} of $\pi_1(X, O)$ into $\pi_1(Y, O_0)$ induced by Φ_0 and Φ_1 are the same. If $\Phi_0(O) \neq \Phi_t(O)$ for some t, $0 < t < 1$, i.e. if $\Phi_t(O) \neq O_0$, then the homomorphisms are the same up to an inner automorphism of $\pi_1(Y, O_0)$.*

Suppose $\Phi_t(O) = O_0$, $0 \leq t \leq 1$, and let ϕ be a closed path in X with origin O. Then $\psi_0 = \Phi_0(\phi)$ and $\psi_1 = \Phi_1(\phi)$ are closed paths in Y with origin O_0. They are homotopic, and the homotopy $\Phi_t(\phi)$ connecting them is stationary at O_0. Consequently the paths ψ_0 and ψ_1 belong to the same class in the group $\pi_1(Y, O_0)$, so that $h_{\Phi_0} = h_{\Phi_1}$.

Now suppose $\Phi_{t_0}(O) \neq O_0$ for some t_0, let $\Phi_1(O) = O_1$, and let $f = \Phi_t(O)$, $0 \leq t \leq 1$, be the path (in general not closed) traced out under the homotopy by the image of O. Denote by f_t the segment of the path f for values of the argument from 0 to t. Then the homotopy $f_t \circ \Phi_t(\phi) \circ f_t^{-1}$ joins the closed path ψ_0 in Y with the path $f_1 \circ \psi_1 \circ f_1^{-1}$ and is stationary at the point O_0. Hence the homomorphism h_{Φ_1} of $\pi_1(X, O)$ into $\pi_1(Y, O_1)$ is the composite of the homomorphism $h_{\Phi_0}: \pi_1(X, O) \to \pi_1(Y, O_0)$ and the isomorphism $H_f: \pi_1(Y, O_0) \to \pi_1(Y, O_1)$ determined by the path f (see Theorem 2.1). If now $O_1 = O_0$, then H_f is an inner automorphism of $\pi_1(Y, O_0)$.

Corollary 1. *Let* $\Phi: X \to Y$ *and* $\Psi: Y \to X$ *be continuous mappings of two arcwise connected spaces, such that* $\Psi\Phi$ *is homotopic to the identity mapping of* X. *Then* h_Φ *is an isomorphism of* $\pi_1(X)$ *into* $\pi_1(Y)$, *and* h_Ψ *is a homomorphism of* $\pi_1(Y)$ *onto* $\pi_1(X)$.

By the corollary to Theorem 2.2 we have $h_{\Psi\Phi} = h_\Psi h_\Phi$. Since $\Psi\Phi$ is homotopic to the identity mapping, it follows by Theorem 2.3 that $h_{\Psi\Phi}$ is an isomorphism of $\pi_1(X)$ onto itself. But this implies both that h_Φ must be an isomorphism and that h_Ψ must be a homomorphism "onto".

A very important role is played in topology by the notion of homotopic equivalence between spaces.

Definition 2.2. The spaces X and Y are called *homotopically equivalent* if there exist continuous mappings $\Phi: X \to Y$ and $\Psi: Y \to X$ such that $\Psi\Phi$ is homotopic to the identity mapping of X and $\Phi\Psi$ is homotopic to the identity mapping of Y.

Corollary 2. *Two arcwise connected homotopically equivalent spaces have the same fundamental group.*

Let $\Phi: X \to Y$, $\Psi: Y \to X$ be continuous mappings, with $\Psi\Phi$ and $\Phi\Psi$ homotopic to the identity mappings of X and Y, respectively. Then our assertion follows at once from Corollary 1, since both h_Φ and h_Ψ must be isomorphisms "onto".

We remark that the converse is in general false. Two arcwise connected spaces may have the same fundamental group but not be homotopically equivalent.

Corollary 3. *The fundamental groups of two homeomorphic arcwise connected spaces are the same.*

This follows from Corollary 2, setting $\Psi = \Phi^{-1}$, where Φ is a homeomorphism. Thus, *the fundamental group is a topological invariant.*

We introduce now a notion which will be useful in considering examples.

Definition 2.3. A subspace X of a space Y is called a *deformation retract* of Y if there exists a homotopy $F_t: Y \to Y$ from the identity mapping $F_0 \equiv 1$ such that $F_1(Y) = X$ and $F_t(X) \subset X$ for every t, $0 \leq t \leq 1$.

Corollary 4. *If* X *is a deformation retract of* Y, *the fundamental groups of* X *and* Y *are isomorphic.*

Let F_t be the homotopy of Y retracting it onto X. Apply Corollary 2, taking for Φ the identity inclusion of X into Y, and setting $\Psi = F_1$. Then the homotopy F_t takes the identity mapping F_0 of Y into $\Phi\Psi$, and the homotopy $F_t\Phi$, which is defined on X because $F_t(X) \subset X$, takes the identity mapping of X into $\Psi\Phi|_X$. Since $\Phi\Psi$ and $\Psi\Phi$ are homotopic to the identity mappings, the groups $\pi_1(X)$ and $\pi_1(Y)$ are isomorphic.

An inclusion, i.e. a homeomorphism of A into B, determines a homomorphism $h_\phi : \pi_1(A) \to \pi_1(B)$, called the *inclusion homomorphism.*

3. Examples of fundamental groups.

1. Example 2.1. The fundamental groups of Euclidean space E^n and of a cube I^n are trivial. Choosing a coordinate system, we may locate the point O at the origin of coordinates. Then a null-homotopy for a path $f(s)$ is given by

$$F_t = (1 - t)f(s), \quad 0 \leqslant s \leqslant 1, \quad 0 \leqslant t \leqslant 1.$$

Example 2.2. The fundamental group of a sphere S^n is trivial if $n > 1$, and is a free cyclic group if $n = 1$.

a) $n = 1$. Let P be a point on S^1. The circle S^1 is the continuous image of the interval I^1, homeomorphic except for the endpoints going into P.

Let a_0 be the class of paths containing this mapping. The group π_1 contains classes of the form $a_0^{\pm k}$, $k \geq 0$. We show that $a_0^k \neq a_0^m$ for $k \neq m$, and that any path f is homotopic to some path in one such class $a_0^{\pm k}$.

Let the length of the circle S^1 be 1. Choose a direction of motion on S^1. The position of any point $f(s)$ on S^1 may be specified by means of a number $l(s)$, the "length" of the segment $f[0, s]$, defined as follows. Since f is uniformly continuous, we may subdivide $[0, 1]$ by points α_i into segments $[\alpha_i, \alpha_{i+1}] = \Delta_i$ such that each $f(\Delta_i)$ lies in a semicircle. Then the directions on S^1 and on $[0, 1]$ allow us to give the length of the arc $(f(\alpha_i), f(\alpha_{i+1}))$ a sign. Let s be a point of subdivision. Then for $l(s)$ we take the algebraic sum of the lengths of the arcs $(f(\alpha_i), f(\alpha_{i+1}))$ for all segments Δ_i to the left of s. Obviously $l(s)$ is independent of the choice of the subdivision, so long as the latter is sufficiently fine and the point s is one of the points of subdivision. Every class a_0^k contains a path f_k defined by the condition

$$l(s) = ks.$$

We assert that no two such paths are homotopic. Suppose there exists a homotopy taking a path f_k into a path f_m. Then there exists a homotopy $f_t(s)$ such that (denoting by $l_t(s)$ the length of $f_t[0, s]$):

$$l_0(s) = ks; \quad l_1(s) = ms, \quad 0 \leqslant s \leqslant 1,$$

and, in particular,

$$l_0(1) = k, \quad l_1(1) = m.$$

But $f_t(1) = P$, so that $l_t(1)$ can take on only integral values, and it follows from the continuity of $l_t(s)$ that $l_t(1) = \text{const}$. Hence $k = m$.

Now let f be an arbitrary path in S^1, defined by the function $l_f(s)$. Since $f(0) = f(1) = P$, we have $l_f(1) = k_f$, $k_f = 0, \pm 1, \cdots$. Consider the homotopy $F_t(s)$ such that

$$l(F_t(s)) = (1 - t)\, l_f(s) + k_f ts.$$

It is readily verified that

$$l(F_t(0)) = 0 \quad \text{and} \quad l(F_t(1)) = k_f, \; 0 \leqslant t \leqslant 1:$$

$$l(F_0(s)) = l_f(s); \quad l(F_1(s)) = k_f s.$$

Consequently, the path f is homotopic to the path $k_f s$, i.e. $f \in a_0^{k_f}$.

b) $n > 1$. In this case, every path f in S^n is null-homotopic.

If the image of f does not cover all of S^n this follows from Example 2.1, since $S^n \backslash \{p\}$, for any point p, is homeomorphic to E^n. Now suppose $f: I_s^1 \to S^n$ is a closed path whose image covers all of S^n. Represent f as a product of several parts:

$$f = f_{0a_1} \circ f_{a_1 a_2} \circ \ldots \circ f_{a_k 1}$$

such that each $f(a_{i-1}, a_i)$, $i = 1, \cdots, k+1$, does not cover all of S^n.

Let w_i, $i = 1, \cdots, k$, be a path in S^n whose image is a simple arc joining the origin O with the point $f(a_i)$. Insert into f, at each point a_i, the null-homotopic path $w_i^{-1} \circ w_i$:

$$f \sim (f_{0a_1} \circ w_1^{-1}) \circ (w_1 \circ f_{a_1 a_2} \circ w_2^{-1}) \circ \ldots \circ (w_k \circ f_{a_k 1}).$$

Since $f(a_{i-1}\, a_i)$ is a compactum which is not all of S^n, $n \geq 2$, and the images of the paths w_i are simple arcs, we have that each $w_{i-1} \circ f_{a_{i-1} a_i} \circ w_i^{-1}$ is a closed path in S^n with origin O, whose image does not cover all of S^n; thus, as we have already seen, $w_{i-1} \circ f_{a_{i-1} a_i} \circ w_i^{-1} \sim 0$, and hence $f \sim 0$.

Thus the group $\pi_1(S^n)$ is trivial for $n > 1$.

Example 2.3. The fundamental group of $E^n \backslash \{p\}$, for any point p, is trivial for $n > 2$, and is a free cyclic group for $n = 2$. (If $n = 1$, $E^1 \backslash \{p\}$ is disconnected.)

The point p may be taken as the origin of coordinates in E^n. Let S^{n-1} be the sphere with center p and radius 1. Then S^{n-1} is a deformation retract of $E^n \backslash \{p\}$. If r_x is the length of the segment $\overline{px}$, the homotopy is given by

$$F_t(x) = \left[1 - t + \frac{t}{r_x}\right] x.$$

Under the homotopy any point x belonging to S^{n-1} remains fixed, since $r_x = 1$. for any such point.

By Corollary 4 of Theorem 2.3 we have $\pi_1(E^n \backslash \{p\}) = \pi_1(S^{n-1})$. Hence by Example 2.2 the fundamental group of $E^n \backslash \{p\}$ is trivial for $n > 2$ and free cyclic for $n = 2$. It is easily seen that the result is unchanged if E^n is replaced by a cube I^n, a topological cube, a half-space, etc., so long as p is in its interior.

The result of Example 2.3 also remains valid, as well as the argument, if p is replaced by a k-dimensional cube I^k or ball Q^k, where $k \leq n$. But I^k cannot be replaced by a k-dimensional topological cube. We shall see in Chapter IV that the fundamental group of $E^3 \backslash l$, where l is a simple arc, or of $E^3 \backslash D^k$, $k = 2, 3$, where D^k is a k-dimensional topological cube, need not be trivial.

Example 2.4. Let R be the closed part of E^3 contained between the faces of a dihedral angle, from which has been removed a line segment l whose endpoints lie on different faces. Then the fundamental group of R is free cyclic.

By a projection along l, R can be deformed homotopically into a single face of the angle (i.e. into a half-plane) with the point belonging to l removed. We then find, just as in Example 2.3, that the fundamental group of R is free cyclic.

It is easily seen that the result is unchanged if R is obtained by removing, not a line segment, but an arc of a circle; or if we replace the dihedral angle by a cube, a trihedral angle, or in general a convex part of E^3.

§3. The van Kampen Theorem and its applications

1. In computing the fundamental group of an arcwise connected space $\mathfrak{S}$, it is often convenient to represent the latter as a union of several arcwise connected spaces $\mathfrak{S}_i$ whose fundamental groups are known. The theorem of van Kampen allows us in certain cases to compute the fundamental group of $\mathfrak{S}$, knowing the fundamental groups of all the $\mathfrak{S}_i$. We first make two preliminary remarks.

Remark 3.1. Let h and h' be two homeomorphisms of the segment I_s^1 onto a simple arc l with endpoints α and β, and let f be a continuous mapping of l into an arcwise connected space X. If $h(0) = h'(0) = \alpha$, $h(1) = h'(1) = \beta$, then $fh \sim fh'$. If $h(0) = \alpha$, $h(1) = \beta$ and $h'(0) = \beta$, $h'(1) = \alpha$, then $fh' \sim (fh)^{-1}$.

$h'^{-1}h$ is a homeomorphism of I_s^1 onto itself. In the first case, the homotopy

$$F_t = fh'\left[(1 - t)\, h'^{-1}h(s) + ts\right]$$

takes the path fh into the path fh'. In the second case, $h'^{-1}h(0) = 1$ and $h'^{-1}h(1) = 0$, and the homotopy

$$\Phi_t = fh' \,[(1 - t)h'^{-1}h\,(s) + t\,(1 - s)]$$

takes the path fh into the path $(fh')^{-1}$.

Remark 3.2. Let $f\colon D \to \mathfrak{S}$ be a continuous mapping of a (not necessarily circular) disk D into a space $\mathfrak{S}$, and κ a continuous mapping of a circle C onto $\operatorname{Fr} D$. Then $f\kappa\colon C \to \mathfrak{S}$ is a closed path in $\mathfrak{S}$ which is homotopic to zero.

It suffices to show that $f\kappa$ can be extended to the circular disk Q with boundary C. Let $h\colon Q \to D$ be a homeomorphism. Then $h^{-1}|_{\operatorname{Fr} D}\kappa\colon C \to C$ is a continuous mapping of the circle onto itself. It can be extended to a mapping $\Phi\colon Q \to Q$ such that the radius to a point $x \in C$ goes into the radius to the point $h^{-1}|_{\operatorname{Fr} D}\kappa(x)$. Then

$$h\Phi : Q \to D, \quad h\Phi\,|_C = \varkappa.$$

Consequently $fh\Phi\colon Q \to \mathfrak{S}$ and $fh\Phi|_C = f\kappa$.

Theorem 3.1. (van Kampen [74]). *Let $\mathfrak{S}$ be the union of two open arcwise connected subsets: $\mathfrak{S} = A \cup B$, with the intersection $A \cap B$ itself arcwise connected and nonvoid. Then the fundamental group $\pi_1(\mathfrak{S})$ is obtained from the free product $\pi_1(A) * \pi_1(B)$ by equating any two classes $a \in \pi_1(A)$ and $b \in \pi_1(B)$ containing the same path in $A \cap B$.*

Proof. 1. Take the origin O in $U = A \cap B$. We show first that a set of generators for the group $\pi_1(\mathfrak{S})$ is the totality of generators of the groups $\pi_1(A)$ and $\pi_1(B)$. To do so, we must show that every closed path in $\mathfrak{S}$ with origin O is homotopic to a product of finitely many closed paths,

$$f \sim f_1 \circ f_2 \circ \cdots \circ f_k,$$

where each f_{2i-1} is a path in A, and each f_{2i} a path in B. Let $f\colon I^1_s \to \mathfrak{S}$ and $K = f(I^1_s)$. The sets $A \setminus B = \mathfrak{S} \setminus B$ and $B \setminus A = \mathfrak{S} \setminus A$ are closed subsets of $\mathfrak{S}$, so that $P_A = f^{-1}[K \cap (A \setminus B)]$ and $P_B = f^{-1}[K \cap (B \setminus A)]$ are compact subsets of I^1_s, with $P_A \cap P_B = \Lambda$. Picking a neighborhood of P_A whose closure does not meet P_B, we can represent I^1_s as a union of finitely many segments,

$$I^1_s = \bigcup_{\rho=1}^{N} \Delta_\rho,$$

such that $\Delta_\rho \cap \Delta_{\rho'} = \Lambda$ for $|\rho - \rho'| > 1$, $\Delta_\rho \cap \Delta_{\rho+1}$ is a point, and

$$P_A \subset \bigcup_i \operatorname{int} \Delta_{2i-1}, \quad P_B \subset \bigcup_i \operatorname{int} \Delta_{2i}.$$

The path f can then be represented as the product of its parts on the segments Δ_ρ: $f = f_{\Delta_1} \circ f_{\Delta_2} \circ \cdots \circ f_{\Delta_N}$, where

$$f(\Delta_{2\lambda-1}) = f_{\Delta_{2\lambda-1}}(I_s) \subset A, \quad f(\Delta_{2\lambda}) = f_{\Delta_{2\lambda}}(I_s) \subset B.$$

If we now insert between the paths f_{Δ_ρ} and $f_{\Delta_{\rho+1}}$ the path $w_\rho \circ w_\rho^{-1}$, where w_ρ is a path in $U = A \cap B$ joining the point $f(\Delta_\rho \cap \Delta_{\rho+1})$ to the point O, we obtain that

$$f \sim (f_{\Delta_1} \circ w_1) \circ (w_1^{-1} \circ f_{\Delta_2} \circ w_2) \circ \ldots \circ (w_{N-1}^{-1} \circ f_{\Delta_N}).$$

Each factor in this product is a closed path with origin O; moreover, $w_{2\lambda-2}^{-1} \circ f_{\Delta_{2\lambda-1}} \circ w_{2\lambda-1}$ is a path $f_{2\lambda-1}$ in A, $w_{2\lambda-1}^{-1} \circ f_{\Delta_{2\lambda}} \circ w_{2\lambda}$ is a path $f_{2\lambda}$ in B, and

$$f \sim f_1 \circ f_2 \circ \ldots \circ f_N.$$

2. Every defining relation for the group $\pi_1(A)$ or $\pi_1(B)$ is obviously also a defining relation for the group $\pi_1(\mathfrak{S})$, but there may appear in $\pi_1(\mathfrak{S})$ new defining relations between elements of the groups $\pi_1(A)$ and $\pi_1(B)$. If a class $a \in \pi_1(A)$ and a class $b \in \pi_1(B)$ both contain a path ϕ lying in U, then in $\pi_1(\mathfrak{S})$ we have the equality $a = b$. We are to show that the group $\pi_1(\mathfrak{S})$ can be specified by adding to the relations for $\pi_1(A)$ and $\pi_1(B)$ only the relations of this type.

Suppose $c = 1$ in the group $\pi_1(\mathfrak{S})$; then from what has just been proved, $c = a_1 b_1 \cdots a_k b_k$, where $a_i \in \pi_1(A)$ and $b_i \in \pi_1(B)$. We must show that the equality $c = 1$ follows from the relations in the groups $\pi_1(A)$ and $\pi_1(B)$ and the equalities of the type $a = b$, where $a \in \pi_1(A)$ and $b \in \pi_1(B)$.

Let f be a path in the class c. Then f can be regarded as a continuous mapping into $\mathfrak{S}$ of the boundary Γ of a square D, where the point $o \in \Gamma$ goes into the origin O. Since f is null-homotopic, the mapping f extends to a mapping $\bar{f}: D \to \mathfrak{S}$. Set $K = \bar{f}(D)$; then $Q_A = \bar{f}^{-1}[K \cap (A \setminus B)]$ and $Q_B = \bar{f}^{-1}[K \cap (B \setminus A)]$ are disjoint compacta in D. Therefore Q_A has a polyhedral neighborhood V_A in D such that

$$Q_A \subset V_A, \quad \bar{V}_A \cap Q_B = \Lambda. \tag{2}$$

By Corollary 2 to Lemma 3.2 of Chapter II, V_A can be chosen so that each component of its closure is a polyhedral open disk int D_i from which have been removed a finite number of disjoint disks whose closures are interior to D_i.[1] It is easily seen that $V_B = D \setminus \bar{V}_A$ is a polyhedral neighborhood of Q_B in D with the same structure. The square D is thus decomposed into a finite number of polyhedral domains $v_{Aj} \subset V_A$ and $v_{Bj} \subset V_B$, each of the form

[1] *Translator's note.* What is meant is: each component of the closure is a disk D_i minus the interiors of a finite number of disjoint disks which lie in the interior of D_i.

$$v_{Aj} = \operatorname{int} D_{Aj} \setminus \bigcup_{i=1}^{s_j} d_{ji}, \quad d_{ji} \subset \operatorname{int} D_{Aj},$$

$$v_{Bj'} = \operatorname{int} D_{Bj'} \setminus \bigcup_{i=1}^{s_j'} d_{j'i}. \quad d_{j'i} \subset \operatorname{int} D_{Bj'},$$

where the D_{Aj}, $D_{Bj'}$, d_{ji} and $d_{j'i}$ are closed polyhedral disks; furthermore

$$\bar{v}_{Aj} \cap \bar{v}_{Aj'} = \Lambda, \quad j \neq j'; \quad \bar{v}_{Bj} \cap \bar{v}_{Bj'} = \Lambda, \ j \neq j';$$

$$v_{Aj} \cap v_{Bj'} = \Lambda.$$

We shall show that each domain $\bar{v}_{Aj}$ and $\bar{v}_{Bj}$ corresponds either to a relation of the type $a = b$ or to a relation in the group $\pi_1(A)$ (or $\pi_1(B)$). For convenience we now drop the index j.

For each domain v_A (and similarly v_B) there are two possible cases.

Case 1). By (2),

$$\bar{v}_A \cap Q_B = \Lambda, \ (\operatorname{Fr} v_A) \cap Q_A = \Lambda.$$

Hence

$$\operatorname{Fr} v_A \subset f^{-1}(U). \tag{3}$$

In case 2), each $\operatorname{Fr} d_i \subset f^{-1}(U)$, but $\operatorname{Fr} D_A$ may intersect $f^{-1}(A \setminus B)$; $\operatorname{Fr} D_A$ consists of a finite number of polygonal lines, of which some lie on Γ and some in the interior of D except for their endpoints. These latter polygonal lines are contained in $f^{-1}(U)$, since they separate off the domain $\bar{v}_A$ in D. Thus $\operatorname{Fr} D_A \cap f^{-1}(U) \neq \Lambda$.[1)]

In either case, pick a point $p_0 \in \operatorname{Fr} D_A \cap f^{-1}(U)$ and a pair of points p_i, p_i' on each $\operatorname{Fr} d_i$, $1 \leq i \leq s$, and, within v_A, join the point p_0 to the point p_1 by a polygonal line l_1, and join each point p_{i-1}' to the point p_i by a polygonal line l_i such that $l_i \cap l_{i'} = \Lambda$, $i \neq i'$. That this can be done is easily shown, using the topology of the plane (see Chapter II). The points p_i and p_i' separate $\operatorname{Fr} d_i$ into two arcs λ_i and μ_i, designated so that in moving along $\operatorname{Fr} d_i$ from p_i to p_i' the disk d_i lies to the right of λ_i and the left of μ_i (Figure 3.2). (In Figure 3.2 read p for o and x for p.)

Set

$$\Theta = \bigcup_{i=1}^{s} l_i \cup \bigcup_{i=1}^{s} \operatorname{Fr} d_i. \tag{4}$$

1) *Translator's note.* If $\operatorname{Fr} D_A \subset \Gamma$, then $\operatorname{Fr} D_A$ is all of Γ and so contains the point $o \in f^{-1}(U)$.

Let Q be a circular disk; its boundary is the union of two semicircles

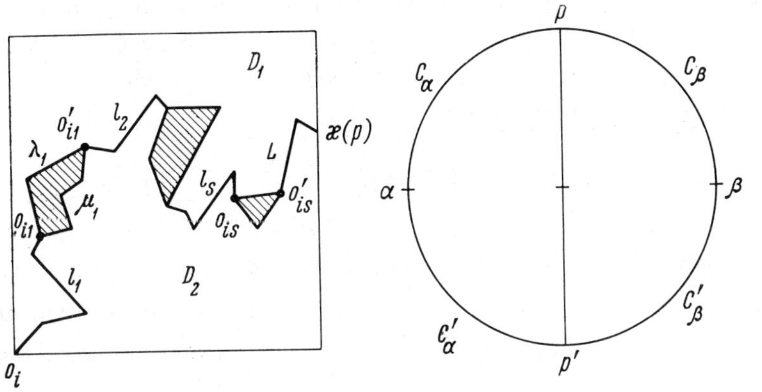

Figure 3.2

$\operatorname{Fr} Q = C \cup C'$, intersecting at their endpoints $C \cap C' = \{\alpha\} \cup \{\beta\}$. We define a continuous mapping κ of Q onto $\bar{v}_A$. We specify it first on $\operatorname{Fr} Q$, setting

$$\varkappa(\alpha) = \varkappa(\beta) = p_0, \quad \varkappa(C) = \operatorname{Fr} D_A,$$

where $\kappa|_C$ is a homeomorphism except for the points α and β.[1] The arc C' is to be mapped continuously onto Θ. Namely, divide it into $4s$ segments

$$C' = \bigcup_{r=1}^{4s} \delta_r$$

and map each segment δ_r homeomorphically onto a simple arc in Θ in the following way:

$$\varkappa(\delta_{2i-1}) = l_i, \quad \varkappa(\delta_{2i}) = \lambda_i,$$

$$\varkappa(\delta_{4s-2i+1}) = \mu_i^{-1}, \quad \varkappa(\delta_{4s-2i+2}) = l_i^{-1}. \tag{5}$$

$$1 \leqslant i \leqslant s$$

The exponent -1 means that the positive direction on δ_ρ is to go into the negative direction on the corresponding arc. Then $\phi = \bar{f}\kappa|_C$ and $\phi' = \bar{f}\kappa|_{C'}$ are closed paths in A with origin $\bar{f}(p_0)$, since $\bar{f}(\bar{v}_A) \subset A$.

We prove next that ϕ and ϕ' are homotopic in A. It suffices to show that the continuous mapping $\kappa|_{C \cup C'}$ extends to a continuous mapping of the disk Q

1) *Translator's note.* Which direction to pick for κ is made evident below.

onto $\overline{v}_A$; for then $\overline{f}\kappa|_Q$ corresponds to a homotopy taking the path ϕ into the path ϕ'. Let x be the point on C diametrically opposite to the point $x' = \kappa^{-1}(p'_s)$. In $v_A \setminus \Theta$ join the point p'_s to the point $p = \kappa(x)$ by a polygonal line L (see Figure 3.2). The mapping κ defined so far on $C \cup C'$ can obviously be extended to the diameter $\overline{x'x}$ so that

$$\varkappa : \overline{x'x} \to L$$

and $\kappa|_{\overline{x'x}}$ is a homeomorphism.

Let C_α and C_β be the segments of the arc C between the points α (β) and x, and let C'_α and C'_β be the segments of C' between the points α (β) and x'. The set $\bigcup_{i=1}^s l_i \cup L$ separates $\overline{v}_A$ into two disks D_1 and D_2, bounded respectively by the polyhedral closed curves

$$\operatorname{Fr} D_1 = \varkappa(C_\alpha) \cup L \cup \bigcup_{i=1}^{s} (l_i \cup \lambda_i),$$

$$\operatorname{Fr} D_2 = \varkappa(C_\beta) \cup L \cup \bigcup_{i=1}^{s} (l_i \cup \mu_i).$$

It is easily verified that κ maps $C_\alpha \cup \overline{xx'} \cup C'_\alpha$ homeomorphically onto $\operatorname{Fr} D_1$, and $C_\beta \cup \overline{xx'} \cup C'_\beta$ homeomorphically onto $\operatorname{Fr} D_2$. By the Schoenflies theorem (Corollary 4 to Theorem 4.3 of Chapter II), κ extends to a homeomorphism of the semidisk $(\alpha x x' \alpha)$ onto D_1 and of the semidisk $(\beta x x' \beta)$ onto D_2. Since the two homeomorphisms agree on $\overline{xx'}$, $\kappa|Q$ is a continuous mapping of the disk Q onto $D_1 \cup D_2 = \overline{v}_A$. And since $\overline{f}(\overline{v}_A) = \overline{f}\kappa(Q) \subset A$, the path ϕ' is homotopic in A to the path ϕ.

Now join the origin O to the point $f(p_0)$ by a path w in $A \cap B$. Then $\Phi = w \circ \phi \circ w^{-1}$ and $\Phi' = w \circ \phi' \circ w^{-1}$ are homotopic paths in A with origin O. Hence Φ and Φ' belong to the same class a of the group $\pi_1(A)$.

In Case 1), the path Φ is contained in $A \cap B$, and so belongs also to a class b of the group $\pi_1(B)$, so that $a = b$ in the group $\pi_1(\mathfrak{S})$.

Thus in Case 1) a domain $\overline{v}_A$ (or $\overline{v}_B$) in the homotopy disk D corresponds to a relation in the group $\pi_1(\mathfrak{S})$ of the type $a = b$, where the classes a and b contain the same path in $A \cap B$; while in Case 2) a domain $\overline{v}_A$ (or $\overline{v}_B$) corresponds to a relation in the group $\pi_1(A)$ (or $\pi_1(B)$).

We now prove the theorem by induction on the number n of domains v_{Aj} and v_{Bj} contained in D.

1) $n = 1$, i.e. the entire disk D is the only domain, say a $\overline{v}_A$. Then $\overline{f}(\Gamma) \subset A$ and $\overline{f}(D) \subset A$, so that f is null-homotopic in A and the relation $c = 1$ holds

within the group $\pi_1(A)$.

2) $n > 1$, and the assertion holds if the homotopy disk is decomposed into $k < n$ domains v_{Aj} and v_{Bj}. There are now two possibilities:

2a) $f(\Gamma) \subset A$ (or $f(\Gamma) \subset B$). In this case, f is a path in A, and $c = a$; it may in particular happen that f is a path in $A \cap B$. Then Γ belongs entirely to a single domain $\overline{v}_A$; and since $n > 1$, the continuum Θ is nonvoid and the domain $v_A \setminus \Theta$ with boundary $\Gamma \cup \Theta$ determines a homotopy of the path $\overline{f}\kappa|_C = \phi$ to the path $\phi' = \overline{f}\kappa|_{C'}$, where κ is as above. The continuum Θ contains $s \leq n - 1$ polyhedral closed curves $\lambda_i \cup \mu_i$, bounding disks d_i, $i = 1, \cdots, s$. The path ϕ' is homotopic to the product of its parts:

$$\varphi' \sim \varphi'|_{\delta_1 \cup \delta_2} \circ \cdots \circ \varphi'|_{\delta_{2s-3} \cup \delta_{2s-2}} \circ \varphi'|_{\bigcup_{\rho=2s-1}^{4s} \delta_\rho}.$$

Insert between these parts, i.e. at the points $x_i = \delta_{2i} \cap \delta_{2i+1}$, $1 \leq i \leq s - 1$, the null-homotopic paths $w_i \circ w_i^{-1}$, where

$$w_i \sim \overline{f}\varkappa|_{\bigcup_{\rho=4s-2i+1}^{4s} \delta_\rho}. \tag{6}$$

By (5) we have $\kappa(x_i) = p_i' \in \lambda_i \cap \mu_i$, and the path w_i corresponds to the mapping $\overline{f}|_{\mu_i^{-1} \cup l_i^{-1} \cup \cdots \cup \mu_1^{-1} \cup l_1^{-1}}$. Then

$$\varphi' \sim (\varphi'|_{\delta_1 \cup \delta_2} \circ w_1) \circ (w_1^{-1} \circ \varphi'|_{\delta_3 \cup \delta_4} \circ w_2) \circ \cdots \circ (w_{s-2}^{-1} \circ \varphi'|_{\delta_{2s-3} \cup \delta_{2s-2}} \circ w_{s-1})$$
$$\circ (w_{s-1}^{-1} \circ \varphi'|_{\bigcup_{\rho=2s-1}^{4s} \delta_\rho}). \tag{7}$$

Each parenthesis is a closed path with origin O. We assert that it is null-homotopic in $\mathfrak{S}$. To see this, observe that from (6) we have

$$w_i \sim \phi'|_{\delta_{4s-2i+1}} \circ \phi'|_{\delta_{4s-2i+2}} \circ w_{i-1}.$$

Hence

$$w_{i-1}^{-1} \circ \varphi'|_{\delta_{2i-1} \cup \delta_{2i}} \circ w_i \sim w_{i-1}^{-1} \circ \varphi'|_{\delta_{2i-1}} \circ \varphi'|_{\delta_{2i}} \circ \varphi'|_{\delta_{4s-2i+1}} \circ \varphi'|_{\delta_{4s-2i+2}} \circ w_{i-1}. \tag{8}$$

By (5), $\kappa(\delta_{2i-1}) = l_i$ and $\kappa(\delta_{4s-2i+2}) = l_i^{-1}$, with κ a homeomorphism on each of these segments. It follows that

$$\varphi'|_{\delta_{4s-2i+2}} \sim (\varphi'|_{\delta_{2i-1}})^{-1}. \tag{9}$$

Indeed, we have

$$\varphi'|_{\delta_{2i-1}} = \overline{f}\varkappa\Phi, \quad \varphi'|_{\delta_{4s-2i+2}} = \overline{f}\varkappa\Psi,$$

where Φ and Ψ are the linear mappings of I^1 onto δ_{2i-1} and $\delta_{4s-2i+2}$. Then $\kappa\Phi$ and $\kappa\Psi$ are homeomorphisms of I^1 onto l_i and l_i^{-1} respectively, and (9) follows from Remark 3.1.

By (5) we have

$$\varkappa(\delta_{2i}) = \lambda_i,\ \varkappa(\delta_{4s-2i+1}) = \mu_i^{-1}.$$

Hence the path $\psi_i = \phi'|_{\delta_{2i}} \circ \phi'|_{\delta_{4s-2i+1}}$ corresponds to the mapping $\bar{f}|_{\mathrm{Fr}\, d_i}$. If in the right-hand side of (8) we now replace the path $\phi'|_{\delta_{4s-2i+2}}$ by the homotopic path $(\phi'|_{\delta_{2i-1}})^{-1}$, we find that to prove what is required, namely that

$$w_{i-1}^{-1}\circ\varphi'|_{\delta_{2i-1}\cup\delta_{2i}}\circ w_i \sim 0, \tag{10}$$

it suffices to show that $\psi_i \sim 0$. But ψ_i is a closed path with origin $\bar{f}(p_i)$, and $\psi_i = \bar{f}k_i$, where k_i is a continuous mapping of the circle onto $\mathrm{Fr}\, d_i$; so that by Remark 3.2 the fact that $\psi_i \sim 0$ follows from the fact that $\bar{f}$ is defined on d_i. On the other hand, ψ_i is a closed path in $A \cap B$, and the number of domains into which d_i is in turn decomposed is less than n; so that by the induction assumption and the fact that the fundamental group is independent of the origin, the homotopy $\psi_i \sim 0$, and so the homotopy (10), follow from the relations in the groups $\pi_1(A)$, $\pi_1(B)$ and the relations of the type $a = b$.

For the first parenthesis in (7), the same thing is proved in a similar fashion from equations (5) for $i = 1$; and for the last, from the fact that the path

$$w_{s-1}^{-1}\circ\varphi'|_{\bigcup_{\rho=2s-1}^{4s}\delta_\rho}$$

is homotopic to the right-hand side of (8) for $i = s$. Consequently the homotopy $\phi' \sim 0$, and so the equality $c = 1$, follow in this case from the relations in the groups $\pi_1(A)$ and $\pi_1(B)$ and the equalities of the type $a = b$.

2b) The path $f \in c$ is a product of several parts $f|_{\Delta_{Aj}}$ and $f|_{\Delta_{Bj}}$, the first of which lie in A, and the others in B. Here Δ_{Aj} and Δ_{Bj} are segments of Γ belonging to domains v_{Aj} and v_{Bj}, respectively. These segments are cut off from Γ by polygonal lines interior to D except for their endpoints. Let $\widehat{pq}$ be one such polygonal line. It decomposes D into two polyhedral disks:

$$D = D_1 \cup D_2.$$

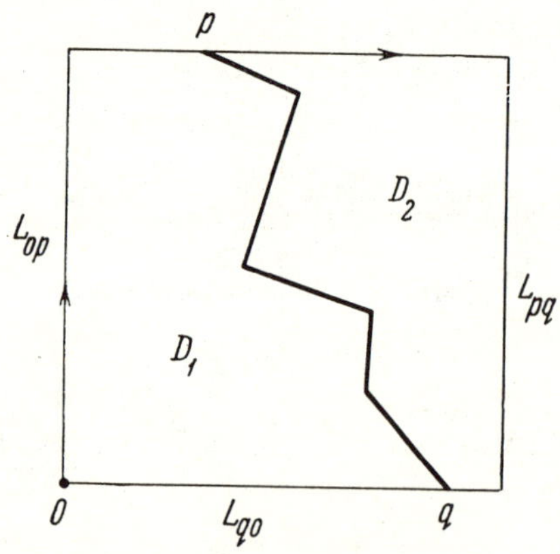

Figure 3.3

Let L_{op}, L_{pq} and L_{qo} be the segments of Γ between the corresponding points (Figure 3.3). Then

$$f \sim (f|_{L_{op}} \circ w) \circ (w^{-1} \circ f|_{L_{pq}} \circ f|_{L_{qo}}),$$

where $w \sim \bar{f}|_{\hat{p}\hat{q} \cup L_{qo}}$. But $f|_{L_{op}} \circ w$ and $\psi = w^{-1} \circ f|_{L_{pq}} \circ f|_{L_{qo}}$ are closed paths in $\mathfrak{S}$, with origin O, circling the boundaries $\operatorname{Fr} D_1$ and $\operatorname{Fr} D_2$, respectively, since

$$\psi \sim f|_{L_{qo}^{-1}} \circ \bar{f}|_{\hat{q}\hat{p} \cup L_{pq}} \circ f|_{L_{qo}},$$

$$L_{op} \cup \hat{p}\hat{q} \cup L_{qo} = \operatorname{Fr} D_1, \quad \hat{q}\hat{p} \cup L_{pq} = \operatorname{Fr} D_2.$$

The paths $f|_{L_{op}} \circ w$ and ψ are null-homotopic, since $\bar{f}$ is defined on D_r, $r = 1, 2$. Each disk D_r contains less than n domains v_{Aj} and v_{Bj}, so that by the induction assumption these homotopies follow from the relations in the groups $\pi_1(A)$ and $\pi_1(B)$ and the equalities $a = b$. But then so does the homotopy $f \sim 0$. This completes the proof of the theorem.

Remark. To obtain all the defining relations in the group $\pi_1(A \cup B)$ of the type $a = b$, it suffices to consider those paths corresponding to generators of $\pi_1(A \cap B)$. Thus a presentation of the group $\pi_1(A \cup B)$ is obtained once one knows presentations of the groups $\pi_1(A)$ and $\pi_1(B)$ and generators of $\pi_1(A \cap B)$.

Corollary 1. *Let $\mathfrak{S}$ be the union of two closed arcwise connected subsets: $\mathfrak{S} = A \cup B$, where $A \cap B$ is arcwise connected and has an arcwise connected neighborhood U such that the inclusion homomorphisms*

$$h_A : \pi_1(A) \to \pi_1(A \cup U),$$
$$h_B : \pi_1(B) \to \pi_1(B \cup U),$$
$$H : \pi_1(A \cap B) \to \pi_1(U)$$

are isomorphisms "onto." Then $\pi_1(\mathfrak{S})$ is obtained from the free product of the groups $\pi_1(A)$ and $\pi_1(B)$ by adjoining the equalities of the type $a = b$, where a and b are classes in $\pi_1(A)$ and $\pi_1(B)$ containing the same path in $A \cap B$.

By the van Kampen Theorem, $\pi_1(\mathfrak{S})$ is obtained from the free product of the groups $\pi_1(A \cup U)$ and $\pi_1(B \cup U)$ by adjoining the equalities of the type $a_U = b_U$, where $a_U \in \pi_1(A \cup U)$, $b_U \in \pi_1(B \cup U)$, and the classes a_U and b_U contain the same path ϕ in U. But ϕ is homotopic in U to a path ϕ' lying in $A \cap B$, and the classes a_U and b_U correspond to classes $a \in \pi_1(A)$ and $b \in \pi_1(B)$ containing the path ϕ'. Thus the equality $a_U = b_U$ implies the equality $a = b$.

Corollary 2. *If $\mathfrak{S} = A \cup B$ is the union of two closed arcwise connected subsets, where the intersection $A \cap B$ is arcwise connected and is a deformation* *retract of an arcwise connected neighborhood U, and A and B are deformation retracts of $A \cup U$ and $B \cup U$, then $\pi_1(\mathfrak{S})$ is computable as in Corollary* 1.

For, by Corollary 4 of Theorem 2.3, h_A, h_B and H then satisfy the conditions of Corollary 1.

Corollary 3. *Under the conditions of the van Kampen Theorem or of Corollaries* 1 *and* 2, *if the group $\pi_1(A \cap B)$ is trivial, then $\pi_1(\mathfrak{S})$ is the free product of the groups $\pi_1(A)$ and $\pi_1(B)$.*

Corollary 4. *Under the conditions of the van Kampen Theorem or of Corollaries* 1 *and* 2, *if the groups $\pi_1(A)$ and $\pi_1(B)$ are trivial, so is the group $\pi_1(A \cup B)$.*

2. Examples

Example 3.1. $\mathfrak{S}$ is the plane E^2 with a finite number of points $p_1, p_2, \cdots, p_k$ removed. Then $\pi_1(\mathfrak{S})$ is the free group on k generators.

Pick the origin O so that any ray from O contains at most one point p_i, and, using half-lines issuing from O and containing no points p_i, divide E^2 into k domains G_i each containing a single point p_i. Then $\mathfrak{S} = \bigcup \mathfrak{S}_i$, where $\mathfrak{S}_i = \overline{G}_i \setminus p_i$. Setting $A_j = \bigcup_{i=1}^{j} \mathfrak{S}_i$, we find that the union $A_j = A_{j-1} \cup \mathfrak{S}_j$ satisfies the conditions of Corollary 2 to the van Kampen Theorem. By Example 2.3, $\pi_1(\mathfrak{S}_j)$ is the free group on one generator a_j; $A_{j-1} \cap \mathfrak{S}_j$ is a half-line (or a line, for $j = k$), so that the group $\pi_1(A_{j-1} \cap \mathfrak{S}_j)$ is trivial. Hence $\pi_1(A_j)$ is the free product of the groups $\pi_1(A_{j-1})$ and $\pi_1(\mathfrak{S}_j)$. Repeated application of the van Kampen Theorem for $j = 2, \cdots, k$ then yields that $\pi_1(\mathfrak{S})$ is the free group on k generators $a_1, \cdots, a_k$, where a_i generates $\pi_1(\mathfrak{S}_i)$ and can be taken as the class of paths in $\mathfrak{S}_i$ that contains a path "circling" the single point p_i just once in a prescribed direction. In other words, a_i contains a path f_i which is a homeomorphism into $\mathfrak{S}_i$ of the circle C, where $O \in f_i(C)$ and p_i is in the interior domain of the simple closed curve $f_i(C)$.

The result is unaltered if in place of points one removes from E^2 line segments, rectangles or disks.

A similar proof shows that for $n > 2$ the complement in E^n of a finite number of points (or parallelepipeds) has trivial fundamental group.

Example 3.2. The fundamental group of $E^3 \setminus C$, where C is a circle, is a free cyclic group.

Using a plane Π perpendicular to the plane containing C and passing through the center of C, taken as the origin, divide $E^3 \setminus C$ into two parts whose closures, A and B, intersect in $\Pi \setminus C$. By Example 2.4 the fundamental group of each part is a free cyclic group, and any path in A or B is homotopic to a path in $A \cap B$.

The generators a of $\pi_1(A)$ and b of $\pi_1(B)$ can be chosen so that the classes a and b contain a path f in $A \cap B$ that circles one of the points in $\Pi \cap C$ in a prescribed direction. Then $a = b$, so that the group $\pi_1(E^3 \setminus C)$ has just one generator.

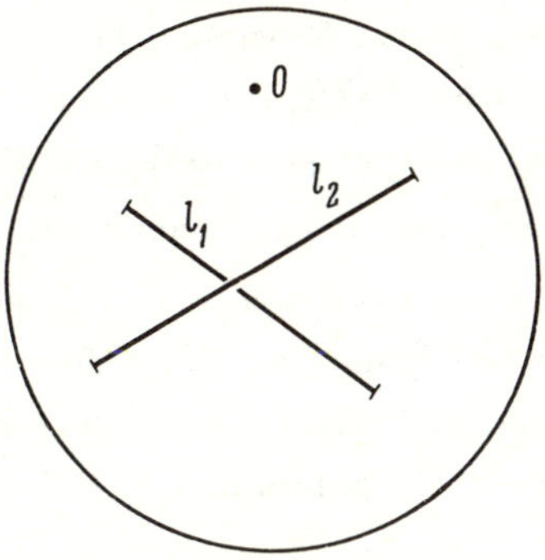

Figure 3.4

Example 3.3. Let R be the closure of a convex domain V in E^3, minus two line segments whose endpoints lie on Fr V. Then $\pi_1(R)$ is a free group on two generators (Figure 3.4).

Let l_1 and l_2 be the given segments. Through a line segment λ parallel and sufficiently close to l_1, pass a plane Π parallel to l_2 and such that Π separates l_1 from l_2. Then Π divides R into two sets A and B, and $R = \overline{A} \cup \overline{B}$, the closures being taken in R.

$\overline{A}$ and $\overline{B}$ are each closed domains in E^3 with a line segment removed. Pick the origin O on the intersection $\Pi \cap \operatorname{Fr} V$. By the remark in Example 2.4, $\pi_1(\overline{A})$ and $\pi_1(\overline{B})$ are free cyclic groups with generators $a \in \pi_1(\overline{A})$ and $b \in \pi_1(\overline{B})$. The intersection $\overline{A} \cap \overline{B}$ is the part of R contained in Π, and the latter, by construction, meets neither l_1 nor l_2. Hence $\overline{A} \cap \overline{B}$ is a planar closed convex domain, and the group $\pi_1(\overline{A} \cap \overline{B})$ is consequently trivial. It is easily seen that $\overline{A} \cup \overline{B}$ satisfies the conditions of Corollary 2 to the van Kampen Theorem. Since $\pi_1(\overline{A} \cap \overline{B})$ is trivial, it follows that $\pi_1(R)$ is the free product of $\pi_1(\overline{A})$ and $\pi_1(\overline{B})$. By Example 2.4, the generator a of $\pi_1(A)$ (and likewise $b \in \pi_1(B)$) can be chosen so that the class a contains a path lying on Fr V and circling one of the points of intersection $l_1 \cap \operatorname{Fr} V$ in a prescribed direction.

In Example 3.3 the line segments can be replaced by arcs of circles. Let us consider one such case.

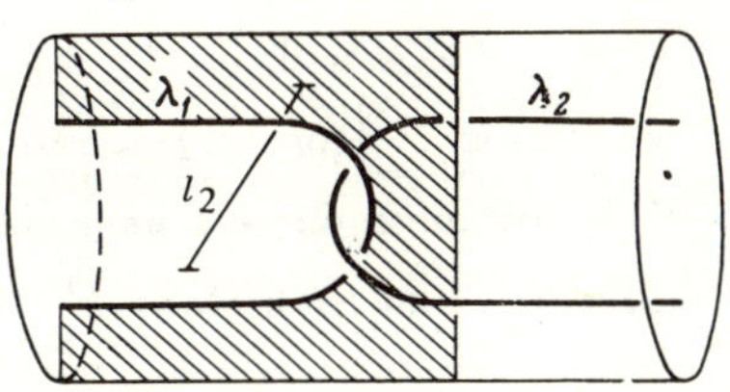

Figure 3.5

Example 3.4. Let R be a bounded circular cylinder X minus two circular arcs λ_1 and λ_2 which are linked in X and lie on perpendicular planes so that the endpoints of each λ_i are on one of the bases of X (Figure 3.5). Then $\pi_1(R)$ is a free group on two generators.

Let Π_1 be the plane containing λ_1, and Π_2 the plane containing λ_2; let P_i, $i = 1, 2$, be the part of $\Pi_i \cap X$ cut off by the arc λ_i and a line segment $l_i \subset \Pi_i \cap X$ perpendicular to Π_j, $j \neq i$; then $P_1 \cap P_2 = \Lambda$. Let U_i be the neighborhood of the P_i in X such that $\overline{U}_1 \cap \overline{U}_2 = \Lambda$. Using the elementary examples described in Chapter II (see Example 2.2), one can construct an isotopy $F_t\colon X \to X$ which is stationary on $X \setminus (U_1 \cup U_2)$ and takes λ_1 into l_1 and λ_2 into l_2. Hence the group $\pi_1(R)$ is isomorphic to the group $\pi_1(X \setminus (l_1 \cup l_2))$ and so, by Example 3.3, is the free group on two generators.

Example 3.5. Let $\mathfrak{S} = \overline{V} \setminus \bigcup_{i=1}^{n} l_i$, where V is a convex domain in E^3, and the l_i, $i = 1, \cdots, n$, are disjoint line segments with endpoints on $\operatorname{Fr} V$. Then $\pi_1(\mathfrak{S})$ is the free group on n generators a_i, $i = 1, \cdots, n$, which may be chosen so that each a_i contains a path $f_i \in \operatorname{Fr} V$ circling one of the points in $l_i \cap \operatorname{Fr} V$ once in a prescribed direction.

We prove this by induction on the number n of segments l_i. For $n \leq 2$ it follows from Example 3.3. Suppose $n > 2$ and the assertion proved for $k < n$ segments. Pick a plane Π separating the segment l_2 from l_1;[1] then Π divides $\mathfrak{S}$ into two parts A and B such that $\mathfrak{S} = A \cup B$, where A and B are of the same form as $\mathfrak{S}$, but the number k_A of segments removed from V_A and the number k_B removed from V_B are each less than n. Pick the origin O on $\Pi \cap \operatorname{Fr} V$.

By the induction assumption $\pi_1(A)$ and $\pi_1(B)$ are free groups with k_A and k_B generators a_i and b_j, respectively, where each a_i contains a path f_i lying on $\operatorname{Fr} V_A$ (and each b_j a path on $\operatorname{Fr} V_B$), circling exactly one point in $l_i \cap \operatorname{Fr} V_A$ (or $l_j \cap \operatorname{Fr} V_B$). The generators of the group $\pi_1(A)$ (and of $\pi_1(B)$) can be chosen so that each segment l_{i_s} that intersects Π corresponds to a generator containing a path $f_{i_s} \subset \Pi \cap V$ that circles the point $l_{i_s} \cap \Pi$ in the positive direction.

The union $\mathfrak{S} = A \cup B$ satisfies the conditions of Corollary 2 to the van Kampen Theorem, and the group $\pi_1(A \cap B)$ is the free group on $r \leq n - 2$ generators, where r is the number of segments l_{i_s} that intersect Π. For each i_s, $s = 1, \cdots, r$, we have $a_{i_s} = b_{i_s}$; thus the group $\pi_1(\mathfrak{S})$ has $k_A + k_B - r = n$ generators, with no relations between them. It remains to show that the generators a_{i_s} can be replaced by new ones containing paths lying on $\operatorname{Fr} V$.[2] Pick new sets of generators for $\pi_1(A)$ and $\pi_1(B)$, by setting $a_i' = a_i$ or $b_i' = b_i$ if $l_i \cap \Pi = \Lambda$ and, for $i = i_s$, $s = 1, \cdots, r$, replacing each a_{i_s} (or b_{i_s}) by a generator a_{i_s}' (or b_{i_s}') containing a path $f_{i_s} \subset \operatorname{Fr} V \setminus \Pi$ circling a point of $l_{i_s} \cap \operatorname{Fr} V$. By the induction assumption, each class a_{i_s} is a product of classes a_i', $1 \leq i \leq k_A$, and each class b_{i_s}' is a

---

1) *Translator's note:* and containing no other l_i.

2) *Translator's note.* The paths for the other generators should also have been chosen to satisfy this requirement.

product of classes b_j, $j \neq i_s$, and of classes $a_{i_s} = b_{i_s}$. Hence each class b'_{i_s} is a product of classes b_j, $j \neq i_s$, and of classes a'_i. Thus we have as generators of $\pi_1(\mathfrak{S})$ the classes a_i or b_i, if $l_i \cap \Pi = \Lambda$, and the classes a'_{i_s} containing paths $f_{i_s} \subset \operatorname{Fr} V_A \setminus \Pi$ for each l_{i_s}; i.e. each generating class of $\pi_1(\mathfrak{S})$ contains a path that lies on $\operatorname{Fr} V$ and circles a point of $l_i \cap \operatorname{Fr} V$.

Example 3.6. Let $S = \overline{V} \setminus \bigcup_{i=1}^n l_i \setminus C$, where $\overline{V}$ is a convex closed domain in E^3, the l_i are line segments with endpoints on $\operatorname{Fr} V$, and C is a two-segment polygonal line with vertex interior to V and endpoints on $\operatorname{Fr} V$.

Construct a plane Π such that $\Pi \cap C$ is the vertex of C and Π contains none of the l_i;[1)] then Π divides S into two parts: $S = A \cup B$, where A and B satisfy the conditions of Example 3.5. Arguing as in Example 3.5, we find that $\pi_1(S)$ is the free product of $n+1$ generators s_i, which may be chosen similarly to those in Example 3.5.

Example 3.7. The fundamental group of a torus is the free abelian group on two generators.

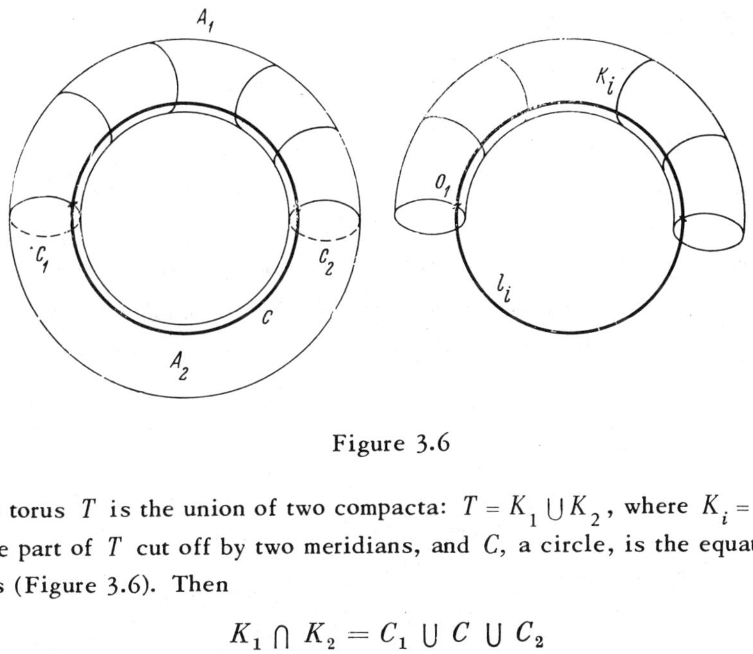

Figure 3.6

The torus T is the union of two compacta: $T = K_1 \cup K_2$, where $K_i = A_i \cup C$; A_i is the part of T cut off by two meridians, and C, a circle, is the equator of the torus (Figure 3.6). Then

$$K_1 \cap K_2 = C_1 \cup C \cup C_2$$

is the union of three circles: the two meridians C_i, $i = 1, 2$, and the equator C, with C_i intersecting C in the point O_i, $i = 1, 2$. The continua K_i and $K_1 \cap K_2$

1) *Translator's note*: and so that the two segments of C are on opposite sides of Π.

are arcwise connected, and $K_1 \cap K_2$ has a neighborhood U in T of which $K_1 \cap K_2$ is a deformation retract; thus the group $\pi_1(U)$ is isomorphic to $\pi_1(K_1 \cap K_2)$, and to compute $\pi_1(T)$ we can apply Corollary 2 of the van Kampen Theorem to the union $K_1 \cup K_2$.

We compute first $\pi_1(K_i)$; we have $K_i = A_i \cup C$, and $A_i \cap C = l_i$ is a circular arc,[1)] while $A_i = C_1 \times l_i$, $i = 1, 2$. Applying Corollary 3 of the van Kampen Theorem to $A_i \cup C$, we find that $\pi_1(K_i)$ is the free product of the groups $\pi_1(A_i)$ and $\pi_1(C)$, since $\pi_1(l_i)$ is trivial. But $A_i = C_1 \times l_i$ and deforms onto C_1, so that $\pi_1(A_i)$ is isomorphic to $\pi_1(C_1)$, i.e. is a free cyclic group with generator γ. Choosing the origin at O_1, we may suppose that γ contains the path that runs over C_1 once, in a prescribed direction. Likewise, $\pi_1(C)$ is free cyclic, with generator Γ containing a path with origin O_1 that runs over C.

Since $K_1 \cap K_2 = C_1 \cup C \cup C_2$, the group $\pi_1(K_1 \cap K_2)$ is easily computed by applying Corollary 3 of the van Kampen Theorem twice: first to the union $C_1 \cup C$, then to the union of the continua $C_1 \cup C$ and $l_1 \cup C_2$, and observing that the groups $\pi_1(C_1 \cap C)$ and $\pi_1[(C_1 \cup C) \cap (l_1 \cup C_2)]$ are trivial, so that the conditions of Corollary 3 are satisfied. Hence $\pi_1(K_1 \cap K_2)$ is the free group on three generators:

$$\gamma, \Gamma \text{ and } w_1 \gamma' w_1^{-1},$$

where γ and Γ are the generators of $\pi_1(C_1)$ and $\pi_1(C)$, γ' is the class of paths with origin O_2 running over the circle C_2 in the same direction as the paths in γ run over C_1 (as viewed from the positive direction on C), and w_i, $i = 1, 2$, is the class of paths running over the arc l_i from the point O_1 to the point O_2.

Since both the groups $\pi_1(K_i)$ are generated by the classes γ and Γ, the classes γ and Γ constitute a complete set of generators for the group $\pi_1(K_1 \cup K_2)$.

Every path in $K_1 \cap K_2$ belonging to the class $w_i \gamma' w_i^{-1}$, $i = 1, 2$, is homotopic in K_i to a path contained in C_1 and belonging to the class γ.

We have, therefore, in the group $\pi_1(K_1 \cup K_2)$,

$$w_1 \gamma' w_1^{-1} = w_2 \gamma' w_2^{-1} = \gamma.$$

Hence, eliminating γ', we obtain

$$w_1^{-1} \gamma w_1 = w_2^{-1} \gamma w_2.$$

Observing that $w_2 = \Gamma^{-1} w_1$, we get

$$w_1^{-1} \gamma w_1 = w_1^{-1} \Gamma \gamma \Gamma^{-1} w_1$$

and so, after cancellation,

---

1) *Translator's note.* In Figure 3.6, l_i should be the label of the other half of the circle.

$$\gamma = \Gamma\gamma\Gamma^{-1}, \quad \gamma\Gamma = \Gamma\gamma.$$

Thus $\pi_1(T)$ is a free abelian group with the generators γ and Γ, containing the paths running over a meridian and an equator of T, respectively.

§4. Computation of the fundamental group of a polyhedral knot in E^3

1. **Definition 4.1.** By a *polyhedral knot* in E^3 is meant a polyhedral closed curve C lying in E^3.

If there exists a homeomorphism $\phi: E^3 \to E^3$ such that $\phi(C)$ is the boundary of a 2-simplex (or is a circle), the knot is called *trivial.* The group $\pi_1(E^3 \setminus C)$ is then isomorphic to the fundamental group of the complement of the circle, i.e. (Example 3.2) it is a free cyclic group. But there are polyhedral knots for which no such homeomorphism exists and for which the group $\pi_1(E^3 \setminus C)$ is not isomorphic to the group of the complement of a circle.[1)] The group $\pi_1(E^3 \setminus C)$ is called the fundamental group of the knot. We shall present here a method of computing this group, based on the van Kampen Theorem.

The polyhedral knot C is the union of finitely many segments and vertices. Pick a plane Π in E^3 that does not intersect C and is not perpendicular to any segment of C, and let ΠC be the orthogonal projection of C onto Π. Obviously, every segment of C projects onto a line segment, and ΠC is the union of these, but they may intersect each other in points other than endpoints. In general, the projection ΠC divides the plane Π into a finite number of domains (unless the projection is a polygonal line).

We call the projection ΠC of the knot C into the plane Π *normal* if it satisfies the following conditions.

a) Every point of ΠC is the image of at most two points of C.

b) The projections of any two segments of C intersect in at most one point. (it follows that ΠC has only finitely many double points, i.e) points which are the images of two points of C).

c) The projection of any vertex of C is a single-valued point (i.e. the image of only one point of C).

Lemma 4.1. *If C is any polyhedral knot, and Π a plane that does not intersect C and is not perpendicular to any segment of C, then there exists an isotopy $F_t: E^3 \to E^3$ taking C into a polyhedral knot C' such that $\Pi C'$ is a normal projection.*

---

1) Conversely, if the fundamental group of the complement is cyclic, the knot is trivial [62].

The isotopy F_t is constructed as the product of a finite number of elementary isotopies ϕ_t^i of the following two types:

1) a rectilinear piece $\overline{p_0p_1}$ of C goes into a two-segment polygonal line $\widehat{p_0p_2p_1}$ intersecting C in the points p_0 and p_1, while $C \setminus \overline{p_0p_1}$ remains stationary, the triangle $p_0p_2p_1$ being such as not to intersect $C \setminus \overline{p_0p_1}$;

2) a piece of C which is a two-segment polygonal line $\widehat{p_0p_2p_1}$ goes into a rectilinear piece $\overline{p_0p_1}$ intersecting C in p_0 and p_1, while $C \setminus \widehat{p_0p_1p_2}$ is unchanged.

Each isotopy ϕ_t^i is, furthermore, stationary outside a small neighborhood u_i of the simplex $p_0p_1p_2$, whose boundary contains the points p_0 and p_1. The isotopy ϕ_t^i is constructed as in Lemma 2.1 of Chapter I, setting $p_0 = q$, $p_1 = q'$, and, for case 1), $p_2 = r$, $x = p$ (where x is a point of $\overline{p_0p_1}$), or, for case 2), $p_2 = p$, $x = r$.

To satisfy condition b) it suffices that every segment of C whose projection intersects that of another segment in more than one point be replaced, via an elementary isotopy 1), by a two-segment polygonal line, in such a way that the projections of the segments of all such polygonal lines are not parallel to the projections of the remaining segments. The knot C_1 thereby obtained is such that its projection ΠC_1 has only finitely many multiple points. If any one of these is the image of a vertex p of C_1, then, applying an elementary isotopy 2), we replace a small piece p_1pp_2 of C_1 by the line segment $\overline{p_1p_2}$, where p_1 and p_2 are single-valued points for C_1; in this manner we take C_1 into a knot C_2 satisfying conditions b) and c).

If, now, there are multiple points in ΠC_2 which are images of more than two points of C_2, consider for each such point x all the points $q_1, \cdots, q_r$ projecting into it, except for the two lowest (closest to Π). Applying the elementary isotopy 1) to appropriate pieces $\overline{p_ip_i'}$ of C_2 containing q_i, $i > 2$, we take each piece $\overline{p_ip_i'}$ into a polygonal line $\widehat{p_it_ip_i'}$ whose projection intersects that of $\overline{p_ip_i'}$ only at the endpoints. C_2 thereby goes into a new polyhedral knot C' satisfying all three conditions a), b) and c), but possibly with more multiple points in $\Pi C'$ than in ΠC.

It is easily seen that this succession of isotopies can be carried out without disturbing, at each step, the conditions already achieved. To do so it suffices to make sure, for each elementary isotopy, that every new segment of the projection is not parallel to any other segment and does not pass through any vertices or multiple points of the projection, and that each new vertex of the knot projects either into the complement of the projection of the original knot or into a single-valued point of the new projection. The polyhedral knot C' obtained as a result of all these operations has a normal projection in Π and is isotopic to C. Hence

$$\pi_1(E^3\setminus C') = \pi_1(E^3\setminus C).$$

In view of Lemma 4.1, to compute the fundamental group of a polyhedral knot it suffices to be able to compute it for one which has a normal projection in some plane.

Let C be such a knot. Pick a point x_0 in one of the interior domains in Π that are complementary to ΠC, such that no ray issuing from x_0 in Π contains more than one point which is either a vertex of ΠC or a double point (this can be done because these latter points lie on only finitely many lines). Through x_0 pass a line l perpendicular to Π, and choose the origin O on l so that the knot C is contained in the slab between the plane Π and the parallel plane Π_0 passing through O. The vertices of C can be of two sorts: of the first sort are those x_i for which the adjoining segments of C lie on opposite sides of the half-plane T which has l as edge and passes through x_i, $x_i \in T$; and of the second sort are those x_i for which both adjoining segments of C lie on the same side of $T \ni x_i$. Using a finite number N of half-planes T_i with edge l, divide E^3 into domains G_i such that:

1) every vertex of C of the first sort lies on a half-plane T_i, while every vertex of the second sort and every double point of ΠC lies within a domain G_i;

2) each domain G_i contains at most one point that is either a double point of ΠC or a vertex of C of the second sort.

Conditions 1) and 2) can be satisfied because of the choice of x_0. Each domain G_i is bounded by a dihedral angle $T_i \cup T_{i+1}$, and the space $\mathfrak{S}_i = \overline{G}_i \setminus C$ is, by 1) and 2), either that of Example 3.5 or that of Example 3.6; in addition, at most two line segments in $C \cap G_i$ can have intersecting projections. Setting $A_k = \bigcup_{i=1}^k \mathfrak{S}_i$, one easily sees that the union $A_{k+1} = A_k \cup \mathfrak{S}_{k+1}$ satisfies the conditions of Corollary 2 to the van Kampen Theorem. Hence, knowing the fundamental group of each $\mathfrak{S}_k$, we can compute in succession the fundamental group of each A_k and, finally, of $A_N = E^3 \setminus C$. Pick a direction on the knot C. It induces a direction on each component l_{iq} of $C \cap G_i$. For the generators of the fundamental group of $\mathfrak{S}_i$ we shall take classes a_{iq}, $q = 1, \cdots, q_i$, such that, if $l_{iq} \cap T_i \neq \Lambda$, then a_{iq} contains a path f_{iq} lying in T_i and circling a point of $l_{iq} \cap T_i$ once clockwise, as viewed from the positive direction on C. Let us describe more precisely the choice of the path f_{iq}.

Since T_i contains no double points of ΠC, every line in T_i parallel to l (the vertical) intersects C in at most one point. Divide T_i by a finite number of verticals into a finite number of strips λ_{iq} such that each λ_{iq} contains just one point $l_{iq} \cap T_i$ of the intersection $T_i \cap C$.[1)] On the line $T_i \cap \Pi_0$ pick a point $O_{iq} \in \lambda_{iq}$, and construct a path $f_{iq} = f_{iq}(s)$, $0 \le s \le 1$, in the following way: for $0 \le s \le 1/3$, $f_{iq}(s)$ is a homeomorphism of the segment $[0, 1/3]$ onto the line segment $[OO_{iq}]$; for $1/3 \le s \le 2/3$, $f_{iq}(s)$ is a closed path (a homeomorphism except for endpoints) with origin O_{iq}, lying in the interior of the strip λ_{iq} and circling the point $l_{iq} \cap T_i$ in the positive direction; finally, for $2/3 \le s \le 1$, $f_{iq}(s)$ is a homeomorphism of the segment $[2/3, 1]$ onto the segment $[O_{iq}O]$. The path so constructed belongs to the generator of the group $\pi_1(T_i \setminus C)$ corresponding to the point $l_{iq} \cap T_i$.

If l_{iq} is a two-segment polygonal line with endpoints on T_{i+1}, then a_{iq} contains a path lying in $T_{i+1} \setminus C$ and circling either one of the points of $l_{iq} \cap T_{i+1}$ in the positive direction.

If l_{iq} is a line segment and its projection into Π contains no double point for which the corresponding point on l_{iq} is the lower one, then the path in T_i circling the point $l_{iq} \cap T_i$ is homotopic in $\overline{G}_i \setminus C$ to a path in T_{i+1} circling the point $l_{iq} \cap T_{i+1}$. Let L be a piece of the knot C whose projection ΠL contains no double points of ΠC for which the corresponding point on L is the lower one, except possibly for the endpoints of L. Then the polygonal line L is of the form

$$L = \bigcup_{r=1}^{k} l'_{i_r q_r},$$

where $l'_{i_r q_r}$ is the piece $L \cap \overline{G}_{i_r}$, $i_{r+1} = i_r \pm 1$, $l'_{i_r q_r} \cap l'_{i_{r+1} q_{r+1}}$ is a point on $\overline{G}_{i_r} \cap \overline{G}_{i_{r+1}}$, and the direction prescribed on C induces on each $l'_{i_r q_r}$ the direction from $l'_{i_{r-1} q_{r-1}} \cap l'_{i_r q_r}$ to $l'_{i_r q_r} \cap l'_{i_{r+1} q_{r+1}}$. If $1 \ne r \ne k$, then $l'_{i_r q_r} = l_{i_r q_r}$ is a component of $C \cap G_{i_r}$; for $r = 1$ and $r = k$, one of the endpoints of $l'_{i_r q_r}$ may lie in the interior of $\overline{G}_{i_r}$. Since ΠL can contain only such double points of ΠC over which the corresponding point in L is the upper one, except for endpoints, it follows that for $1 \ne r \ne k$ the path in $\mathfrak{G}_{i_r}$ corresponding to $l'_{i_r q_r}$ is homotopic, on the one hand, to the path in T_{i_r} circling the point $l'_{i_r q_r} \cap T_{i_r}$ in the positive direction and, on the other hand, to the path in T_{i_r+1} circling the point $l'_{i_r q_r} \cap T_{i_r+1}$ in the positive direction. It is contained in the class $a_{i_r q_r}$, where $a_{i_r q_r}$ is the generator of the group $\pi_1(\mathfrak{G}_{i_r})$ corresponding to the segment $l_{i_r q_r}$. Hence

---

1) *Translator's note.* This needs rephrasing if a two-segment l_{iq} meets T_i; in that case, f_{iq} is to circle either one of the two points of $l_{iq} \cap T_i$. Similar modifications are needed elsewhere in this argument to take case of such "re-entrant" components l_{iq}.

$$a_{i_1q_1} = a_{i_2q_2} = \ldots = a_{i_{k-1}q_{k-1}} = a_{i_kq_k}, \tag{11}$$

where $a'_{i_1q_1}$ is the class in $\pi_1(\mathfrak{S}_{i_1})$ that contains a path in $\mathfrak{S}_{i_1} \cap \mathfrak{S}_{i_2}$ circling the point $l'_{i_1q_1} \cap \overline{G}_{i_1} \cap \overline{G}_{i_2}$ in the positive direction.

If one endpoint of the segment $l'_{i_1q_1}$ is a lower point of C over a double point of ΠC, and $i_2 = i_1 + 1$, then the class $a'_{i_1q_1}$ may not be the same as the class $a_{i_1q_1}$. Similarly, the class $a'_{i_kq_k}$ may not be the same as the class $a_{i_kq_k}$ if $i_k = i_{k-1} - 1$. Let us compute the class $a'_{i_1q_1}$ in $\pi_1(\mathfrak{S}_{i_1})$.

Let y_1 be the endpoint of $l'_{i_1q_1}$, and suppose the direction prescribed on C induces on $l'_{i_1q_1}$ the direction from y_1 to $l'_{i_1q_1} \cap \overline{G}_{i_1} \cap \overline{G}_{i_2}$. If $i_2 = i_1 - 1$, then $\overline{G}_{i_1} \cap \overline{G}_{i_2} = T_{i_1}$ and $a_{i_1q_1} = a'_{i_1q_1}$. Suppose now that $i_2 = i_1 + 1$; then $\overline{G}_{i_1} \cap \overline{G}_{i_2} = T_{i_1+1} = T_{i_2}$.

Suppose $x \in \overline{G}_{i_1}$ is a double point of ΠC which is the image of the points $y_1 \in l_{i_1q_1}$, where $l'_{i_1q_1} \subset l_{i_1q_1}$, and $y_2 \in l_{i_1s}$, with y_2 lying "over" y_1, and $l_{i_1q_1}$ and l_{i_1s} rectilinear components of $\overline{G}_{i_1} \cap C$. The direction prescribed on C induces on $l_{i_1q_1}$ the direction from T_{i_1} to T_{i_1+1} (since $i_2 = i_1 + 1$).

Let $Q \times \lambda$ be the topological product of the planar quadrilateral Q, with vertices $\Pi l_{i_1q_1} \cap T_{i_1}$, $\Pi l_{i_1s} \cap T_{i_1}$, $\Pi l_{i_1q_1} \cap T_{i_2}$ and $\Pi l_{i_1s} \cap T_{i_2}$, and a line λ perpendicular to the plane Π. By Condition 2), G_{i_1} contains no vertex of C of the second sort, or any other double point of ΠC, and therefore $(Q \times \lambda) \cap C = l_{i_1q_1} \cup l_{i_1s}$. This is true also for $U \times \lambda$, where U is a sufficiently small quadrilateral neighborhood of Q in $\Pi \cap \overline{G}_{i_1}$. The planes Π and Π_0 cut off from $U \times \lambda$ a cylinder Z containing $l_{i_1q_1} \cup l_{i_1s}$; obviously $Z \cap C = l_{i_1q_1} \cup l_{i_1s}$. The two side faces of Z are rectangles: Q_1 with boundary ν_1, lying in T_{i_1}, and Q_2 with boundary ν_2, lying in T_{i_2} (Figure 3.7).[1)]

Let $O_1 \in \nu_1$ and $O_2 \in \nu_2$ be the vertices of Z lying on the edge μ closest to O in the plane Π_0.

The rectangle Q_1 contains the points $l_{i_1q_1} \cap T_{i_1}$ and $l_{i_1s} \cap T_{i_1}$; the rectangle Q_2, the points $l_{i_1q_1} \cap T_{i_2}$ and $l_{i_1s} \cap T_{i_2}$.

Consider two paths ϕ_1 and ϕ_2 in $\mathfrak{S}_{i_1}$, where ϕ_j, $j = 1, 2$, is the following continuous mapping of the segment I_s^1 into $\mathfrak{S}_{i_1} \cap T_{i_j}$: the segment $[0 \le s \le 1/3]$ is mapped homeomorphically onto the segment $[OO_j] \subset \mathfrak{S}_{i_1} \cap T_{i_j}$; the segment

1) *Translator's note.* Directions in Figure 3.7 seem to be at odds with those specified in the text.

$[1/3 \le s \le 2/3]$ is mapped homeomorphically, except for endpoints, onto the curve ν_j, in the positive direction around the point $l_{i_1 q_1} \cap T_{i_j}$; finally, the segment $[2/3 \le s \le 1]$ is mapped homeomorphically onto the segment $[O_j O]$. Then ϕ_j is a closed path in $\mathfrak{E}_{i_1} \cap T_{i_j}$, with origin O, circling the point $l_{i_1 q_1} \cap T_{i_j}$ once in the positive direction, and circling the point $l_{i_1 s} \cap T_{i_j}$ in the positive or negative direction according as the segment $l_{i_1 s}$ is directed from T_{i_1} to T_{i_2} or from T_{i_2} to T_{i_1}.

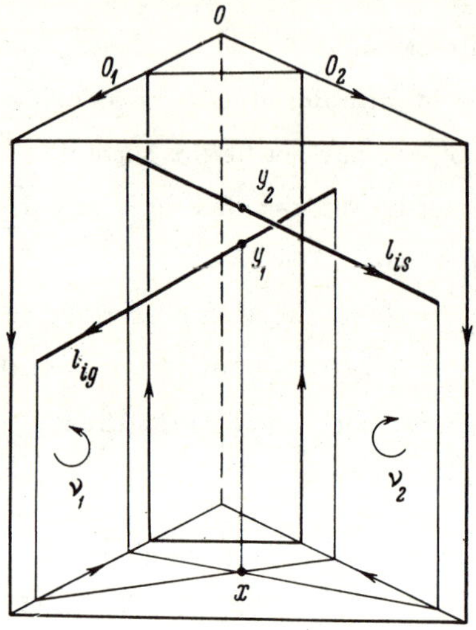

Figure 3.7

If now, in the rectangle Q_1, we separate the points $l_{i_1 s} \cap T_{i_1}$ and $l_{i_1 q_1} \cap T_{i_1}$ by a polygonal line with endpoints O_1 and $\alpha \in \nu_1$, we find that the path ϕ_1 belongs to the class in $\pi_1(\mathfrak{E}_{i_1})$ which is the product of the classes $a_{i_1 q_1}$ and $a^{\epsilon}_{i_1 s}$, where $\epsilon = \pm 1$ depending on the direction in which ϕ_1 circles the point $l_{i_1 s} \cap T_{i_1}$. Thus[1)]

$$\varphi_1 \in a_{i_1 q_1} a^{\varepsilon}_{i_1 s}. \tag{12}$$

In exactly the same way, ϕ_2 belongs to the class in $\pi_1(\mathfrak{E}_{i_1})$ which is the product of the classes $a'_{i_1 q_1}$ and $a^{\epsilon}_{i_1 s}$, except that, since the segments $l_{i_1 q_1}$ and $l_{i_1 s}$ cross each other, the path ϕ_2 circles the points of intersection of these segments with T_{i_2} in the opposite order from path ϕ_1. Hence

$$\varphi_2 \in a^{\varepsilon}_{i_1 s} a'_{i_1 q_1}. \tag{13}$$

We now show that ϕ_2 is homotopic in $\mathfrak{E}_{i_1}$ to ϕ_1. Since Π, Π_0, and the fore and aft faces of the cylinder Z do not intersect C, the homotopy F_t taking ϕ_1 into ϕ_2 can be constructed as follows: for each t, $0 \le t \le 1$, the point O_1 goes into the point O_t of the edge $[O_1 O_2]$ defined by

$$F_t(O_1) = O_1(1 - t) + tO_2,$$

and the path ϕ_1 goes into the path $\phi_t = F_t$, where ϕ_t maps the segment $[0 \le s \le 1/3]$ of I^1_s homeomorphically onto the line segment OO_t in Π_0, the

1) *Translator's note.* The product may be in the opposite order from that in (12).

segment $[1/3 \leq s \leq 2/3]$ homeomorphically, except for endpoints, onto the closed curve $(\mathrm{Fr}\, Z) \cap T_t$, where T_t is the half-plane with edge l that contains the point O_t, $\phi_t(1/3) = \phi_t(2/3) = O_t$, and finally the segment $[2/3 \leq s \leq 1]$ homeomorphically onto the segment $O_t O$. It follows, by (12) and (13), that

$$a_{i_1 q_1} a^{\epsilon}_{i_1 s} = a^{\epsilon}_{i_1 s} a'_{i_1 q_1}. \tag{14}$$

And since $a'_{i_1 q_1} = a_{i_2 q_2}$, by (11) we have

$$a_{i_1 q_1} a^{\epsilon}_{i_1 s} = a^{\epsilon}_{i_1 s} a_{i_2 q_2}. \tag{15}$$

Thus the segment $l_{i_1 q_1}$, which over the double point of ΠC passes under the segment $l_{i_1 s}$, corresponds to two classes of paths: $a_{i_1 q_1}$ and $a'_{i_1 q_1} = a_{i_2 q_2}$, connected by relation (15). On the other hand, the class $a_{i_1 s}$ corresponding to the upper segment $l_{i_1 s}$ contains both the path in T_{i_1} circling the point $l_{i_1 s} \cap T_{i_1}$ and the path in T_{i_2} circling the point $l_{i_1 s} \cap T_{i_2}$, so that

$$a_{i_1 s} = a_{i_2 s}.$$

One computes in the same way $a'_{i_k q_k}$.

We now use these considerations to indicate the procedure for computing the fundamental group of the knot C. C is the union of a finite number of polygonal lines L_j:

$$C = \bigcup_{j=1}^{k} L_j,$$

where L_j is the piece of C between two successive points y_j and y_{j+1} that project into double points x_j and x_{j+1} of ΠC and are the lower points of C over x_j and x_{j+1}. Aside from y_j and y_{j+1}, ΠL may contain only such double points of ΠC for which the corresponding points on L_j are the upper ones on C. By (11), each piece L_j of C corresponds to a single generator c_j of the group $\pi_1(E^3 \setminus C)$, where c_j contains, in every half-plane T_l intersecting L_j and containing no vertex of L_j of the second sort, a path circling a point of $L_j \cap T_l$ in the positive direction. The set of all c_j, $j = 1, \cdots, k$, constitutes a generating set for the group $\pi_1(E^3 \setminus C)$, since by the van Kampen Theorem such a set is given by all the generators a_{iq} of the groups $\pi_1(\mathcal{E}_i)$, and these satisfy the relations (11). The point $y_j \in L_{j-1} \cap L_j$ has above it a point $z_j \in L_s$, where L_s is another such piece of C. The pieces L_{j-1}, L_j and L_s correspond to classes c_{j-1}, c_j and c_s; and if $c_s \in \pi_1(\mathcal{E}_{i_1})$, then $c_s = a_{i_1 s}$, $c_{j-1} = a_{i_1 q_1}$ and $c_j = a_{i_2 q_2}$. Consequently the classes c_s, c_{j-1}, c_j are connected by relations (15), so that we have obtained defining relations for the fundamental group of the knot:

$$c_{j-1}c_s^{\varepsilon} = c_s^{\varepsilon}c_j, \quad \varepsilon = \pm 1, \tag{16}$$

where ϵ is $+1$ or -1 according as the positive direction on L_s projects onto the positive or the negative direction on L_j.[1] Since the relations (11) and (15) constitute a complete set of relations between the elements of the free groups $\pi_1(\mathfrak{S}_i)$, $i = 1, 2, \cdots, N$, the relations (16), taken for all double points of ΠC, constitute a complete set of defining relations for the group $\pi_1(E^3 \setminus C)$.

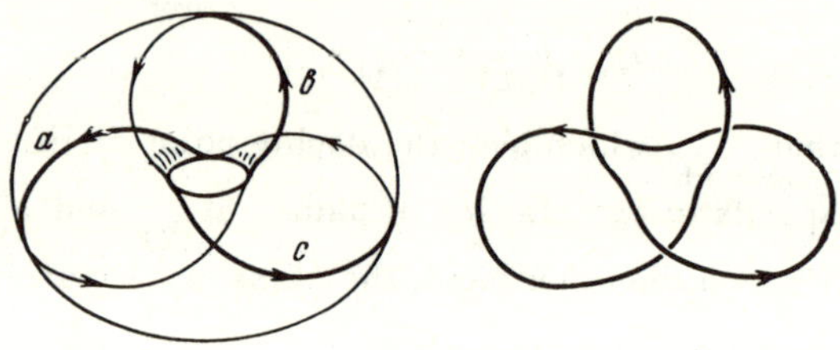

Figure 3.8

Example 4.1. The fundamental group of the simplest knot.

The simplest knot C lies on a torus and circles the meridian of the torus 3 times and the equator 2 times (Figure 3.8). It is a special case of a knot on the torus that circles the meridian p times and the equator q times, where p and q are relatively prime. The projection of the knot into the plane perpendicular to the planes of all the meridians is a normal projection, with three double points. The knot C breaks up into three simple arcs L_a, L_b and L_c, which correspond to classes a, b, c of paths in $E^3 \setminus C$. The three defining relations corresponding to the double points are of the form

$$ab = bc, \quad bc = ca, \quad ca = ab.$$

Each of these follows from the other two; moreover, from the first and third relation, c can be expressed in terms of a and b:

$$c = b^{-1}ab = aba^{-1}. \tag{17}$$

Thus the group $\pi_1(E^3 \setminus C)$ is a group with two generators a and b and one defining relation:

$$aba = bab.$$

The group $\pi_1(E^3 \setminus C)$ can be presented in another way, replacing the generators a and b by new generators A and B, where A contains a path circling the

[1] *Translator's note.* This prescription for the sign of ϵ seems unclear. Essentially, the sign of ϵ depends on the orientation induced in the plane Π by the ordered pair of directions $(\Pi L_s, \Pi L_j)$ at the point of crossing.

equator of the torus in the positive direction, and B a path circling the meridian of the torus in the positive direction. It is easily seen that $B = ab$. Expressing A in terms of a, b and c, we find that $A = bca$. Substituting the value of c from (17), we obtain $A = aba = bab$, whence $A^2 = (aba)(bab) = B^3$.

Thus the fundamental group of the simplest knot is the group with two generators A and B and the one defining relation

$$A^2 = B^3.$$

To prove that the knot C is nontrivial, it suffices to show that the fundamental group of the space $E^3\backslash C$ is not isomorphic to that of the complement of the standard circle, i.e. is not a free cyclic group. This follows from the fact that the group $\pi_1(E^3\backslash C)$ has the following representation into the noncommutative group S_3 of permutations of three integers (i.e. a homomorphism onto this group):

$$A \to (12)(3), \quad B \to (123).$$

§5. The fundamental group of a domain in E^n

1. There are classical methods for computing the fundamental groups of finite polyhedra (see for example [80]). It is therefore appropriate to reduce the computation of the fundamental group of a domain in E^n to that of the fundamental groups of finite polyhedra. To do this we shall need the notion of the limit of a direct spectrum of groups. Let us recall it.

Definition 5.1. By a *direct spectrum* of groups $\{G_\alpha\}$ we mean a family of groups G_α, where the index set $A = \{\alpha\}$ is a directed set and there is defined, for each pair of indices $\alpha < \beta$, a homomorphism

$$h^\alpha_\beta\colon G_\alpha \to G_\beta,$$

where, for every set of three indices such that $\alpha < \beta < \gamma$, one has

$$h^\alpha_\gamma = h^\beta_\gamma h^\alpha_\beta. \tag{18}$$

Divide the set of all elements g_α of all the groups G_α into classes such that two elements $g_\alpha \in G_\alpha$ and $g_\beta \in G_\beta$ belong to the same class if and only if there exists a γ, $\alpha < \gamma$ and $\beta < \gamma$, such that

$$h^\alpha_\gamma(g_\alpha) = h^\beta_\gamma(g_\beta).$$

It is easily seen that every element g_α belongs to only one class g and completely determines it. The element g_α is called a representative of the class g in the group G_α. The class g need not have a representative in every group G_β and may have several representatives in the same group G_α.

We define a multiplication in the set $G = \{g\}$ of all classes in the following way. Let g_α be a representative of the class g in the group G_α, and g'_β a representative of the class g' in the group G_β. Pick an index γ such that $\alpha < \gamma$ and $\beta < \gamma$, and define as the product $g \cdot g'$ the class with the representative $h^\alpha_\gamma(g_\alpha) \cdot h^\beta_\gamma(g_\beta)$. in the group G_γ.

It is easily verified at once that G is a group. The identity of G is the class of all elements g_α of the groups G_α for which there exists a $\beta > \alpha$ such that $h^\alpha_\beta(g_\alpha)$ is the identity element of the group G_β. The group G is called the limit of the direct spectrum of groups G_α.

Let A' be a subset of A cofinal in A, i.e. such that for every $\alpha \in A$ there exists an $\alpha' \in A'$ for which $\alpha < \alpha'$. Then the set of groups $\{G_{\alpha'}\}$, $\alpha' \in A'$, is a direct spectrum, whose limit G' is isomorphic to the group G. Indeed, for any class $g \in G$ with representative $g_\alpha \in G_\alpha$, there exists an $\alpha' \in A'$ such that $\alpha' > \alpha$; then $g_{\alpha'} = h^\alpha_{\alpha'}(g_\alpha)$ is a representative of g in the group $G_{\alpha'}$, and also a representative of a class g' in the group $G'_{\alpha'}$. Different classes g_1 and g_2 in G determine different classes g'_1 and g'_2 in G', since for any representatives $g_{1\alpha}$ and $g_{2\beta}$ of g_1 and g_2 and any $\alpha' \in A'$ such that $\alpha < \alpha'$ and $\beta < \alpha'$, we have $h^\alpha_{\alpha'}(g_{1\alpha}) \neq h^\beta_{\alpha'}(g_{2\beta})$. By (18), the isomorphism so defined from G into G' is onto. Thus to compute the limit of a direct spectrum $\{G_\alpha\}$ it suffices to compute the limit of the direct spectrum determined by a cofinal subset of the set A.

2. **Computation of $\pi_1(U)$, where U is a domain in E^n.**

Theorem 5.1. *The fundamental group of a domain U in E^n is the limit of the direct spectrum of the fundamental groups of all connected finite polyhedra in U containing a given point O.*

Any point $x \in U$ belongs to such a polyhedron, since it can be joined by a polygonal line in U to the point O. The set $\{\alpha\}$ that indexes the polyhedra $P_\alpha \subset U$ is a directed set if we put $\alpha < \beta$ whenever $P_\alpha \subset P_\beta$. Let π^α_1 be the fundamental group of P_α. We show first of all that the set of groups π^α_1 constitutes a direct spectrum. If $\alpha < \beta$, then $P_\alpha \subset P_\beta$, so that there exists an inclusion homomorphism for the fundamental group,

$$h^\alpha_\beta : \pi^\alpha_1 \to \pi^\beta_1;$$

and if $\alpha < \beta < \gamma$, then $P_\alpha \subset P_\beta \subset P_\gamma$, and by the corollary to Theorem 2.2 the inclusion homomorphisms are related as in (18).

We show now that $\pi_1(U) = \lim_\alpha \pi^\alpha_1$. A representative of an element $\sigma \in \lim_\alpha \pi^\alpha_1$ in the group π^α_1 is a class of paths with origin O which are homotopic to each other in P_α and therefore also in U. These classes are connected by the relations

$$h^\alpha_\gamma(\sigma_\alpha) = h^\beta_\gamma(\sigma_\beta), \quad \alpha < \gamma, \ \beta < \gamma, \tag{19}$$

defining the class σ. Since h^α_β is an inclusion homomorphism, σ is a class of homotopic paths in U. Different classes σ and σ' contain paths which are not homotopic in any P_α, and consequently not homotopic in U. Indeed, a homotopy between two paths f and f' in U determines a continuous mapping F into U of the homotopy rectangle D. Then $F(D)$ is a continuum, and for a sufficiently fine triangulation T of E^n the star $\mathrm{St}_T F(D)$ is a finite polyhedron P contained in U, so that the paths f and f' are homotopic in P. Thus $\lim_\alpha \pi_1^\alpha$ maps isomorphically into the group $\pi_1(U)$. We now show that this isomorphism is onto.

Let $\tau \in \pi_1(U)$ be a homotopy class of paths in U with origin O, and take $\phi \in \tau$. Since ϕ is a continuous mapping of the segment I^1_s into U, its image $\phi(I^1_s)$ is a continuum, and for a sufficiently fine triangulation T of E^n we have

$$\mathrm{St}_T\, \phi(I^1_s) = P_\alpha \subset U,$$

so that ϕ is a path in the finite polyhedron P_α. Thus $\phi \in \tau_\alpha$, where τ_α is an element of the group π_1^α. If $\phi(I^1_s) \subset P_\alpha$ and $\phi(I^1_s) \subset P_\beta$, then $\phi \in \tau_\alpha$ and $\phi \in \tau_\beta$, and for every γ such that $\gamma > \alpha$ and $\gamma > \beta$ we have $\phi \in \tau_\gamma$. Consequently

$$\tau_\gamma = h^\alpha_\gamma(\tau_\alpha) = h^\beta_\gamma(\tau_\beta), \tag{20}$$

And if $\alpha < \beta < \gamma$, then

$$\tau_\gamma = h^\alpha_\gamma(\tau_\alpha) = h^\beta_\gamma h^\alpha_\beta(\tau_\alpha). \tag{21}$$

Thus, to the element τ of $\pi_1(U)$ there correspond classes τ_α in certain groups π_1^α. By (20) and (21) the classes τ_α are representatives in the groups π_1^α of some element σ of the group $\lim_\alpha \pi_1^\alpha$. Consequently the elements $\tau \in \pi_1(U)$ correspond in a one-to-one fashion to the elements $\sigma \in \lim_\alpha \pi_1^\alpha$, and the group $\pi_1(U)$ is isomorphic to the group $\lim_\alpha \pi_1^\alpha$.

We now indicate a procedure for computing $\pi_1(U)$ when we know the groups $\pi_1(U_k)$ for an increasing sequence of domains $U_k \ni O$ which are finite polyhedra such that $U = \bigcup_{k=1}^\infty U_k$.

Lemma 5.1. *Any domain U in E^n is the countable union of an increasing sequence of closed domains $\overline{U}_k$, each of which is a finite polyhedron.*

Let O be a point in U, and T_1 a triangulation of E^n sufficiently fine to have $\mathrm{St}_{T_1} O \subset U \cap Q_1$, where by O_k is meant the ball of radius k and center O. Let U_k be the component of O in the interior of the union of all closed n-simplices of the triangulation T_k that are contained in $U \cap Q_k$, where T_k is the $(k-1)$th barycentric subdivision of the triangulation T_1. It is easily seen that

$$\overline{U}_k \subset U, \quad U = \bigcup_{k=1}^\infty U_k, \tag{22}$$

since any point $x \in U$ can be joined to O by a polygonal line l in U, and for sufficiently large k we have $\mathrm{St}_{T_k} l \subset U_k$, so that $x \in U_k \in \overline{U}_k$.

Thus to determine the group $\pi_1(U)$ it suffices to consider the countable direct spectrum of the groups π_1^k for the increasing sequence of domains (22). By choosing the sets U_k in the most convenient way, we can compute the group $\pi_1(U)$, knowing the groups $\pi_1(U_k)$ for all U_k and the inclusion homomorphisms h_{k+1}^k. Since the representatives, in the groups π_1^α, of an element $\tau \in \pi_1(U)$ are classes of paths homotopic to each other (for the various α) in U, the group $\pi_1(U)$ can be presented in the following way:

1) the set of generators of $\pi_1(U)$ consists of all the generators of all the groups $\pi_1(U_k)$;

2) the set of defining relations for $\pi_1(U)$ consists of all the defining relations for all the groups $\pi_1(U_k)$, together with new defining relations of the form

$$g_{k+1} = h_{k+1}^k(g_k) = h_{k+1}^k(g_k'), \tag{23}$$

which follow from the inclusion homomorphism determining the spectrum.

Remark 5.1. If the domain U is the unbounded component of the complement of a compactum $K \subset E^n$, then $U \supset E^n \backslash Q^n$, where Q^n is an n-ball with K in its interior. Since $Q^n \cap U$ is a deformation retract of U, it follows that $\pi_1(U) = \pi_1(U \cap Q^n)$.

Example 5.1. The fundamental group of the complement of a zero-dimensional tame compactum K in E^n is trivial for $n \geq 3$.

Since for a zero-dimensional tame compactum K there exists a homeomorphism $H: E^n \to E^n$ such that $H(K)$ lies on a line, it suffices to consider the case that K lies on a line. As polyhedral neighborhoods of K we may take finite unions of parallelepipeds $p_{i_1 \cdots i_k}$:

$$K \subset P_k = \bigcup_{i_1 i_2 \ldots i_k} p_{i_1 \ldots i_k}; \ \overline{p}_{i_1 \ldots i_k} \cap \overline{p}_{j_1 \ldots j_k} = \Lambda, \tag{24}$$

$$(j_1 \ldots j_k) \neq (i_1 \ldots i_k),$$

where the indices i_r take on the values 0 and 1.

By the remark in Example 3.1, each $U_k = E^n \backslash \overline{P}_k$ has trivial fundamental group π_1^k. But then the group $\pi_1(E^n \backslash K)$, being the limit of the direct spectrum of groups π_1^k, is likewise trivial.

Example 5.2. The fundamental group of $E^2 \backslash K$, where K is an arbitrary zero-dimensional compactum in the plane without isolated points.

By Corollary 2 to Theorem 3.2 of Chapter II, there exists a homeomorphism

$H\colon E^2 \to E^2$ taking K into the Cantor perfect set. Consequently it suffices to consider the case that K is the Cantor perfect set, which can be given as in (24), with all the $p_{i_1 \cdots i_k}$ rectangles and

$$\bar{p}_{i_1 \ldots i_k i_{k+1}} \subset p_{i_1 \ldots i_k}, \ i_{k+1} = 0, 1.$$

By Example 3.1 the fundamental group of $U_k = E^2 \setminus \bar{P}_k$ is the free group on 2^k generators $a_{i_1 \cdots i_k}$, where $a_{i_1 \cdots i_k}$ contains a path circling the rectangle $\bar{p}_{i_1 \cdots i_k}$ in the positive direction.

The homomorphism h^k_{k+1} of $\pi_1(U_k)$ into $\pi_1(U_{k+1})$ is defined by the relations

$$a_{i_1 \ldots i_k} = a_{i_1 \ldots i_k 0} \cdot a_{i_1 \ldots i_k 1};\ i_r = 0, 1,\ r = 1, \ldots, k. \tag{25}$$

Thus the fundamental group of the complement $E^2 \setminus K$ of a zero-dimensional perfect compactum in the plane is the group with a countable set of generators $a_{i_1 \cdots i_k}$, $i_r = 0, 1$, connected by the defining relations (25).

No class $a_{i_1 \cdots i_k}$ in the group $\pi_1(E^2 \setminus K)$ can be contained in the identity class, i.e. can consist of paths null-homotopic in $E^2 \setminus K$. Indeed, if a path $\phi \in a_{i_1 \cdots i_k}$ were null-homotopic in $E^2 \setminus K$, it would be null-homotopic in U_p for some $p \geq k$, which is impossible, since

$$a_{i_1 \ldots i_k} = \prod_{j_r = 0, 1,\ r > k} a_{i_1 \ldots i_k j_{k+1} \cdots j_p},$$

and $\pi_1(U_p)$ is freely generated by the $a_{j_1 \cdots j_p}$, $j_r = 0, 1$.

We remark that if the zero-dimensional compactum $K \subset E^2$ has isolated points, the fundamental group $\pi_1(E^2 \setminus K)$ can still be given in the same way, except that one must set $a_{i_1 \cdots i_k 1} = 1$ whenever the rectangle $p_{i_1 \cdots i_k}$ contains just one point of K.

CHAPTER IV

WILD IMBEDDINGS IN E^3

§1. Definitions

We shall consider in this chapter a number of examples of simple arcs, disks, spheres and zero-dimensional compacta wildly imbedded in E^3. Recall that a set P homeomorphic to a polyhedron is said to be *wildly imbedded*, or simply *wild*, in the space E^n if there exists no homeomorphism of E^n onto itself taking P into a rectilinear polyhedron. Otherwise P is called *tame* in E^n. These notions were introduced in studying imbeddings in E^3. For that case they are justified by the theorem of Alexander [17] to the effect that every polyhedral sphere S^2 divides E^3 into two domains, whose closures are semilinearly homeomorphic to the closures of the complementary domains of the boundary of a 3-simplex (if the imbedding is into a polyhedral sphere S^3, the closures of both complementary domains are 3-cells). For imbeddings into E^n, $n > 3$, it is unknown whether the corresponding theorem is true, so that in this case, for imbeddings of disks and spheres, it is more expedient to consider, instead of tame imbeddings, another notion, that of "trivial imbedding" (see Chapter 5), for which the requirement of a homeomorphism onto a polyhedron under a homeomorphism of all of E^n is replaced by a more stringent condition: homeomorphism onto a simplex, in the case of disks D^k, $k \leq n$, and onto the boundary of a $(k+1)$-simplex in the case of spheres S^k, $k < n$. Observe that any polyhedral knot in E^3 is tame, by definition, but need not be trivial.

We shall describe first some examples of simple arcs, disks and spheres wildly imbedded in E^3, which were constructed by Alexander [18] and Artin and Fox [21], as well as some wild imbeddings that can be obtained by using these as a starting point. We shall then study the wild zero-dimensional compactum of Antoine [20] and the wild imbeddings of simple arcs, disks and spheres in E^3 obtainable by its means.

Let P be a subset of E^n homeomorphic to a polyhedron, and take $x \in P$. P is said to be *locally tame* at x if there exists a neighborhood u_x of x in E^n and a homeomorphism of its closure $\overline{u}_x$ into E^n, $h\colon \overline{u}_x \to E^n$, such that $h(\overline{u}_x \cap P)$ is a rectilinear polyhedron. Obviously, if P is tame, then P is locally tame at every point. Bing and Moise have shown that if P is locally tame in E^3 at every point, then P is tame in E^3 [23,56]. We shall not present the proof of this theorem.

For imbeddings in E^n, $n > 3$, instead of a locally tame imbedding of a manifold M^k in E^n, $k < n$, we consider a *locally flat* imbedding, for which the condition that $h(\bar{u}_x \cap M^k)$ be a polyhedron is replaced by the condition $h(\bar{u}_x \cap M^k) \subset E^k$, where E^k is a k-dimensional hyperplane of E^n. In Chapter V we shall prove Brown's theorem that for a sphere $S^{n-1} \subset S^n$, for any n, local flatness of the imbedding implies triviality. For $n = 3$ the Bing-Moise theorem implies that a locally tame imbedding of S^2 is tame and therefore, by the Alexander theorem [17], trivial. We shall now show that *for a simple arc* $l \subset E^n$, *if* l *is locally tame, then* l *is tame.*

Observe that, by Theorem 2.3 of Chapter I, if an arc is locally tame, it is locally flat. We show that if l is locally flat, then l is trivially imbedded, and consequently l is tame in E^n.

We prove first that l is imbedded in E^n together with a neighborhood, i.e. that it has a neighborhood V of which there exists a homeomorphism F into E^n such that $F(l)$ is a line segment. By the local flatness condition, for every $x \in l$ there exists a neighborhood u_x in E^n and a homeomorphism ϕ_x: $u_x \to E^n$ such that $\phi_x(u_x \cap l)$ is a rectilinear (open) interval, or a half-open interval if x is an endpoint. From the family $\{u_x\}$ take a finite number of neighborhoods $u_1, \cdots, u_N$ such that

$$l \subset \bigcup_{i=1}^{N} u_i; \quad u_i \cap u_j = \Lambda, \quad \text{if} \quad |i-j| > 1,$$

and suppose ϕ_i: $u_i \to E^n$, where $\phi_i(u_i \cap l)$ is a rectilinear interval. The arc l can be represented in the form $l = \bigcup_{i=1}^N l_i$, where the l_i are closed segments of l, $l_i \cap l_{i+1}$ is a point, and $l_i \subset u_i$.

By induction on the number of summands, we show that for the arc $\lambda_k = \bigcup_{i=1}^k l_i$ there exists a neighborhood U_k and a homeomorphism f_k: $U_k \to E^n$ such that $f_k(\lambda_k)$ is a rectilinear segment. By assumption this holds for $k = 1$. Suppose it holds for $k - 1 < N$ summands; we show it holds for k summands. By Theorem 2.3 of Chapter I, ϕ_k may be chosen so that

$$\varphi_k(\lambda_{k-1} \cap \bar{u}_k) = f_{k-1}(\lambda_{k-1} \cap \bar{u}_k) \tag{1}$$

and the segment $\phi_k(l_k)$ is a prolongation of the segment $f_{k-1}(\lambda_{k-1})$. In $f_{k-1}(U_{k-1})$ we take a parallelepiped $\Delta \supset f_{k-1}(\lambda_{k-1})$, and in $\phi_k(u_k)$ a parallelepiped $\delta \supset \phi_k(l_k)$, such that

$$\varphi_k^{-1}(\delta \cap \Delta) \subset U_{k-1}. \tag{2}$$

This is possible because of (1) and the fact that $\lambda_{k-1} \cap l_k \in U_{k-1} \cap u_k$; set

$$V_{k-1} = f_{k-1}^{-1}(\Delta); \quad v_k = \varphi_k^{-1}(\delta).$$

There exists a homeomorphism h of the parallelepiped δ onto itself which is stationary on the boundary and takes the line segment $\phi_k(\lambda_k \cap v_k)$ into the segment $\phi_k(\lambda_{k-1} \cap v_k)$:

$$\underset{y \in \operatorname{Fr}\delta}{h(y) = y}; \quad h\varphi_k(\lambda_k \cap v_k) = \varphi_k(\lambda_{k-1} \cap v_k) \subset \Delta \cap \delta. \tag{3}$$

Define a homeomorphism Φ of the domain v_k onto itself by $\Phi = \phi_k^{-1} h \phi_k$. By (2) and (3),

$$\underset{x \in \operatorname{Fr} v_k}{\Phi(x) = x}; \quad \Phi(\lambda_k \cap v_k) \subset \varphi_k^{-1}(\delta \cap \Delta) \subset U_{k-1}. \tag{4}$$

Consequently Φ extends to a homeomorphism $\overline{\Phi}$ of the domain $U_{k-1} \cup v_k$ onto itself, given by

$$\underset{x \in v_k}{\overline{\Phi}(x) = \Phi(x)}; \quad \underset{x \in U_{k-1} \setminus v_k}{\overline{\Phi}(x) = x}.$$

It follows from (4) that

$$\overline{\Phi}(\lambda_k) = \lambda_{k-1} \subset U_{k-1}. \tag{5}$$

There is in v_k a neighborhood w_k of the arc l_k such that $\overline{\Phi}(w_k) \subset U_{k-1}$. Set

$$U_k = U_{k-1} \cup w_k; \quad f_k = f_{k-1}\overline{\Phi}|_{U_k}.$$

By (5), f_k is a homeomorphism of the domain U_k into E^n, such that $f_k(\lambda_k)$ is a line segment. Thus in a finite number of steps we construct a domain $V = U_N$ containing l, and a homeomorphism $F\colon V \to E^n$ taking l into a line segment.

Consequently the arc l is imbedded in E^* together with a neighborhood.[1] We shall prove in Chapter V (Theorem 2.2) that if a simple arc is imbedded in E^n together with a neighborhood, it is imbedded trivially in E^n, and hence is tame.

From the foregoing it follows that every wild simple arc, disk or sphere in E^3 has points of *local wildness*, or *wild points*, i.e. points at which the imbedding is not locally tame. We shall consider in §2 examples of imbeddings in E^3 with finitely many wild points, and in §§3 and 4 examples of imbeddings with infinitely many such points.

---

1) A similar, but technically more complicated, proof shows that a locally flat imbedding is a cube is an imbedding together with a neighborhood [14].

§2. Examples of simple arcs, disks and spheres in E^3 with finitely many wild points

1. **Examples 2.1** ([21], Example 1.1). *A simple arc with two knotted ends.* This arc L is pictured in Figure 4.1. It can be

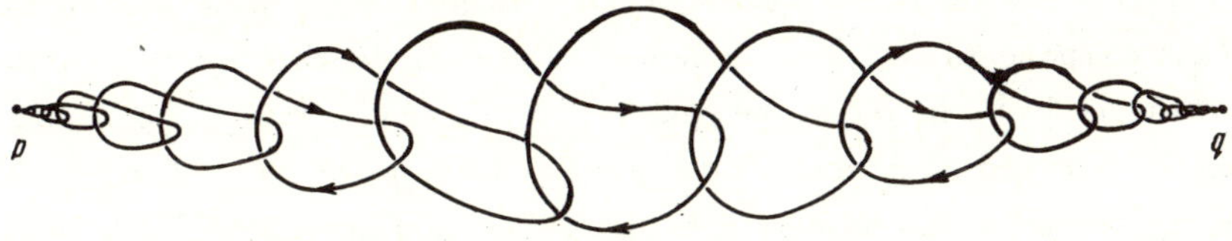

Figure 4.1

regarded as locally polyhedral, and therefore locally tame, everywhere except at the endpoints p and q. We show that the simple arc L is widly imbedded in E^3 by showing that the fundamental group of $E^3\setminus L$ is nontrivial. L is the union of three infinite families of polyhedral simple arcs; any one of these arcs has no

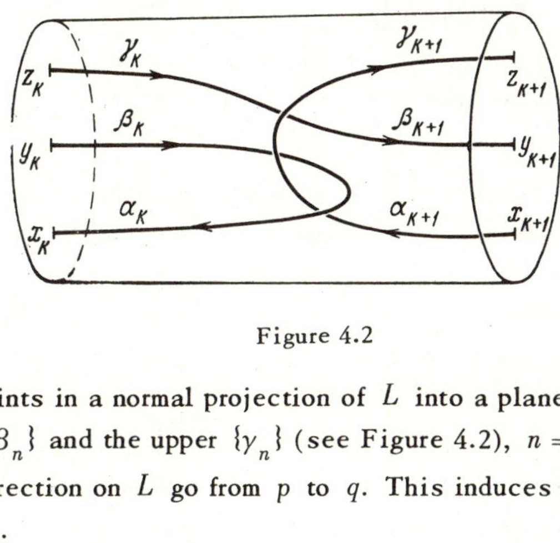

Figure 4.2

"lower" double points in a normal projection of L into a plane: the lower family $\{\alpha_n\}$, the middle $\{\beta_n\}$ and the upper $\{\gamma_n\}$ (see Figure 4.2), $n = 0, \pm 1, \pm 2, \cdots$. Let the positive direction on L go from p to q. This induces a direction on each arc α_n, β_n and γ_n.

Outside a convex domain V containing L, take a line R in the plane perpendicular to $\overline{pq}$ and passing through the center of $\overline{pq}$. Through R, take a set of half-planes Π_n, $n = 0, \pm 1, \pm 2, \cdots$, such that $L \cap \Pi_n$ consists of three points: $x_n \in \alpha_n$, $y_n \in \beta_n$ $z_n \in \gamma_n$ (the proportions of L can be so chosen that it is possible), arranged from below to above on Π_n.

We shall compute $\pi_1(E^3\setminus L)$ as the direct limit of groups $\pi_1(E^3\setminus \overline{U}_n)$, where each $\overline{U}_n$ is the following closed neighborhood of L: $\overline{U}_n$ is the topological product of a sufficiently small disk with the part of L lying in the domain bounded by $\Pi_{-n} \cup \Pi_n$ that does not contain the points p and q, plus two truncated cones with bases on Π_{-n} and Π_n containing the parts of L adjacent to p and q. The domains

U_n can be chosen so that $\overline{U}_{n+1} \subset U_n$. We compute $\pi_1(E^3 \setminus \overline{U}_n)$ by means of the van Kampen Theorem. Choose the origin O on the line R, and divide $E^3 \setminus \overline{U}_n$ by the half-planes Π_k, $-n \leq k \leq n$, into domains V_k between Π_k and Π_{k+1}, and an "outside" domain V_{+n} bounded by $\Pi_{-n} \cup \Pi_{+n}$ and containing the points p and q. Each V_k, $-n \leq k < +n$, is the domain O_k of a dihedral angle with three tubes around the components of $L \cap O_k$ removed. These components are three arcs: the first intersects $\operatorname{Fr} V_k$ in Π_k at the points $x_k \in \alpha_k$ and $y_k \in \beta_k$; the second crosses the first and intersects $\operatorname{Fr} V_k$ in Π_{k+1} at the points $x_{k+1} \in \alpha_{k+1}$ and $z_{k+1} \in \gamma_{k+1}$; finally, the third crosses the second and intersects Π_k at the point $z_k \in \gamma_k$ and Π_{k+1} at the point $y_{k+1} \in \beta_{k+1}$ (Figure 4.2).

Let $X_k = \bigcup_{i=-n}^{k} V_i$; then for $k < +n$, $X_k \cap V_{k+1}$ is the part of Π_{k+1} obtained by removing three disks with centers x_{k+1}, y_{k+1}, z_{k+1}, while $X_{n-1} \cap V_{+n}$ is obtained by removing one disk each from Π_{+n} and Π_{-n}. Each union $X_k \cup V_{k+1}$ satisfies the conditions of Corollary 2 to the van Kampen Theorem, so that, knowing the fundamental group of each V_k, $-n \leq k \leq +n$, we can compute $\pi_1(E^3 \setminus \overline{U}_n)$. It is easily seen that the group $\pi_1(V_{+n})$ is trivial. Observe that a path in V_k circling one of the arcs α_{k+1}, β_{k+1}, γ_{k+1}, and a path in V_{k+1}, $k < +n$, circling the same arc in the same direction, are both homotopic to a path in Π_{k+1} circling the corresponding point x_{k+1}, y_{k+1} or z_{k+1}.

To express the equality of the corresponding elements in the group $\pi_1(X_k \cup V_{k+1})$, it suffices to designate by the same symbol, viz., a_{k+1}, b_{k+1} or c_{k+1}, the element $\pi_1(X_k)$ and the element of $\pi_1(V_{k+1})$ that contain the path in Π_{k+1} circling the point x_{k+1}, y_{k+1} or z_{k+1}, respectively. These elements constitute a complete set of generators for the free group $\pi_1(X_k \cap V_{k+1})$ (Example 3.1 of Chapter III).

To compute the group $\pi_1(X_{k+1})$, it suffices to write down the defining relations for all the groups $\pi_1(V_i)$, $i \leq k+1$. Let us compute the group $\pi_1(V_i)$. Each V_i is the closed domain of a dihedral angle with three tubes, the neighborhoods of the arcs, removed. As in Example 3.4 of Chapter III, it is easily shown that V_i is homeomorphic to a convex domain with three line segments removed, i.e. $\pi_1(V_i)$ is a free group on three generators, which may be taken as a_i, b_i and c_i (see Figure 4.2 and Example 3.5 of Chapter III).[1)] The classes a_k, b_k, c_k in $\pi_1(X_{k+1})$

---

1) *Translator's note*, There is some confusion here and in the next paragraph: a_i and b_i, being conjugate, cannot belong to the same generating set of $\pi_1(V_i)$. The proper statement is that $\pi_1(V_k)$ has the six generators $a_k, \cdots, c_{k+1}$ with the defining relations (6) (and so, of course, is free on three generators). Similarly, the group $\pi_1(X_l)$, $l<n$, has generators a_k, b_k, c_k, $-n \leq k \leq l+1$, and defining relations (6), $-n \leq k \leq l$.

are connected by certain relations with the classes a_{k+1}, b_{k+1}, c_{k+1}. These can be computed in the same way as for the group of a knot (see Chapter III), by considering the relations for all double points in V_k of the normal projection of L. There are three such points in V_k (see Figure 4.2), for which we find the following relations, taking each class as containing a path circling the corresponding arc clockwise, as viewed from the positive direction on L:

$$a_k^{-1}a_{k+1} = c_{k+1}a_k^{-1} = b_k^{-1}c_{k+1}; \qquad (6)$$
$$c_kc_{k+1} = c_{k+1}b_{k+1}.$$

For the group $\pi_1(X_l)$, $l < +n$, a complete set of generators consists of the elements a_k, b_k, c_k, $-n \le k \le l$; and the set of relations (6), $-n \le k \le l$, is a complete set of defining relations. For the group $\pi_1(E^3\setminus\overline{U}_n)$ a complete set of generators consists of all the elements a_k, b_k, c_k, $-n \le k \le +n$. All the relations (6), $-n \le k \le +n$, are defining relations for $\pi_1(E^3\setminus\overline{U}_n)$. It is easily seen that

$$c_kb_ka_k^{-1} = c_{k+1}b_{k+1}a_{k+1}^{-1},$$

so that, since $\pi_1(V_{+n})$ is trivial, we obtain

$$c_kb_ka_k^{-1} = 1, \qquad (7)$$

where $-n \le k \le +n$.

Thus the group $\pi_1(E^3\setminus\overline{U}_n)$ is determined by generators $\{a_k, b_k, c_k\}$, $-n \le k \le +n$, and the set of defining relations (6) and (7).

To pass from the groups $\pi_1(E^3\setminus\overline{U}_n)$ to their direct limit $\pi_1(E^3\setminus L)$, observe that the inclusion homomorphisms

$$h_n : \pi_1(E^3\setminus\overline{U}_n) \to \pi_1(E^3\setminus\overline{U}_{n+1})$$

are isomorphisms into. Thus each generator a_k, b_k, c_k of $\pi_1(E^3\setminus\overline{U}_n)$ goes into the same generator of $\pi_1(E^3\setminus\overline{U}_{n+1})$, six new generators being added with indices $\pm(n+1)$. Therefore, in accordance with the description in §5 of Chapter III, the group $\pi_1(E^3\setminus L)$ can be determined in the following way.

A complete set of generators for $\pi_1(E^3\setminus L)$ consists of all the a_k, b_k, c_k, and a complete set of defining relations consists of all the relations (6) and (7), $-\infty < k < +\infty$.

Let us now perform some transformations on the group. First of all, from (6), the generators a and b can be expressed in terms of c:

$$b_k = c_k^{-1}c_{k-1}c_k; \qquad (8)$$
$$a_k = c_{k+1}^{-1}c_k^{-1}c_{k-1}c_kc_{k+1}.$$

Substituting these values of a_k and b_k into (7), we find, after cancellation, that

$$c_{k-1}c_kc_{k+1}^{-1}c_k^{-1}c_{k-1}^{-1}c_kc_{k+1} = 1$$

or, transforming this,

$$c_{k-1}c_kc_{k+1} = c_kc_{k+1}c_{k-1}c_k. \quad -\infty < k < +\infty. \tag{9}$$

Substituting the values in (8) into the unused relation (6) $a_k^{-1}a_{k+1} = c_{k+1}a_k^{-1}$, we find that the latter follows from (9).

Thus a complete set of generators for $\pi_1(E^3\backslash L)$ is given by all the classes c_k; and a complete set of defining relations is given by (9). The group $\pi_1(E^3\backslash L)$ is nontrivial, since it has the following representation into the permutation group on five elements:

$$c_k \rightarrow (12345) \text{ if } k \text{ is even,}$$
$$c_k \rightarrow (14235) \text{ if } k \text{ is odd.}$$

Consequently the arc L is wildly imbedded in E^3. The only wild points, as we have seen, can be p and q. The proof that L is locally wild at p and q can also be carried out in the same way as in Example 2.4 below. The fact that $\pi_1(E^3\backslash L)$ is here nontrivial implies that $E^3\backslash L$ is not homeomorphic to the complement of a line segment.

2. Using Example 2.1 as a starting point, a number of wild imbeddings of arcs, disks and spheres can be constructed in E^3. We shall now consider some of them.

Example 2.1a. *A disk $D \subset E^3$ with two wild points on the boundary.*

The boundary of the disk D is a pair of simple arcs L_1 and L_2, imbedded like L, close to one another and intersecting only at the endpoints: $L_1 \cap L_2 = \{p\} \cup \{q\}$; while the part of D in the domain V_n of the dihedral angle $\Pi_{-n} \cup \Pi_{+n}$ is the topological product of $L_1 \cap V_n$ by a line segment (which contracts to a point as one approaches p and q). It is easily seen that the group $\pi_1(E^3\backslash D)$ is isomorphic to $\pi_1(E^3\backslash L)$, since $E^3\backslash D$ is a deformation retract of $E^3\backslash L$, i.e. the group $\pi_1(E^3\backslash D)$ is nontrivial; so that D is wildly imbedded in E^3, and $E^3\backslash D$ is not homeomorphic to $L^3\backslash\sigma^2$, where σ^2 is a 2-simplex. The disk has two wild points, which lie on the boundary.

Example 2.1b. *A sphere $S^2 \subset E^3$, whose exterior complementary domain is not homeomorphic to the complement of a ball (or of a point).*

The sphere S^2 can be obtained as the union of two disks D_1 and D_2 with common boundary $L_1 \cup L_2$, imbedded like the disk D of Example 2.1a, such that

$$D_1 \cap D_2 = L_1 \cup L_2.$$

The closure $\overline{U}$ of the interior domain of S^2 is, if we exclude the points p and q, the topological product of $L_1 \setminus (\{p\} \cup \{q\})$ by a disk d. Hence $\overline{U}$ is a topological ball (a 3-cell), and the domain U is homeomorphic to E^3. The part of S^2 that lies in V_n (see Example 2.1) is the topological product of $L_1 \cap V_n$ by a circle C. The exterior complementary domain U_c of S^2 is not simply connected, since it is homeomorphic to $E^3 \setminus L$ and so has fundamental group isomorphic to $\pi_1(E^3 \setminus L)$, i.e. nontrivial. If the space E^3 is enlarged to a sphere S^3 by adding a point, then S^2 divides S^3 into two domains, one of which is not homeomorphic to E^3.

Example 2.1c. *A disk D' with two interior wild points* can be obtained from the sphere S^2 of Example 2.1b if we remove from the latter a small disk d with center in the middle arc $\gamma_0 \subset L_1$. The disk D' contains the arc L_2, and any path f in $E^3 \setminus D'$ that belongs to any one of the classes c_n^2 of the group $\pi_1(E^3 \setminus L_2)$ is not null-homotopic in $E^3 \setminus L_2$, and so also not in $E^3 \setminus D'$. Consequently the group $\pi_1(E^3 \setminus D')$ is nontrivial, and D' is wildly imbedded in E^3. It is easily verified that $E^3 \setminus D'$ is homeomorphic to $E^3 \setminus L_2$.

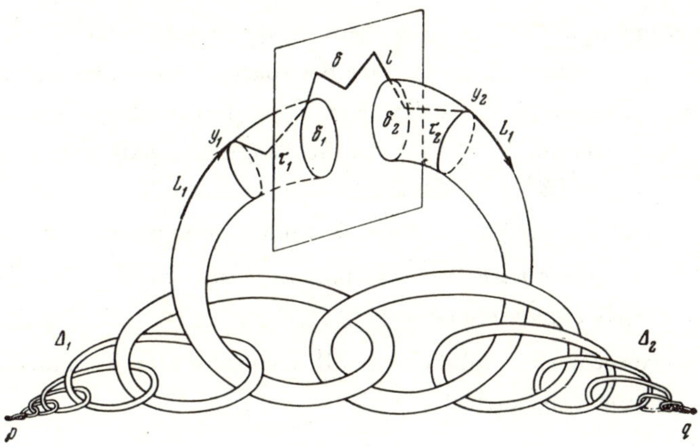

Figure 4.3

Example 2.2. *A disk $\Delta \subset E^3$ that does not lie on any sphere contained in E^3.* To construct Δ, remove from the sphere S^2 of Example 2.1b a small cylinder τ that contains the centers x_1 and x_2 of the arcs $\gamma_{10} \subset L_1$ and $\gamma_{20} \subset L_2$, and is cut out from S^2 by two planar disks, such that

$$\tau \cap (L_1 \cup L_2) \subset \gamma_{10} \cup \gamma_{20}.$$

Next, consider a disk δ in the plane perpendicular to the axis of the cylinder τ (the segments of the arcs γ_{10} and γ_{20} lying on τ can be supposed rectilinear) such that $\delta \cap S^2 = \delta \cap \tau$ is a circle C in the *interior* of δ and containing the points x_1 and x_2. The cylinder τ separates S^2 into two disks: Δ_1 containing the point p, and Δ_2, containing q. In the interior of δ, take two disjoint disks δ_1 and δ_2 with centers x_1 and x_2; join the boundary of Δ_1 with the boundary of δ_1 by a cylinder τ_1 lying in a small neighborhood W of the cylinder τ on one side of δ, and join the boundary of Δ_2 with the boundary of δ_2 by a cylinder τ_2 lying on the other side of δ (Figure 4.3). Set

$$\Delta = \delta\setminus\delta_1 \cup (\tau_1 \cup \Delta_1)\setminus\delta_2 \cup (\tau_2 \cup \Delta_2).$$

Δ is the disk obtained from δ by replacing each disk δ_i, $i = 1, 2$, by the disk $\tau_1 \cup \Delta_i$.

We now show that the disk Δ does not lie on any sphere contained in E^3.

Consider the points $y_i = L_1 \cap \operatorname{Fr}\Delta_i$, and join y_1 to y_2 by a polygonal line l in $(\delta\setminus\delta_1\setminus\delta_2 \cup \tau_1 \cup \tau_2) \cap W$ that does not meet the boundary of δ. Then

$$L = (L_1 \cap \Delta_1) \cup l \cup (L_1 \cap \Delta_2)$$

is a simple arc in $\Delta \cup W$. Since the endpoints of l are the same as those of $L_1 \cap \tau$, and $l \subset W$, there exists an isotopy $F_t\colon E^3 \to E^3$ which is stationary on $\Delta_1 \cup \Delta_2$ and outside W and takes l into $L_1 \cap \tau$, and therefore $\lambda = L_1\setminus\tau \cup l$ into L_1. Hence the group $\pi_1(E^3\setminus\lambda)$ is isomorphic to the group $\pi_1(E^3\setminus L_1)$. Now, if Δ were contained in a sphere $\Sigma^2 \subset E^3$, its boundary Γ would be null-homotopic in $\Sigma^2\setminus\operatorname{int}\Delta$, and therefore in $E^3\setminus\operatorname{int}\Delta$. But Γ is also the boundary of δ, and if the origin O is chosen on Γ, then Γ is the image of a path belonging to the class c_0 of the group $\pi_1(E^3\setminus\lambda)$. Since this path is not null-homotopic in $E^3\setminus\lambda$, neither is it, a fortiori, in $\Sigma^2\setminus\operatorname{int}\Delta$; thus Δ cannot be contained in a sphere.

Example 2.3. *A sphere with four wild points, both of whose complementary domains are not simply connected* (*Figure* 4.4).

To make the construction, take two small disjoint disks δ_1 and δ_2 on the sphere S^2 of Example 2.1b, such that

$$(\delta_1 \cup \delta_2) \cap L_1 \subset \gamma_{10}; \quad (\delta_1 \cup \delta_2) \cap L_2 = \Lambda.$$

Join their boundaries by a segment l of the arc γ_{10}. Add to l two disjoint arcs λ_1 and λ_2, each having just one endpoint in common with l, and, except for this endpoint, lying in the interior domain U of S^2:

$$\lambda_i \cap l = \{x_i\} \in \delta_i; \quad \lambda_i \setminus \{x_i\} \subset U; \quad \lambda_1 \cap \lambda_2 = \Lambda.$$

Furthermore, the arcs λ_i are to be so chosen that the union

$$\lambda = \lambda_1 \cup l \cup \lambda_2$$

is a simple arc imbedded in E^3 like the arc L of Example 2.1. Thus the arcs λ_1 and λ_2 have each a knotted end, with corresponding endpoints r_1 and r_2 in U.

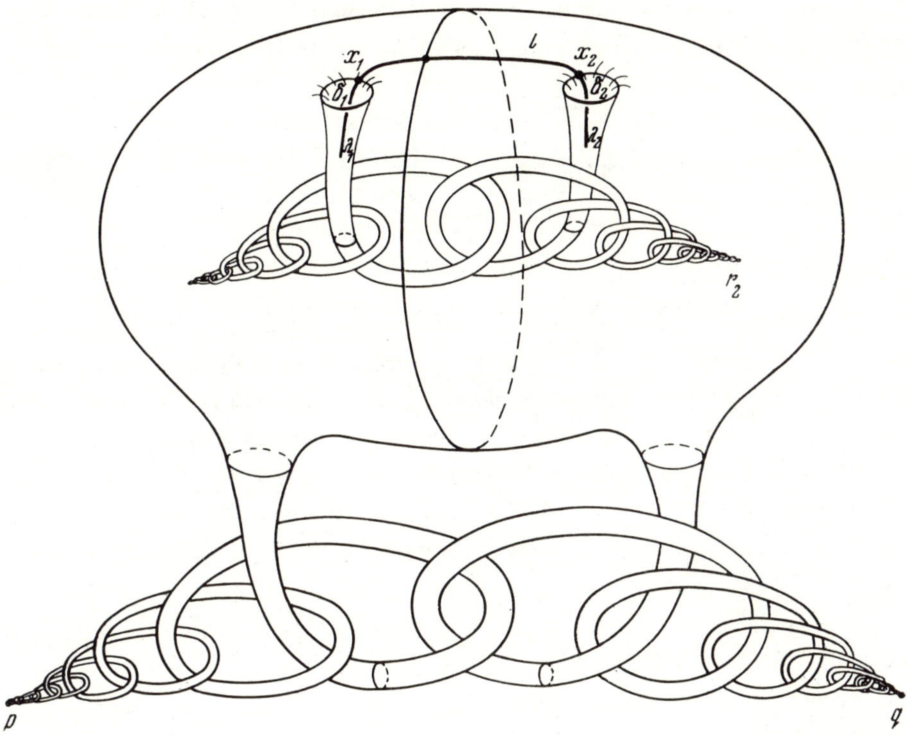

Figure 4.4

To each arc λ_i attach a disk Δ_i, $i = 1, 2$, lying in U except for its boundary, which coincides with the boundary of δ_i, such that $\Delta_1 \cap \Delta_2 = \Lambda$ and $\Delta_i \setminus \{r_i\}$ is the topological product of the arc $\lambda_i \setminus \{r_i\}$ by a circle (λ_i is situated on Δ_i in the same way as, in Example 2.2, $L_1 \cap \Delta_i$ is on Δ_i). Now set

$$\Sigma^2 = S^2 \setminus \delta_1 \setminus \delta_2 \cup \Delta_1 \cup \Delta_2.$$

Then Σ^2 is a sphere satisfying the required conditions. The exterior domain V' of Σ^2 is homeomorphic to the exterior domain V of S^2. Indeed, the spheres $\delta_i \cup \Delta_i$, $i = 1, 2$, bound topological cones K_i, with bases δ_i and vertices r_i, so that there exists a homeomorphism of V onto $V' = V \cup \bigcup_{i=1,2}(K_i \setminus \Delta_i)$, stationary outside a small neighborhood of $K_1 \cup K_2$. Since the domain V is not simply connected, neither is V'. Likewise, the interior domain U' of Σ^2 is not simply connected. To see this, consider a path f in U' that belongs to some class c_n^λ of the group $\pi_1(E^3 \setminus \lambda)$ (see Example 2.2). Since f is not null-homotopic in $E^3 \setminus \lambda$, neither is it, a fortiori, in U'. Consequently the group $\pi_1(U')$ is nontrivial and U' is not simply connected. It is easily determined, in fact, that the group $\pi_1(U')$ is isomorphic to $\pi_1(E^3 \setminus \lambda)$.

3. We shall now consider examples constructed by Artin and Fox [21] of wild imbeddings with a single wild point.

Example 2.4. *A wild simple arc L with a single "knotted" end, whose complement is homeomorphic to that of a point (Figure 4.5).*

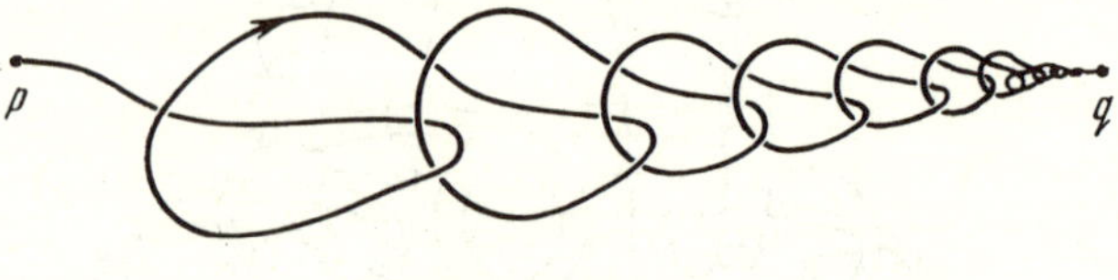

Figure 4.5

The arc L is the part of the arc L of Example 2.1 from the point $z_0 \in \gamma_0$ to the point q (or from p to z_0). L can be assumed to lie in the half-space cut off by the plane π_0 passing through the point z_0. An easy iterative computation shows that the group $\pi_1(E^3 \setminus L)$ is trivial (by using the considerations of Example 2.1 and observing that here $c_0 = 1$). But this does not imply that $E^3 \setminus L$ is homeomorphic to the complement of a point. To prove the latter, we shall construct a continuous mapping F of E^3 onto itself, under which $F(L) = q$ and which is a homeomorphism on $E^3 \setminus L$.

At the point z_0 we adjoin to L a line segment $\overline{z_1 z_0}$ such that $\overline{z_1 z_0} \cup L$ is a simple arc. Then $\nu = \overline{z_1 z_0} \cup L \setminus q$ has a neighborhood W in E^3 which is the topological product of ν by a disk whose center lies on ν (and whose diameter approaches zero as one approaches q). $\overline{W}$ is a topological cone with vertex q: there exists a homeomorphism $h\colon \overline{W} \to K$, where K is a circular cone with vertex O, and $h(q) = O$, $h(\overline{z_1 z_0} \cup L)$ is the altitude of K, and the base D of K is the image of the disk whose center lies at the point z_1. There exists a continuous mapping ϕ of K onto itself, under which the segment $h(L)$ of the altitude goes into the vertex O, and which is a homeomorphsim outside $h(L)$ and is stationary on the boundary.

To construct ϕ, consider three circular cones contained in K: cone K_1 with base D and vertex $h(z_0)$; a truncated cone K_2 with lower base D, whose upper base D' is parallel to D, is contained in the *interior* of K, and has its center on $h(L)$; and, finally, cone K_3 with base D' and vertex $h(q)$. Let C, C_i, $1 \le i \le 3$, be the lateral surfaces of these cones. Every line intersecting D and perpendicular to D contains points

$$a \in D;\; b \in C_1;\; c \in C_2 \cup C_3;\; d \in C.$$

The continuous mapping ϕ is to take the segment $\overline{ab}$ linearly into the segment $\overline{ac}$, and $\overline{bd}$ linearly into $\overline{cd}$ (ϕ can be given by formulas (8) of Chapter I, setting $t = 1$). Setting $F = h^{-1}\phi h$, we obtain a continuous mapping of $\overline{W}$ onto itself, stationary on the boundary, such that $F(L) = q$ and F is a homeomorphism outside L. Setting

$$F(x) = x, \quad x \in E^3 \setminus W,$$

we obtain the required continuous mapping taking $F(E^3 \setminus L)$ homeomorphically into $E^3 \setminus \{q\}$. If E^3 is enlarged to a sphere S^3 by adding a point, then $S^3 \setminus L$ is homeomorphic to E^3.[1)]

To prove that the imbedding of L into E^3 is wild, we make use of the following observation.

Let l be a tame simple arc in E^3, and let

$$V_1 \supset V_2 \supset \ldots \supset V_n \supset \ldots$$

be a sequence of neighborhoods of an endpoint q of l in E^3 such that $q = \bigcap_{n=1}^{\infty} V_n$. Then there exists an integer n_0 such that for $n > n_0$ the inclusion homomorphism $H_n\colon \pi_1(V_n \setminus l) \to \pi_1(V_1 \setminus l)$ is trivial (i.e. the group $\pi_1(V_n \setminus l)$ maps into the identity element of the group $\pi_1(V_1 \setminus l)$).

Since l is tame, there exists a homeomorphism $h\colon E^3 \to E^3$ taking l into a line segment $[A, B]$. Every neighborhood V_n goes into a neighborhood $W_n = h(V_n)$ of the point B. The group $\pi_1(W_n \setminus [A, B])$ is isomorphic to the group $\pi_1(V_n \setminus l)$, $n = 1, 2, \cdots$, and the inclusion homomorphism of $\pi_1(W_n \setminus [A, B])$ into $\pi_1(W_1 \setminus [A, B])$ is isomorphic to the homomorphism H_n. Hence it suffices to prove the assertion in the case of a segment. But for sufficiently large n, W_1 contains a ball δ with center B such that

$$W_n \subset \delta \subset W_1.$$

The inclusion homomorphism of the group $\pi_1(W_n \setminus [A, B])$ into the group $\pi_1(W_1 \setminus [A, B])$ is the product of the inclusion homomorphism of $\pi_1(W_n \setminus [A, B])$ into $\pi_1(\delta \setminus [A, B])$ and the inclusion homomorphism of $\pi_1(\delta \setminus [A, B])$ into $\pi_1(W_1 \setminus [A, B])$. But the group $\pi_1(\delta \setminus [A, B])$ is trivial; consequently it, and so also $\pi_1(W_n \setminus [A, B])$, is mapped under the inclusion homomorphism into the identity element of $\pi_1(W_1 \setminus [A, B])$, so that the latter homomorphism is trivial. Hence, so also is the homomorphism $H_n\colon \pi_1(V_n \setminus l) \to \pi_1(V_1 \setminus l)$ trivial for sufficiently large n.

Now let $\{V_n\}$ be a sequence of open parallelepipeds with center at the point q of the arc L, shrinking down to q, with $\operatorname{Fr} V_n$ intersecting L in the three points x_n, y_n and z_n (see Example 2.1). In the same way as in Example 2.1 it can be shown that the group $\pi_1(V_n \setminus l)$ has the set of generators a_n, c_k^n, $k \geq n$, where c_k^n is a class of null-homotopic paths in $V_n \setminus l$ circling the arc γ_k.

Under the inclusion homomorphism H_n each class c_k^n, $k \geq n$, goes into the

1) This proof goes through for an arbitrary simple arc in E^3 with an endpoint as its only wild point.

class c_k^1 containing all the paths in c_k^n.

Since $c_k^1 \neq 1$, H_n is nontrivial for every n; consequently the imbedding of L in E^3 is wild.

As in Examples 2.1a, b and c, we can use the arc L to construct a disk D_L with a single wild point q, either on its boundary or in its interior, and a sphere S_L^2 with a single wild point. Just as for L, one proves that $E^3 \setminus D_L$, as well as the exterior component of $E^3 \setminus S_L^2$, is homeomorphic to $E^3 \setminus \{q\}$.

It is easily verified that the interior domain U of S_L^2 is homeomorphic to E^3 (since it is homeomorphic to the interior of a circular cone). Thus both complementary domains of $S_L^2 \subset S^3$ are homeomorphic to E^3, and $\overline{U}$ is an n-cell. But $S^3 \setminus U$ is *not* an n-cell.

Artin and Fox have shown [21] that the simple arc which is the union of L with its mirror image L' in the plane Π_0 is a wild simple arc in E^3, whose complement has trivial fundamental group but is not homeomorphic to E^3.

§3. Imbeddings in E^3 with infinitely many wild points

1. Observe that it follows at once from the definition that the set A of wild points of a topological polyhedron P is closed. Indeed, any neighborhood v of a limit point x of A contains a wild point $y \in A$, so that no homeomorphism h: $v \to E^n$ exists such that $h(v \cap P)$ is a rectilinear polyhedron; in other words x is also a wild point. Likewise, the set of nonflat points on a manifold is closed.

Example 3.1 *A simple arc with infinitely many wild points.*

In the line segment $[0, 1]$, take an infinite set of disjoint intervals (a_i, b_i) and replace each segment $[a_i, b_i]$ by a simple arc L_i like that of Example 2.1, with endpoints a_i and b_i, such that $\operatorname{diam} L_i < 2(b_i - a_i)$ and L_i lies in the slab between the planes perpendicular to $[0, 1]$ at a_i and b_i. Then

$$L = [0, 1] \setminus \bigcup_{i=1}^{\infty} (a_i, b_i) \cup \bigcup_{i=1}^{\infty} L_i$$

is a simple arc. Each point a_i and b_i is wild on L_i, and so also on L. The set X of all wild points on L is the closure of the set $\{a_i, b_i\}$ of all endpoints of the arcs (a_i, b_i).

Example 3.2. *A disk and a sphere with infinitely many wild points.*

On a circular disk C (or on the surface of a ball), take a countable set of disjoint disks δ_i. Replace each δ_i by a disk D_i with a single interior wild point a_i, where the boundary of D_i coincides with that of δ_i, $\operatorname{diam} D_i < 2 \operatorname{diam} \delta_i$, and $D_i \cap D_j = \Lambda$, $i \neq j$. Set

$$D = C \setminus \bigcup_{i=1}^{\infty} \delta_i \cup \bigcup_{i=1}^{\infty} D_i.$$

The set X of wild points on D is the closure of the set $\{a_i\}$. If for the δ_i we take the squares whose interiors are removed in constructing a Sierpiński curve Z, then $X = Z \cup \{a_i\}$. The set of wild points on the disk is in that case one-dimensional. A similar construction yields a sphere with infinitely many wild points.

2. **Example 3.3.** *A disk Δ which cannot be enlarged in E^3 to any disk D containing Δ in its interior.*

In a square C, take a countable sequence of disjoint squares Δ_i converging to a point x_0 on the boundary of C, such that the boundary of Δ_i consists of a segment l_i of the boundary of C and a polygonal line in the interior of C. In the interior of each Δ_i take two disjoint disks δ_i and δ_i', and, as in Example 2.2, replace δ_i and δ_i' by disks D_i and D_i' whose boundaries coincide with those of δ_i and δ_i', such that the disk

$$\Delta_i' = \Delta_i \setminus \delta_i \setminus \delta_i' \cup D_i \cup D_i'$$

cannot be extended to a sphere in E^3. Since the boundary of Δ_i contains a segment of the boundary of C, the construction can be performed so that

$$\Delta_i' \cap (C \setminus \Delta_i) = \Lambda; \qquad \Delta_i' \cap \Delta_j' = \Lambda, \quad i \neq j;$$

$$\operatorname{diam} \Delta_i' < 2 \operatorname{diam} \Delta_i.$$

Then

$$\Delta = C \setminus \bigcup_{i=1}^{\infty} \Delta_i \cup \bigcup_{i=1}^{\infty} \Delta_i'$$

is a disk (Figure 4.6). We assert that Δ satisfies the required condition. Each disk Δ_i' contains an arc L_i situated in E^3 like the arc L of Example 2.1, with

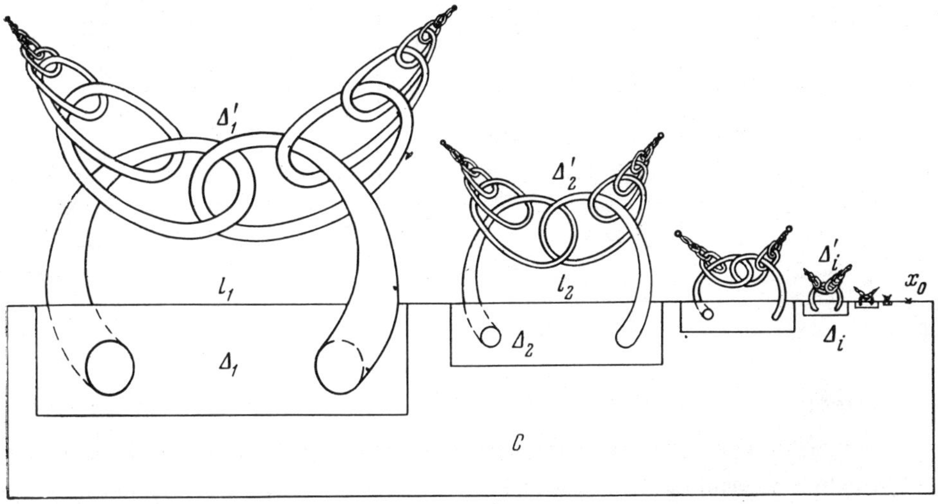

Figure 4.6

$L_i \cap \Delta_i$ a segment of the arc γ_{i0}. Suppose there exists a disk $D \subset E^3$ containing Δ in its interior. Then $\overline{D\setminus\Delta}$ contains a disk σ whose boundary consists of a segment $[a, b]$ of the boundary of C containing x_0, together with a simple arc in $D\setminus\Delta$. Pick i so large that Δ_i' is contained in a ball Q with center x_0 such that

$$Q \cap \operatorname{Fr} \sigma \subset [a, b].$$

Then

$$\Delta_i' \cap \operatorname{Fr} C \subset [a, b].$$

Since $L_i \subset \Delta_i'$ and $L_i \cap \operatorname{Fr} C = \Lambda$, we have $L_i \cap \sigma = \Lambda$. Take x_0 as the origin in $E^3\setminus L_i$. Then a path f circling the boundary of σ is homotopic in $E^3\setminus L_i$ to a path ϕ circling $\operatorname{Fr} C$, since $L_i \subset Q$ and the part of f in $E^3\setminus Q$ between the points of $[a, b] \cap \operatorname{Fr} Q$ is homotopic to the part of ϕ in $E^3\setminus Q$. But ϕ is in the class c_0^i of the group $\pi_1(E^3\setminus L_i)$ (Example 2.1), and so is not null-homotopic in $E^3\setminus L_i$; hence neither is f. On the other hand, f is a homeomorphism of a circle onto the boundary of σ, so that f is null-homotopic in σ and therefore also in $E^3\setminus L_i$; so that we have arrived at a contradiction. Thus there is no disk D in E_3 containing Δ in its interior.

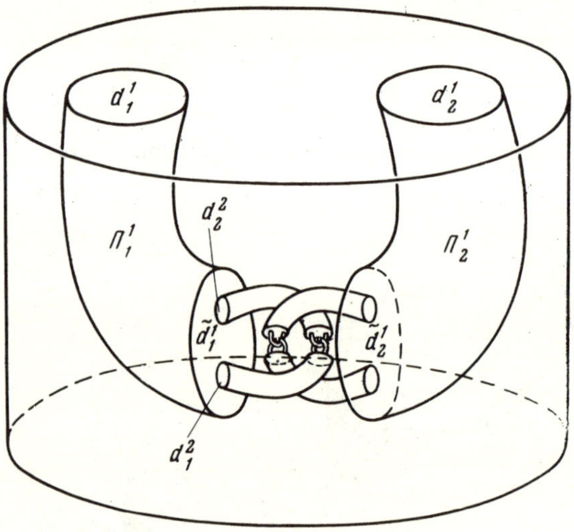

Figure 4.7

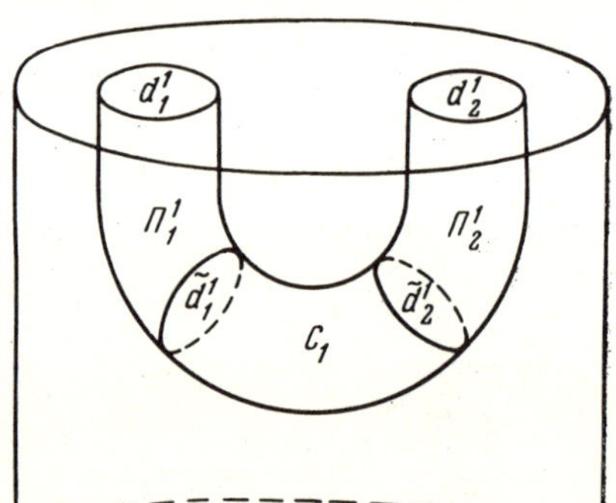

Figure 4.8

Example 3.4. *Alexander's "horned sphere"* [18] (*Figure* 4.7).

This is the first example in the literature of a sphere wildly imbedded in E^3 and bounding a domain not homeomorphic to E^3. The sphere S_A is constructed as the limit of a sequence of tame spheres S_k. To construct S_1 take two disjoint disks d_1^1 and d_2^1, of diameter $<\frac{1}{4}$, on the standard sphere Σ, and in the interior domain V of Σ construct a two-dimensional handle R^1 (a topological cylinder) of diameter <1, with boundary $\operatorname{Fr} d_1^1 \cup \operatorname{Fr} d_2^1$ (Figure 4.8). In the closed domain $\overline{V}^1$ bounded by $d_1^1 \cup R^1 \cup d_2^1$, take two disks $\tilde{d}_1^1$ and $\tilde{d}_2^1$ slicing through $\overline{V}^1$, such that the cylinders Π_i^1, $i=1, 2$, cut off from $\overline{V}^1$, with bases d_i^1 and $\tilde{d}_i^1$, are disjoint, and $\operatorname{diam} \overline{V^1 \setminus \Pi_1^1 \setminus \Pi_2^1} < \frac{1}{2}$. Let λ_i^1 be the lateral surface of the cylinder Π_i^1, and set

$$S_1 = \Sigma \setminus \bigcup_{i=1,2} d_i^1 \cup \bigcup_{i=1,2} (\lambda_i^1 \cup \tilde{d}_i^1).$$

Obviously S_1 is a sphere, since each disk d_i^1 has been replaced by a disk $\lambda_i \cup \tilde{d}_i^1$. Now set

$$C^1 = \overline{V^1 \setminus \Pi_1^1 \setminus \Pi_2^1}.$$

C^1 is a three-dimensional cylinder with bases $\tilde{d}_1^1$ and $\tilde{d}_2^1$. In each disk $\tilde{d}_i^1$, $i=1, 2$, remove two disks d_{2i-1}^2 and d_{2i}^2, and attach to their boundaries a handle R_i^2 lying in the interior of C^1. The handles R_1^2 and R_2^2 are to be chosen so that they link in C^1 like the simple arcs in Example 3.4 of Chapter III. Next, as in the construction of S_1, in each closed domain $\overline{V}_i^2$, $i=1, 2$, with boundary $d_{2i-1}^2 \cup R_i^2 \cup d_{2i}^2$, take two disks $\tilde{d}_{2i-1}^2$ and $\tilde{d}_{2i}^2$, denote by λ_j^2 the lateral surfaces of the cylinders cut off by these disks from $\overline{V}_i^2$, and set

$$S_2 = S_1 \setminus \bigcup_{i=1}^{4} d_i^2 \cup \bigcup_{i=1}^{4} (\lambda_i^2 \cup \tilde{d}_i^2).$$

Obviously S_2 is a sphere, and $\operatorname{diam}(\lambda_i^2 \cup \tilde{d}_i^2) < \frac{1}{2}$ since $\operatorname{diam} C^1 < \frac{1}{2}$.

The construction now proceeds in a similar fashion. Suppose the sphere S_k has been constructed, with 2^k disks $\tilde{d}_j^k$ singled out on it, where the pair $\tilde{d}_{2i-1}^k$ and $\tilde{d}_{2i}^k$ lie in a domain $\overline{V}_i^k$ of diameter less than $\frac{1}{2}^{k-1}$ and bounded by a handle R_i^k and disks d_{2i-1}^k and d_{2i}^k. Suppose also that the diameter of the cylinder C_i^k cut out from $\overline{V}_i^k$ by the disks $\tilde{d}_{2i-1}^k$ and $\tilde{d}_{2i}^k$ is no greater than $\frac{1}{2}^k$. In the cylinder C_i^k, construct handles R_{2i-1}^{k+1} and R_{2i}^{k+1}, and cylinders Π_j^{k+1}, $4i-3 \leq j \leq 4i$, with boundaries $d_j^{k+1} \cup \lambda_j^{k+1} \cup \tilde{d}_j^{k+1}$, in the same way as in the construction of S_2, choosing the disks d_j^{k+1} sufficiently small; and let

$$S_{k+1} = S_k \setminus \bigcup_{j=1}^{2^{k+1}} d_j^{k+1} \cup \bigcup_{j=1}^{2^{k+1}} (\lambda_j^{k+1} \cup \tilde{d}_j^{k+1}).$$

Since $\operatorname{diam} C_i^k < \frac{1}{2}^k$, we have $\operatorname{diam}(\lambda_j^{k+1} \cup \tilde{d}_j^{k+1}) < \frac{1}{2}^k$. The construction can be performed, moreover, so that $\operatorname{diam} C_j^{k+1} < \frac{1}{2}^{k+1}$, where C_j^{k+1} is the cylinder cut out from the domain $\overline{V}_j^{k+1}$ by the disks $\tilde{d}_{2j+1}^{k+1}$ and $\tilde{d}_{2j}^{k+1}$.

Continuing this procedure, we construct an infinite sequence of spheres

$S_1, S_2, \cdots, S_k, \cdots$. For each sphere S_k, there exists a homeomorphism ϕ_k: $S_k \to S_{k+1}$ such that

$$\varphi_k|_{S_k \setminus \bigcup_i d_i^{k+1}} = 1; \qquad \varphi_k(d_i^{k+1}) = \lambda_i^{k+1} \cup \tilde{d}_i^{k+1}.$$

Consider the sequence of homeomorphisms f_k: $S_1 \to S_{k+1}$, where $f_k = \phi_k \phi_{k-1} \cdots \phi_1$. The disks d_i^k can be chosen in succession so that[1] $\operatorname{diam} f_{k-1}^{-1}(d_i^k) < ½^k$. Obviously $\rho(x, \phi_k(x)) < ½^k$, so that $\rho[f_k(x), f_{k+1}(x)] < ½^{k-1}$ and the sequence of homeomorphisms f_k converges uniformly to a continuous mapping f.

We assert that f is a homeomorphism, i.e. that $f(x) \neq f(x')$ for $x \neq x'$. If there exists a k such that $f_k(x) \in S_k \setminus \bigcup d_i^k$ and $f_k(x') \in S_k \setminus \bigcup d_i^k$, then $f(x) = f_k(x)$ and $f(x') = f_k(x')$, so that $f(x) \neq f(x')$. Now suppose that $f_k(x) \notin S_k \setminus \bigcup d_i^k$ for every k. Then there exists a decreasing sequence of cylinders C_j^k such that $f(x) = \bigcap_{k=1}^{\infty} C_i^k$, with $C_{i_{k+1}}^{k+1} \subset \operatorname{int} C_{i_k}^k$. Since $x' \neq x$, we have that for some k either $f_k(x') \in S_k \setminus \bigcup d_i^k \subset S_k \setminus \bigcup C_i^k$, or else $f_k(x') \in C_j^k$, $j \neq i$, since the diameters of the $f_{k-1}^{-1}(d_i^k)$ approach zero with increasing k; thus in the latter case $C_i^k \cap C_j^k = \Lambda$. In either case $f(x') \neq f(x)$. Hence the sequence of spheres S_k converges uniformly to a sphere S_A, the "horned sphere" of Alexander.

We shall now show that the interior domain U bounded by the sphere S_A is not homeomorphic to the open ball. Indeed, its fundamental group is nontrivial.

Let us compute $\pi_1(U)$. U is representable in the form

$$U = (V \setminus \overline{V}^1) \cup \bigcup_{k=1}^{\infty} \bigcup_{i=1}^{2^{k-1}} (\tilde{C}_i^k \setminus \overline{V}_{2i-1}^{k+1} \setminus \overline{V}_{2i}^{k+1}), \tag{10}$$

where V is the interior domain of the standard sphere, and $\tilde{C}_i^k$ is obtained from the closed cylinder C_i^k by removing its two bases. It is convenient to replace the disjoint summands in (10) by nondisjoint summands, so as to be able to apply the van Kampen Theorem. First of all, observe that $\pi_1(V \setminus \overline{V}^1) = \pi_1(V \setminus \overline{V}^1 \cup \sigma^1)$, where σ^1 is the lateral surface of the cylinder $\overline{C}^1$. This follows from the fact that $V \setminus \overline{V}^1$ can be deformed continuously onto a subset homeomorphic to $V \setminus \overline{V}^1 \cup \sigma^1$ In the same way we have, for each C_i^k,

$$\pi_1(\tilde{C}_i^k \setminus \overline{V}_{2i-1}^{k+1} \setminus \overline{V}_{2i}^{k+1} \cup \sigma_{2i-1}^{k+1} \cup \sigma_{2i}^{k+1}) = \pi_1(\tilde{C}_i^k \setminus \overline{V}_{2i-1}^{k+1} \setminus \overline{V}_{2i}^{k+1}), \tag{11}$$

where σ_j^{k+1} is the lateral surface of the cylinder $C_j^{k+1} \subset C_i^k$.

Now pick the origin O in $V \setminus \overline{V}^1$, and on each σ_i^k a point O_i^k; construct a polygonal line $l^1 \subset V \setminus \overline{V}^1$ joining O to O^1, and in each $\tilde{C}_i^k \setminus \overline{V}_{2i-1}^{k+1} \setminus \overline{V}_{2i}^{k+1}$ two

1) *Translator's note.* In this and the following paragraph there seems to be an (easily corrected) inconsistency with the preceding notation.

polygonal lines κ^{k+1}_{2i-1} and κ^{k+1}_{2i}, intersecting only at the point O^k_i and joining O^k_i to O^{k+1}_{2i-1} and O^{k+1}_{2i}, respectively. Set

$$l^k_i = l^1 \cup \varkappa^2_{i_2} \cup \ldots \cup \varkappa^k_{i_k}, \quad \text{if} \quad O^k_{i_k} \in C^r_{i_r}, \ r \leqslant k;$$

$$A^k_i = l^k_i \cup \tilde{C}^k_i \setminus \overline{V}^{k+1}_{2i-1} \setminus \overline{V}^{k+1}_{2i} \cup \sigma^{k+1}_{2i-1} \cup \sigma^{k+1}_{2i},$$

$$k \geqslant 1, \ 1 \leqslant i \leqslant 2^{k-1}.$$

Since l^k_i intersects $\overline{A^k_i \setminus l^k_i}$ in a single point, and the group $\pi_1(l^k_i)$ is trivial, it follows from (11) that the group $\pi_1(A^k_i)$ is isomorphic to the group $\pi_1(\tilde{C}^k_i \setminus \overline{V}^{k+1}_{2i-1} \setminus \overline{V}^{k+1}_{2i})$.

Since $l^k_i \subset U$, equality (10) may be replaced by the equality

$$U = (V \setminus \overline{V}^1) \cup \bigcup_{k=1}^{\infty} \bigcup_{i=1}^{2^{k-1}} A^k_i. \tag{12}$$

It is easily seen that $V \setminus \overline{V}^1$ is homeomorphic to a convex domain with one line segment removed, while $\overline{C}^k_i \setminus \overline{V}^{k+1}_{2i-1} \setminus \overline{V}^{k+1}_{2i}$ is homeomorphic to the space of Example 3.4 of Chapter III. Hence the group $\pi_1(V \setminus \overline{V}^1 \cup \sigma^1)$, being isomorphic to $\pi_1(V \setminus \overline{V}^1)$, is free cyclic with generator a_0, and each $\pi_1(A^k_i)$ is a free group on two generators a^k_{2i} and a^k_{2i-1}.

On the handle $\overline{V}^1$ we take the positive direction as running from the disk d^1_1 to the disk d^1_2 and pick the class a_0 so as to contain a path f circling $\overline{V}^1$ clockwise as viewed from this positive direction. Pick each class a^k_i in succession so as to contain a path $w\phi w^{-1}$, where the image of the path w is a simple arc, and ϕ is a path in $\tilde{d}^k_i$ circling the disk d^{k+1}_{2i-1} in the positive direction. Here the direction on $\overline{V}^{k+1}_i$ is to be chosen so that on the portion containing the disk d^{k+1}_{2i-1} it agrees with the direction on the axis of the cylinder $C^k_p \supset \overline{V}^{k+1}_i$ (Figure 4.9).

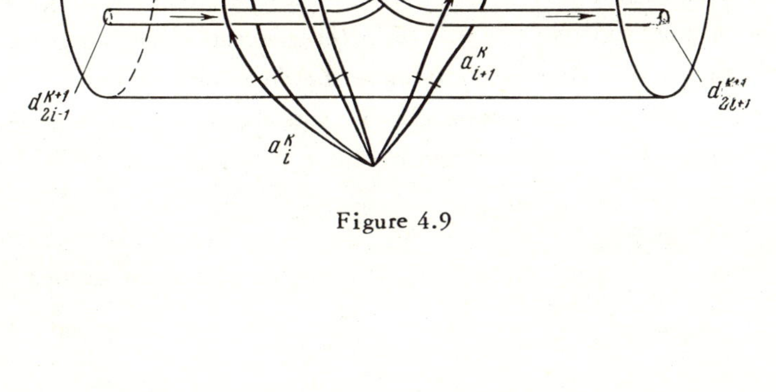

Figure 4.9

We compute, to start off with, the fundamental group of $K^1 = (V\backslash\overline{V}^1 \cup \sigma^1) \cup A_1$, where $A_1 = A_1^1$. It is easily seen that

$$(V\backslash\overline{V}^1 \cup \sigma^1) \cap A_1 = l^1 \cup \sigma^1$$

and that $l^1 \cup \sigma^1$ is a deformation retract of a suitable neighborhood G containing it in K_1, while $V\backslash\overline{V}^1 \cup \sigma^1$ and A_1 are retracts of $V\backslash\overline{V}^1 \cup \sigma^1 \cup G$ and $A_1 \cup G$, respectively, so that Corollary 2 to the van Kampen Theorem applies.

Since σ^1 deforms to a circle, it follows that $\pi_1(l^1 \cup \sigma^1)$ is free cyclic, and the class a_0 of the group $\pi_1(V\backslash\overline{V}^1 \cup \sigma^1)$ contains a path f^0 that lies in $l^1 \cup \sigma^1$ and is contained in a generator of the group $\pi_1(l^1 \cup \sigma^1)$. This path is homotopic in A_1 to the product of paths $\tilde{f}^{-1} \in \tilde{a}_1^{-1}$ and $f \in a_1$ circling the two parts of the handle R_1^2 (see Figure 4.9).

Therefore in the group $\pi_1(K_1)$ we have

$$a_0 = \tilde{a}_1^{-1} a_1. \tag{13}$$

To put this in terms of the generators of the groups $\pi_1(V\backslash\overline{V}_1 \cup \sigma^1)$ and $\pi_1(A_1)$, we write out, as in Example 2.1, the relations between the classes a_1, $\tilde{a}_1$, a_2 and $\tilde{a}_2$:

$$a_1 a_2^{-1} = a_2^{-1} \tilde{a}_1 = \tilde{a}_1 \tilde{a}_2^{-1}.$$

Using this to determine $\tilde{a}_1$, and substituting into (13), we get

$$a_0 = a_2 a_1^{-1} a_2^{-1} a_1. \tag{14}$$

This relation expresses the inclusion homomorphism of the group $\pi_1(V\backslash\overline{V}^1 \cup \sigma^1)$ into the group $\pi_1(K_1)$.

Now set

$$K_r = \bigcup_{k=1}^{r} \bigcup_{i=1}^{2^{k-1}} A_i^k; \quad D_r = \bigcup_{i=1}^{2^{r-1}} A_i^r.$$

Since $A_i^r \cap A_{i'}^r$, $i \neq i'$, is a simple arc containing the point O, it follows that $\pi_1(D_r)$ is the free product of the groups $\pi_1(A_i^r)$, i.e. the free group on the generators a_j^r, $1 \leq j \leq 2^r$. Moreover, $K_{r-1} \cap D_r = \bigcup_{i=1}^{2^{r-1}} (l_i^r \cup \sigma_i^r)$, $\sigma_i^r \cap \sigma_{i'}^r = \Lambda$, $i \neq i'$, and we find in the same way that $\pi_1(K_{r-1} \cap D_r)$ is a free group on the generators a_i^{r-1}. Applying induction, suppose the a_i^{r-1} are free generators for the group $\pi_1(K_{r-1})$ (this is true for $r-1=1$); then, arguing in the same way as for K_1, we find that $\pi_1(K_r)$ is a free group on the generators a_j^r, and that the inclusion homomorphism of $\pi_1(K_{r-1})$ into $\pi_1(K_r)$ is determined by the 2^{r-1} equalities

$$a_i^{r-1} = a_{2i}^r (a_{2i-1}^r)^{-1} (a_{2i}^r)^{-1} a_{2i-1}^r. \tag{15}$$

The group $\pi_1(u)$ is the direct limit of the spectrum of the groups $\pi_1(K_r)$, and so, applying the rule given in §5 of Chapter III, it can be presented in the following way: the generators of $\pi_1(U)$ are the elements

$$a_0,\ a_i^r,\ 1 \leqslant r < \infty,\ 1 \leqslant i \leqslant 2^r;$$

and they are connected by the defining relations (14) and (15).

We assert that the group $\pi_1(U)$ is nontrivial. Indeed, none of the elements a_0, a_i^r, $1 \le r < +\infty$, is the identity element. To prove this, it suffices to show that in every group $\pi_1(K_n)$, $n \geq r$, we have $a_i^r \neq 1$. But this follows from the facts that $\pi_1(K_n)$ is freely generated by the a_j^n and, as is easily verified by formulas (15), that a_i^r is an irreducible product of the elements $(a_j^n)^{\pm 1}$. Thus in the group $\pi_1(U)$ we have

$$a_0 \neq 1;\ a_i^r \neq 1,\ 1 \leqslant r < +\infty,\ i \leqslant 2^r.$$

§4. The wild zero-dimensional compactum of Antoine

1. We shall now describe a zero-dimensional compactum A which is wildly imbedded in E^3, constructed by Antoine [20] in 1921. A is the intersection of a decreasing sequence of sets A_k, each of which is the union of finitely many disjoint solid toruses whose diameters approach zero with $1/k$. Let T_0 be a solid torus of diameter 1. In the interior of T_0, take a closed chain of four solid toruses T_i, all of the same diameter (where T_i is the topological product of a *planar* closed curve by a disk), such that each T_i is linked with the adjoining T_{i-1} and T_{i+1}, T_4 is linked with T_1, and the chain $\bigcup_{i=1}^4 T_i$ cannot be shrunk to a point in T_0 (Figure 4.10).

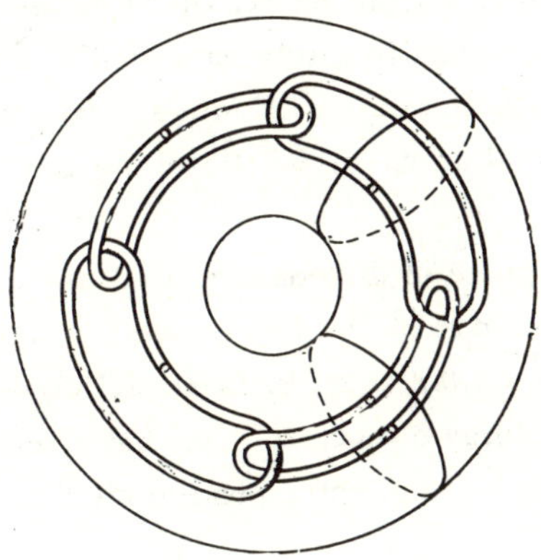

Figure 4.10

Suppose we have constructed A_k, as the union of $n_k = 4^k$ solid toruses $T_{i_1 \cdots i_k}$ all of the same diameter:

$$A_k = \bigcup_{i_1 \ldots i_k} T_{i_1 \ldots i_k},\ 1 \leqslant i_\rho \leqslant 4,$$

such that

$$T_{i_1 \ldots i_k} \cap T_{j_1 \ldots j_k} = \Lambda, \quad \text{if} \quad (i_1 \ldots i_k) \neq (j_1 \ldots j_k).$$

Denote the sequence of indices $i_1 \cdots i_k$ by ν_k. To construct A_{k+1}, take, in the interior of each T_{ν_k}, in the same way as for T_0, a chain of four disjoint solid toruses $T_{\nu_k i}$, all of the same diameter, such that each $T_{\nu_k i}$ is linked with $T_{\nu_k(i-1)}$ and $T_{\nu_k(i+1)}$, and $\bigcup_{i=1}^4 T_{\nu_k i}$ cannot be shrunk to a point in T_{ν_k}. It is easily verified that this can be done so that diam $T_{\nu_k i} < \frac{3}{4}$ diam T_{ν_k}. Set

$$A_{k+1} = \bigcup_{\nu_k} \bigcup_{i_{k+1}} T_{\nu_k i_{k+1}}.$$

Then A_{k+1} is the union of disjoint solid toruses $T_{\nu_{k+1}}$, $\nu_{k+1} = (\nu_k i_{k+1})$, and the diameters of the T_{ν_k} approach zero with $1/k$. Now set

$$A = \bigcap_{k=0}^{\infty} A_k = \bigcap_{k=0}^{\infty} \bigcup_{i_1 \dots i_k} T_{i_1 \dots i_k}, \quad 1 \leqslant i_p \leqslant 4.$$

Obviously A is a compactum, since each A_k is a compactum. By construction each $T_{\nu_{k+1}} = T_{\nu_k i}$ is contained in the interior of T_{ν_k}; hence

$$A \subset \bigcup_{i_1 \dots i_k} \operatorname{int} T_{i_1 \dots i_k},$$

so that A is the union of disjoint open-closed subsets $A \cap T_{i_1 \cdots i_k}$ of arbitrarily small diameter. Thus $\dim A = 0$.

2. We have seen that the complement of a tame zero-dimensional compactum in E^3 has trivial fundamental group (see Example 5.1 of Chapter III). Consequently, to prove that our zero-dimensional compactum is wildly imbedded in E^3, i.e. that no homeomorphism exists of E^3 onto itself carrying A into a compactum on a line, it suffices to show that the fundamental group $\pi_1(E^3 \setminus A)$ of the complement of A is nontrivial.

Pick the origin O in $E^3 \setminus T_0$, outside the plane of the equator. Pick the direction on the equator of T_0 clockwise (as viewed from O); and on the meridian, clockwise as viewed from the positive direction on the equator. Next, define the positive direction on the equator and on the meridian of each T_{ν_k} so that directions on all the T_{ν_k} agree with the direction on T_0 and correspond to clockwise rotations (see Figure 4.10).

The group $\pi_1(E^3 \setminus A)$ is the direct limit of the groups $\pi_1(E^3 \setminus A_k)$, $k = 0, 1, \cdots$. Since $E^3 \setminus A_0 = E^3 \setminus T_0$ is homeomorphic to the complement of a circle, the group $\pi_1(E^3 \setminus A_0)$ is free cyclic with generator a_0, where a_0 contains a path circling the meridian of T_0 in the positive direction. Observe that $\pi_1(E^3 \setminus A_k) = \pi_1(E^3 \setminus A_k)$, since the complement $E^3 \setminus O_\epsilon(A_k)$ of a sufficiently small ϵ-neighborhood of A_k is homeomorphic to $E^3 \setminus A_k$ and is a deformation retract of $E^3 \setminus A_k$.

We shall determine the group $\pi_1(E^3 \setminus A_k)$ by means of the van Kampen Theorem, setting

$$E^3 \setminus A_{k+1} = \overline{E^3 \setminus A_k} \cup K_k,$$

where

$$K_k = \bigcup_{\nu_k} (T_{\nu_k} \setminus A_{k+1} \cup l_{\nu_k}), \tag{16}$$

l_{ν_k} being a polygonal line lying in $\overline{E^3 \setminus A_k}$ and joining the origin O to a point on T_{ν_k}, such that each $l_{\nu_k i}$ is the union of l_{ν_k} and a polygonal line $\lambda_{\nu_k i}$, where $\lambda_{\nu_k i}$ passes through the interior of $T_{\nu_k} \setminus \bigcup_{j=1}^4 T_{\nu_k j}$ and joins a point $O_{\nu_k} \in \operatorname{Fr} T_{\nu_k}$ to a point $O_{\nu_k i} \in \operatorname{Fr} T_{\nu_k i}$.

K_k is obviously connected and arcwise connected, since each $K_{\nu_k} = T_{\nu_k} \setminus A_{k+1} \cup l_{\nu_k}$ is, and $K_{\nu_k} \cap K_{\nu_k'}$, for $\nu_k' \neq \nu_k$, is a polygonal line.

Hence the fundamental group $\pi_1(K_k)$ is the free product of the groups $\pi_1(K_{\nu_k})$. The intersection

$$\overline{E^3 \setminus A_k} \cap K_k = \bigcup_{\nu_k} (\operatorname{Fr} T_{\nu_k} \cup l_{\nu_k}) \tag{17}$$

is an arcwise connected continuum, and has an arbitrarily close neighborhood in $E^3 \setminus A_{k+1}$ satisfying the conditions of Corollary 2 to the van Kampen Theorem; thus we may determine $\pi_1(E^3 \setminus A_{k+1})$ from $\pi_1(E^3 \setminus A_k)$ and $\pi_1(K_k)$.

The group $\pi_1(E^3 \setminus A_1)$ is easily found directly. But we shall determine it by means of the procedure just suggested, since this fits the subsequent computation. Thus

$$X_i = (C_i \cap K_0) \cup \mathscr{E}_0,$$

where l_0 is a polygonal line joining the origin O to a point $O_0 \in \operatorname{Fr} T_0$, and $l_0 \setminus \{O_0\} \subset E^3 \setminus T_0$. Let us compute $\pi_1(K_0)$. Observe that $\pi_1(K_0)$ is isomorphic to $\pi_1(K_0')$, where $K_0' = \overline{K_0 \setminus l_0}$.

Divide T_0 into four cylinders C_i by half-planes orthogonal to the equator of T_0, such that each half-plane intersects just one T_i, and each cylinder C_i intersects T_i and T_{i+1} (mod 4). Set

$$K_0 = l_0 \cup T_0 \setminus A_1 = l_0 \cup T_0 \setminus \bigcup_{i=1}^4 T_i,$$

where $\mathfrak{S}_0$ is the equator of T_0. Then

$$K_0' = \bigcup_{i=1}^4 X_i.$$

Pick the point O_0 on $C_1 \cap \mathfrak{S}_0$. Since $(C_i \cap K_0) \cap \mathfrak{S}_0$ is a simple arc, we have

$$\pi_1(X_i) = \pi_1(C_i \cap K_0) * \pi_1(\mathscr{E}_0).$$

But $C_i \cap K_0$ is homeomorphic to the space of Example 3.4 of Chapter III, so that $\pi_1(C_i \cap K_0)$ is a free group on two generators a_i and a_{i+1}; take for these the paths with origin O_0 that circle, in the positive direction, those meridians of T_i

and T_{i+1} which lie on the parts of T_i and T_{i+1} on which the positive direction agrees with that on T_0 (Figure 4.11).[1)] The group $\pi_1(\mathcal{E}_0)$ is free cyclic with generator Γ. Hence $\pi_1(X_i)$ is a free group on the generators a_i, a_{i+1}, Γ. Set

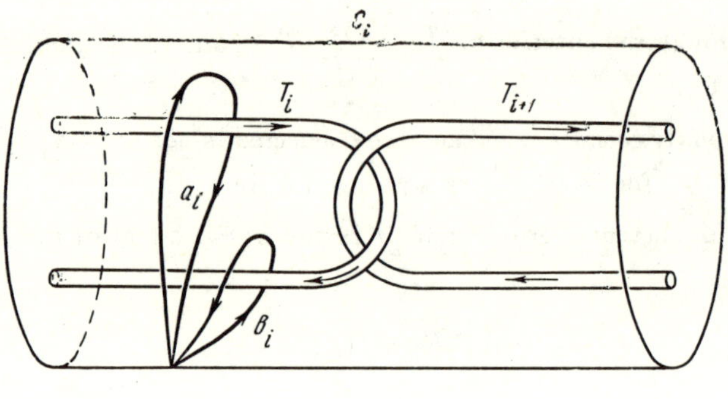

Figure 4.11

$Y_r = \bigcup_{i=1}^r X_i$; then

$$Y_{r-1} \cap X_r = (C_{r-1} \cap C_r)\setminus T_r \cup \mathcal{E}_0, \quad r < 4,$$
$$Y_3 \cap X_4 = [C_3 \cap C_4 \setminus T_4] \cup [C_4 \cap C_1 \setminus T_1] \cup \mathcal{E}_0.$$

$C_{r-1} \cap C_r \setminus T_r$ is a disk minus two small disks in T_r, and $(C_{r-1} \cap C_r \setminus T_r) \cap \mathcal{E}_0$ is a point.

It is easily verified that the union $Y_{r-1} \cup X_r$ satisfies the condition of the van Kampen Theorem. Thus we can compute in succession the groups $\pi_1(Y_r)$, $r = 1, \cdots, 4$. This requires that we write down the equalities between classes in $\pi_1(Y_{r-1})$ and $\pi_1(X_r)$ containing the same path in $Y_{r-1} \cap X_r$. We denote such classes by the same symbols in both groups. These are the classes Γ, a_r and b_r, where b_r is the class in $\pi_1(Y_{r-1})$ and $\pi_1(X_r)$ containing a path in $Y_{r-1} \cap X_r$ that circles the second component of the intersection $C_{r-1} \cap C_r \cap T_r$.

In the same way as in Example 3.4, we find that

$$a_i^{-1}a_{i+1} = a_{i+1}b_i^{-1} = b_i^{-1}b_{i+1},$$
$$i = 1, \ldots, 4 \pmod 4. \tag{18}$$

Also, as in Example 3.7 of Chapter III,

$$a_ib_i^{-1}\Gamma = \Gamma a_ib_i^{-1}. \tag{19}$$

Eliminating b_i and b_{i+1} from (18), we obtain

$$a_ia_{i+1}^{-1}a_i^{-1}a_{i+1} = a_{i-1}\,a_i^{-1}a_{i-1}^{-1}a_i,$$
$$i = 1, \ldots, 4 \pmod 4). \tag{20}$$

---

1) *Translator's note.* Properly speaking, what is being described is the image of $\pi_1(C_i \cap K_0)$ in $\pi_1(X_i)$, since O_0 need not be in $C_i \cap K_0$.

Applying the van Kampen Theorem to the union of the spaces $\overline{E^3\setminus T_0}$ and K_0 and observing that $\Gamma = 1$ in the group $\pi_1(\overline{E^3\setminus T_0})$ and that

$$a_0 = a_i b_i^{-1} = a_i a_{i+1}^{-1} a_i^{-1} a_{i+1}, \tag{21}$$

we find that the group $\pi_1(E^3\setminus A_1)$ has the generators $a_1, \cdots, a_4$, connected by the relations (20) (mod 4). To show that $\pi_1(E^3\setminus A_1)$ is nontrivial and, indeed, that $a_0 \neq 1$, it suffices to verify that the group has the following representation into the permutation group on six integers:

$$a_0 \to (1, 2)(3, 4)(5)(6);$$
$$a_1 \to (1, 2)(3)(4)(5, 6); \quad a_3 \to (1)(2)(3, 4)(5, 6);$$
$$a_2 \to (1, 3)(2, 4)(5)(6); \quad a_4 \to (1, 4)(2, 3)(5)(6).$$

3. Assuming the group $\pi_1(\overline{E^3\setminus A_k})$ determined, we compute the group $\pi_1(E^3\setminus A_{k+1})$.

We suppose the group $\pi_1(E^3\setminus A_k)$ has generators

$$a_{i_1\ldots i_k} = a_{\nu_k}, \ 1 \leqslant i_\rho \leqslant 4, \ \rho = 1, \ldots, k.$$

and elements

$$a_0, \ a_{i_1\ldots i_r} = a_{\nu_r}, \ r < k.$$

These are all connected by the defining relations

$$a_{\nu_r} = a_{\nu_r i} a_{\nu_r (i+1)}^{-1} a_{\nu_r i}^{-1} a_{\nu_r (i+1)},$$
$$\nu_r = i_1 \ldots i_r, \ i_\rho = 1, \ldots, 4, \ 1 \leqslant r < k, \ \nu_0 = 0. \tag{22}$$

By (20) and (21) the group $\pi_1(\overline{E^3\setminus A_1})$ is of this form. Let $\Gamma_{\nu_{k-1}i}$ be the element of $\pi_1(\overline{E^3\setminus A_k})$ containing a path $l_{\nu_{k-1}i} w_{\nu_{k-1}i} l_{\nu_{k-1}i}^{-1}$, where $w_{\nu_{k-1}i}$ is a path in $T_{\nu_{k-1}i}$ circling the equator of $T_{\nu_{k-1}i}$ in the positive direction, with origin $O_{\nu_{k-1}i}$. Then

$$\Gamma_{\nu_{k-1}i} = a_{\nu_{k-1}(i+1)} b_{\nu_{k-1}(i-1)} = a_{\nu_{k-1}(i+1)} a_{\nu_{k-1}i}^{-1} a_{\nu_{k-1}(i-1)} a_{\nu_{k-1}i}, \tag{23}$$

where the classes $b_{\nu_{k-1}i}$ and $a_{\nu_{k-1}i}$ are connected, in view of (22) and the choice of the arcs $l_{\nu_{k-1}i}$, by relations similar to (18).

It follows from (22) that

$$a_{\nu_{k-1}i} \Gamma_{\nu_{k-1}i} = \Gamma_{\nu_{k-1}i} a_{\nu_{k-1}i}, \tag{24}$$

since

$$a_{\nu_{k-1}i} a_{\nu_{k-1}(i+1)} a_{\nu_{k-1}i}^{-1} a_{\nu_{k-1}(i-1)} = a_{\nu_{k-1}(i+1)} a_{\nu_{k-1}i}^{-1} a_{\nu_{k-1}(i-1)} a_{\nu_{k-1}i}.$$

Now, $E^3\backslash A_{k+1} = \overline{E^3\backslash A_k} \cup K_k$, and the van Kampen Theorem applies to this union. By (16) the group $\pi_1(K_k)$ is the free product of the groups $\pi_1(K_{\nu_k})$, where $K_{\nu_k} = T_{\nu_k}\backslash A_{k+1} \cup l_{\nu_k}$. But K_{ν_k} is homeomorphic to K_0, so that $\pi_1(K_{\nu_k})$ is isomorphic to $\pi_1(K_0)$; thus the group $\pi_1(K_k)$ has the complete set of generators

$$a_{\nu_k i_{k+1}},\ 1 \leqslant i_r \leqslant 4,\ r = 1, 2, \ldots, k+1,\ \nu_k = i_1 \ldots i_k;\ \Gamma_{\nu_k},$$

connected by defining relations which are obtained from (19) and (20) by replacing a_i by $a_{\nu_k i}$ and Γ by $\Gamma_{\nu_k i}$, with

$$b_{\nu_k i} = a^{-1}_{\nu_k (i+1)} a_{\nu_k i} a_{\nu_k (i+1)}.$$

Applying the van Kampen Theorem to the union $\overline{E^3\backslash A_k} \cup K_k$, we observe that, by (17)

$$a_{\nu_k} = a_{\nu_k i} b^{-1}_{\nu_k i} = a_{\nu_k i}\, a^{-1}_{\nu_k (i+1)}\, a^{-1}_{\nu_k i}\, a_{\nu_k (i+1)},\quad i = 1, \ldots, 4 \,(\mathrm{mod}\, 4). \tag{25}$$

$\Gamma_{\nu_k i}$ is defined by (23), and, using (25), the defining relations of the group $\pi_1(K_k)$ that contain $\Gamma_{\nu_k i}$ then follow from (24), and so from (22). Thus the passage from the group $\pi_1(E^3\backslash A_k)$ to the group $\pi_1(E^3\backslash A_{k+1})$ amounts to adjoining generators $a_{\nu_k i}$, $1 \leq i \leq 4$, and the defining relations (25); i.e. the group $\pi_1(E^3\backslash A_{k+1})$ has generators $a_{i_1 \cdots i_k i_{k+1}}$, $1 \leq i_r \leq 4$, related among themselves and to the classes $a_{i_1 \cdots i_r}$, $r \leq k$, by relations (22) for $r \leq k$.

4. The group $\pi_1(E^3\backslash A)$ is the direct limit of the groups $\pi_1(E^3\backslash A_k)$. Its generators are the elements defined by the representatives $a_{\nu_k} = a_{i_1 \cdots i_k}$ in the groups $\pi_1(E^3\backslash A_k)$.

To designate the element of $\pi_1(E^3\backslash A)$ represented by a_{ν_k}, we retain the notation for it in $\pi_1(E^3\backslash A_k)$. Then the elements of $\pi_1(E^3\backslash A)$ are connected by relations (22), $k < +\infty$, which constitute a complete set of defining relations for $\pi_1(E^3\backslash A)$. To prove that $\pi_1(E^3\backslash A)$ is nontrivial, we show that, in it, $a_0 \neq 1$.

Suppose the converse. Then by definition of the identity element in the direct limit of a sequence of groups, there exists an integer p such that $a_0^p \neq 1$, where $a^p_{\nu_k}$ is the representative of a_{ν_k} in the group $\pi_1(E^3\backslash A_p)$, $p \geq k$. Let p be the smallest integer such that $a_0^{p-1} \neq 1$. By (21),

$$a_0^p = a_i^p (a^p_{i+1})^{-1} (a_i^p)^{-1} a^p_{i+1}. \tag{26}$$

But each element a_i^p, $1 \leq i \leq 4$, is a product of elements $a^p_{i i_2 \cdots i_p}$, *all with the same first index.* The subgroup G_i generated by the generators of $\pi_1(E^3\backslash A_p)$ which have first index equal to i, is isomorphic in a natural fashion with the group $\pi_1(E^3\backslash A_{p-1})$, the element $a_i^p \in G_i$ corresponding to the element

$a_0^{p-1} \in \pi_1(E^3\setminus A_{p-1})$. By assumption we have $a_0^{p-1} \neq 1$, so that $a_i^p \neq 1$ in the group G_i. But in the group $\pi_1(E^3\setminus A_p)$ the elements a_i^p are connected only by relations (26), since by (22) the remaining relations contain only elements with the same first index. Consequently the subgroup generated by the a_i^p, $1 \leq i \leq 4$, is isomorphic to the group $\pi_1(E^3\setminus A_1)$, which, as we have seen, is nontrivial, and $a_0^p \neq 1$, contrary to assumption. Thus the assumption that $a_0 = 1$ in the group $\pi_1(E^3\setminus A)$ leads to a contradiction; consequently $a_0 \neq 1$ and $\pi_1(E^3\setminus A)$ is nontrivial. We remark that a similar proof shows, for any element $a_{i_1\cdots i_k}$, that $a_{i_1\cdots i_k} \neq 1$.

We have thereby proved that the zero-dimensional compactum A is wildly imbedded in E^3.

5. Wildly imbedded zero-dimensional compacta exist in any E^n, $n \geq 3$. Examples of such have been constructed by Ivanov [3] and Blankinship [31], their constructions being generalizations of that of Antoine. There is also an example of a wild zero-dimensional compactum in E^3 for which the fundamental group of the complement is trivial [46].

By means of zero-dimensional wild compacta, examples of wildly imbedded k-cells and k-spheres in E^n, $k < n$, for any $n \geq 3$, can be constructed on the basis of the following theorem (see [2]).

Theorem 4.1. *For any zero-dimensional compactum A in E^n there exists in E^n (or S^n) a k-sphere S^k containing it, of any dimension $k < n$.*

It suffices to prove the theorem for $k = n - 1$. Indeed, assuming the latter, suppose that for some $k < n$ we have constructed a sphere S^k in E^n (or S^n) containing A, and let $h: S^k \to \Sigma^k$ be a homeomorphism of S^k onto the standard sphere Σ^k of dimension k. Since the theorem holds for $k - 1$, there exists in Σ^k a $(k-1)$-sphere S^{k-1} containing the zero-dimensional compactum $h(A)$. Then the sphere $s^{k-1} = h^{-1}(S^{k-1})$ is contained in S^k, so also in E^n, and contains A.

Thus, once we can pass from k to $k - 1$, we can construct in succession spheres S^k containing A, for any $k \leq n - 1$.

Let us now construct a sphere S^{n-1} containing A. Represent A in the form

$$A = \bigcap_{k=1}^{\infty} \bigcup_{i=1}^{n_k} \bar{u}_i^k;$$

$$\bar{u}_i^k \cap \bar{u}_j^k = \Lambda,\ i \neq j,\ \operatorname{diam} \bar{u}_i^k < \frac{1}{k},\ \bar{u}_i^k \subset u_j^{k-1} \quad \text{for some } j,$$

where $\bar{u}_i^k$ is a closed polyhedral domain in E^n with connected boundary. Take a closed n-simplex in $E^n\setminus \bigcup_{i=1}^{n_1} \bar{u}_i^1$; we shall regard its boundary Σ_0 as the zeroth approximation to S^{n-1}. We now construct a countable sequence of polyhedral $(n-1)$-spheres Σ_k converging uniformly to the desired sphere S^{n-1}. To construct the sphere Σ_1, pick n_1 disjoint closed $(n-1)$-simplices d_i^1 on Σ_0, and one $(n-1)$-simplex $\widetilde{d}_i^1$, similar to d_i^1, on the boundary of each $\overline{u}_i^1$. Next, join the center of d_i^1 to the center of $\widetilde{d}_i^1$ by a polygonal line l_i^1 in $E^n \setminus \bigcup_{i=1}^{n_1} \overline{u}_i$, in such a way that $l_i^1 \cap l_j^1 = \Lambda$ for $i \neq j$. For each polygonal line, construct a tubular neighborhood $\overline{V}_i^1$, this being a cylinder with bases d_i^1 and $\widetilde{d}_i^1$ and axis l_i^1. The construction can be performed so that $\overline{V}_i^1 \cap \overline{V}_j^1 = \Lambda$ for $i \neq j$. Let λ_i^1 be the lateral surface of $\overline{V}_i^1$. The union $\lambda_i^1 \cup \widetilde{d}_i^1$ is obviously a topological $(n-1)$-dimensional disk with boundary $\operatorname{Fr} d_i^1$, since λ_i^1 is homeomorphic to the product of $\operatorname{Fr} d_i^1$ by a line segment. Now set

$$\Sigma_1 = \Sigma_0 \setminus \bigcup_{i=1}^{n_1} d_i^1 \cup \bigcup_{i=1}^{n_1} (\lambda_i^1 \cup \widetilde{d}_i^1).$$

Suppose the sphere Σ_k has been constructed, so as to intersect each $\overline{u}_i^k$ in an $(n-1)$-simplex $\widetilde{d}_i^k$ on the boundary of $\overline{u}_i^k$. The subsequent construction takes place inside each domain u_i^k. On each $\widetilde{d}_i^k$, pick as many disjoint $(n-1)$-simplices d_j^{k+1}, of sufficiently small diameter, as there are domains $\overline{u}_j^{k+1}$ in u_i^k; and on the boundary of each domain u_j^{k+1} pick an $(n-1)$-simplex $\widetilde{d}_j^{k+1}$ similar to the corresponding d_j^{k+1}. Next, as in the construction of Σ_1, construct simple arcs l_j^{k+1} and cylinders $\overline{V}_j^{k+1}$ with bases d^{k+1} and $\widetilde{d}_j^{k+1}$, such that if $d_j^{k+1} \subset \operatorname{Fr} u_i^k$, then

$$V_j^{k+1} \subset u_i^k \setminus \bigcup_{\rho} \overline{u}_\rho^{k+1}, \ \overline{V}_j^{k+1} \cap \overline{u}_j^{k+1} = \widetilde{d}_j^{k+1};$$

and

$$\overline{V}_j^{k+1} \cap \overline{V}_{j'}^{k+1} = \Lambda, \ j \neq j'.$$

Denoting by λ_j^{k+1} the lateral surface of $\overline{V}_j^{k+1}$, and observing that $\lambda_j^{k+1} \cup \widetilde{d}_j^{k+1}$ is an $(n-1)$-disk, set

$$\Sigma_{k+1} = \Sigma_k \setminus \bigcup_{j=1}^{n_{k+1}} d_j^{k+1} \cup \bigcup_{j=1}^{n_{k+1}} (\lambda_j^{k+1} \cup \widetilde{d}_j^{k+1}).$$

Consider a homeomorphism $\phi_k\colon \Sigma_k \to \Sigma_{k+1}$ such that

$$\varphi_k |_{\Sigma_k \setminus \bigcup_j d_j^{k+1}} = 1; \ \varphi_k (d_j^{k+1}) = \lambda_j^{k+1} \cup \widetilde{d}_j^{k+1},$$

$$j = 1, 2, \ldots, n_{k+1}.$$

If we set $f_k = \phi_k \phi_{k-1} \cdots \phi_1 \phi_0$, and pick the sequence of simplices d_j^k such that $\operatorname{diam} f_{k-1}^{-1}(d_j^k) < 1/k$, we can show, just as in Example 3.4,[1] that the sequence of homeomorphisms f_k converges uniformly to a homeomorphism of Σ_0 onto a certain sphere S^{n-1}, the limit of the spheres Σ_k. It remains only to show

---

[1] *Translator's note.* As in that example, there seems to be a notational inconsistency here. One way to correct it is to replace d_j^k in the inequality by $\widetilde{d}_j^k$.

that $A \subset S^{n-1}$. But this follows from the fact that for any k and $r \geq 0$, the sphere Σ_{k+r} intersects each $\bar{u}_i^k$. This implies that S^{n-1} intersects every $\bar{u}_i^k$ for all k, and hence that S^{n-1} contains A.

Corollary. *For any zero-dimensional compactum A in E^n (or S^n), and any $k < n$, there exists a k-cell $S^k \subset E^n$ containing A.*

To prove this, consider a homeomorphism $h: S^k \to \Sigma^k$, of a sphere $S^k \subset E^n$ containing A, onto the standard sphere Σ^k. Pick in Σ^k a sufficiently small standard $(k-1)$-sphere σ^{k-1} separating Σ^k into two k-cells, such that $h(A)$ is contained in the interior of one of them, say Δ^k. Then $h^{-1}(\Delta^k)$ is a k-cell in S^k, and therefore in E^n, that contains A. Observe that for $k = 1$, i.e. for a simple arc, the corollary is a classical result.

It follows from the definition that *if a simple arc $l \subset E^n$ contains a wild zero-dimensional compactum A, it is itself wildly imbedded in E^n.* The same is easily proved for a 2-sphere S^2 or for a two-dimensional disk.

Indeed, suppose there exists a homeomorphism $\phi: E^n \to E^n$ taking S^2 into the boundary $\mathrm{Fr}\,\sigma^3$ of a 3-simplex. By Corollary 2 to Theorem 3.2 of Chapter II, there exists a homeomorphism $\Psi: \mathrm{Fr}\,\sigma^3 \to \mathrm{Fr}\,\sigma^3$ taking $\phi(A)$ into a subset of an edge of σ^3. The mapping Ψ, defined on $\mathrm{Fr}\,\sigma^3$, extends to a homeomorphism $\Psi: E^n \to E^n$. To see this, first extend Ψ to a homeomorphism Ψ' of the three-dimensional hyperplane containing σ^3—which can be taken as the (x_1, x_2, x_3) hyperplane of the coordinate system $(x_1, x_2, \cdots, x_n)$—so that the ray from the center O of σ^3 passing through a point $p \in \mathrm{Fr}\,\sigma^3$ goes into the ray from O passing through the point $\Psi(p)$. Then let

$$\Psi(x_1, x_2, \ldots, x_n) = [\Psi'(x_1, x_2, x_3), x_4, \ldots, x_n).$$

The mapping $\Psi\phi$ is now a homeomorphism of E^n onto itself, taking A into a subset of an edge of σ^3, i.e. into a subset of a line segment; so that A is tame, contradicting the hypothesis.

It can, in fact, be proved that for any $k < n$, where $n \geq 3$, if a k-sphere S^k (resp. k-cell Q^k) in E^n contains a wild zero-dimensional compactum, then S^k (or Q^k) is wildly imbedded in E^n, since a zero-dimensional compactum A that lies in a hyperplane of E^n is always tame in E^n [2].

Thus the existence in E^n, $n \geq 3$, of a wildly imbedded zero-dimensional compactum allows one to construct, for any $k < n$, spheres S^k and k-cells wildly imbedded in E^n.

Observe that the "horned sphere" S_A of Alexander (Example 3.4) has a perfect set A of wild points, which is a *tame* zero-dimensional compactum, since

$$A = \bigcap_{k=1}^{\infty} \bigcup_{i=1}^{n_k} C_i^k, \quad C_i^k \cap C_j^k = \Lambda, \quad i \neq j,$$

where the C_i^k are topological n-cells, so that A is cellularly separated.

It has been proved recently [38, 51, 15], that if $n \geq 5$ and $k \neq n-2$, the set of wild points for a sphere S^k in E^n cannot have isolated points. *Hence the set of wild points for a sphere S^k in E^n, for $n \geq 5$ and $k \neq n-2$, is a perfect set.*

Thus we have here an essential difference from the case $n = 3$; we have seen in §1 that isolated wild points can exist on a sphere S^k both for $k = 1$ and $k = 2$. For $k = n - 2$, isolated wild points can exist for any n; in particular, a sphere S^{n-2} wildly imbedded in E^n can have just a single wild point [10].

CHAPTER V

TOPOLOGICAL IMBEDDINGS OF A SPHERE S^{n-1} INTO THE SPHERE S^n AND INTO E^n

§1. The generalized Schoenflies theorem

1. We have considered in Chapter IV a number of examples which show that the problem of the imbedding of manifolds, in particular of spheres and cells, into the space E^n or into the spheres S^n is for $n \geq 3$ considerably more complicated than for $n = 2$. A number of questions arise in this connection, concerning the characterization of "nice" imbeddings and, indeed, the precise definition of the notion of a nice imbedding.

Once the question of imbedding in the plane had been studied, and the first examples constructed of "wild" imbeddings in E^3, notably the "horned sphere" of Alexander [18], bounding a domain in E^3 which is not homeomorphic to E^3 (see Chapter IV, Example 3.4), the question arose of determining criteria for an imbedding to be "nice."

It is natural to regard as "nice" an imbedding of an $(n - 1)$-sphere S^{n-1} into the sphere S^n if the closures of the domains into which S^{n-1} separates S^n are n-cells.

It is natural, also, to study first of all the imbedding of a polyhedral sphere. A topological sphere $S^k \subset E^n$ is called *polyhedral* if it is a rectilinear polyhedron in some triagulation of E^n.

A polyhedral sphere S^k in S^n, $k < n$, is defined in the same way, regarding S^n as the boundary of an $(n + 1)$-simplex. If we take for S^n the boundary of an $(n + 1)$-ball in E^{n+1}, we shall mean by a spherical polyhedron in S^n a polyhedron in the spherical triangulation of S^n obtained by projecting, from the center of S^n, a rectilinear triangulation of the boundary of an $(n + 1)$-simplex inside S^n and containing the center of S^n. The simplices of such a triangulation will be called *spherical.*

A spherical triangulation of S^n is obviously isomorphic to a rectilinear triangulation of the boundary of an $(n + 1)$-simplex.

Alexander proved in 1922 [17] that *every polyhedral 2-sphere in S^3 divides S^3 into domains whose closures are 3-cells* (topological cubes). This gave rise to the notion of a tame imbedding of a topological polyhedron (i.e. a set homeomorphic to a polyhedron) in E^n or in S^n.

Definition 1.1. A subset P of E^n homeomorphic to a polyhedron is called *tame* in E^n (S^n) if there exists a homeomorphism h of E^n (S^n) onto itself taking

P into a rectilinear polyhedron.

This notion turned out to be very useful in studying 3-space E^3 (or S^3). A number of papers in the fifties, mainly those of Bing [22–29] and Moise [53–60], as well as others, studied the polyhedral structure of 3-space E^3 and topological imbeddings in E^3, and obtained a number of important results concerning three-dimensional manifolds. We do not present these results, since they are obtained by means of a very cumbersome polyhedral technique, specialized for E^3, whereas there are at present sufficiently general results (though considerably less general than for E^3) on imbeddings of spheres and cells in E^n (or S^n) for arbitrary $n \geq 3$.

The notion of "tame" imbedding is unsuited for studying imbeddings in E^n with $n > 3$, since it is unknown up to the present whether the closures of the domains into which a polyhedral sphere Σ^{n-1} divides S^n, where $n \geq 4$, are topological cubes.

This question was quite recently answered affirmatively by Newman [61] in the case of the so-called *combinatorial spheres* (see §5), but it is unknown for $n > 3$ whether every polyhedral sphere is combinatorial. Indeed, this is one of the fundamental classical problems of topology, and is known as the "Hauptvermutung" for spheres.

Accordingly, as we have indicated in Chapter IV, in studying the question of the imbedding, in E^n (or S^n), of spheres S^k or cells I^k, the notion used is that of *trivial imbedding*. Let us call *standard* the sphere Σ^k in E^n given by the equalities

$$\sum_{i=1}^{k+1} x_i^2 = 1; \; x_{k+2} = \cdots = x_n = 0.$$

The standard cell I^k is defined similarly.

Definition 1.2. An imbedding of a sphere S^k (or cell Q^k) in E^n (or S^n) is called *trivial* if there exists a homeomorphism h of E^n (or S^n) onto itself taking S^k (or Q^k) into the standard k-sphere (or k-cube).

An intensive development is being carried out of the topology of imbeddings in E^n or S^n, having started in 1959 with the papers of Mazur [49, 50] giving a sufficiency criterion for an imbedding of a topological sphere S^{n-1} in S^n to be trivial, for any n. This criterion is known as the "generalized Schoenflies theorem." Mazur proved it under a certain restriction, which we shall describe in §3.5. In 1960 Brown [32] gave a second proof of the Schoenflies theorem with no restrictions; we shall present his proof in this chapter.

The two proofs, Mazur's and Brown's, are quite different, very original in concept, and technically not complicated. Moreover, the Schoenflies theorem, once established, initiated a large number of important new results in the theory of n-dimensional topological manifolds. We shall indicate some of them in this chapter.

In 1963 appeared the very interesting papers of Stallings [73] and Zeeman [77], giving sufficiency criteria for an imbedding of a sphere S^k in E^n to be trivial, where $k < n - 2$. The study of these cases is considerably more complicated than the case $k = n - 1$, and new methods were required. We shall not present these newer papers in the present monograph. We merely state the theorems of Stallings and Zeeman at the end of the chapter.

1. It will be convenient in this chapter to speak of the imbedding of a sphere S^{n-1} into a sphere S^n. To obtain the criterion for triviality of S^{n-1} in E^n from that for S^{n-1} in S^n, it suffices to enlarge E^n to a sphere by adding a point.

Theorem 1.1. *If the closures of both complementary domains of a sphere S^{n-1} in the sphere S^n are topological cubes, then there exists a homeomorphism of S^n onto itself taking the equator of S^n into S^{n-1}.*

Let $f\colon \Sigma^{n-1} \to S^n$ be a homeomorphism into S^n of the standard sphere Σ^{n-1}, which we may regard as the equator of S^n, and let V_i be the domains in S^n complementary to S^{n-1}. By hypothesis, their closures $\overline{V}_i$ are n-cells. For each $\overline{V}_i$, $i = 1, 2$, there exists a homeomorphism

$$\varphi_i : I_i^n \to \overline{V}_i, \quad i = 1, 2,$$

of the standard ball I_i^n bounding the equator.

Then there is defined on the equator Σ^{n-1} the homeomorphism

$$\varphi_i^{-1} f : \Sigma^{n-1} \to \Sigma^{n-1}.$$

This can be extended to a homeomorphism $\Psi_i\colon I_i^n \to I_i^n$, $i = 1, 2$, taking the radius to the point $x \in \operatorname{Fr} I_i^n$ into the radius to the point $\phi_i^{-1} f(x)$. Then

$$\varphi_i \Psi_i : I_i^n \to \overline{V}_i;$$

moreover,

$$\varphi_i \Psi_i |_{\Sigma^{n-1}} = \varphi_i \varphi_i^{-1} f = f.$$

Define a mapping $F\colon S^n \to S^n$ by setting

$$F(x) = \varphi_i \Psi_i (x), \quad x \in I_i^n, \quad i = 1, 2.$$

Then since $\operatorname{Fr} I_1^n = \operatorname{Fr} I_2^n$ is the equator Σ^{n-1}, we have

$$F(\Sigma^{n-1}) = f(\Sigma^{n-1}) = S^n \text{ and } F|_{\Sigma^{n-1}} = f.$$

Corollary 1. *If f is a homeomorphism of the equator of the sphere S^n onto S^{n-1}, and the closures of the complementary domains of S^{n-1} in S^n are topological cubes, then f extends to a homeomorphism of S^n onto itself.*

This is the homeomorphism F.

Corollary 2. *The union of two topological balls with boundaries identified is a sphere.*[1)]

Let $M = \overline{V}_1 \cup \overline{V}_2$, where the $\overline{V}_i$ are n-cells and $\overline{V}_1 \cap \overline{V}_2 = \mathrm{Fr}\,\overline{V}_1 \cap \mathrm{Fr}\,\overline{V}_2$. The sphere Σ^n is separated by the equator Σ^{n-1} into two hemispheres Q^n_1 and Q^n_2. Since the $\overline{V}_i$ are n-cells, there exist homeomorphisms

$$\varphi_i : Q^n_i \to \overline{V}_i.$$

As in the proof of the theorem, construct a homeomorphism $\Psi: Q^n_2 \to Q^n_2$ agreeing on Σ^{n-1} with $\phi_2^{-1}\phi_1$. Then the homeomorphism

$$\varphi_2\Psi : Q^n_2 \to \overline{V}_2$$

agrees on Σ^{n-1} with ϕ_1, so that, setting

$$\Phi|_{Q^n_1} = \varphi_1, \quad \Phi|_{Q^n_2} = \varphi_2\Psi,$$

we obtain a homeomorphism of Σ^n onto M.

2. **Brown's proof of the Schoenflies theorem** [32].

Lemma 1.1. *Let f be a continuous mapping of a cube Q^n into the sphere S^n which has only finitely many nondegenerate point-inverses, these lying in the interior of Q^n. Then $f(Q^n)$ is a closed domain in S^n bounded by a sphere $S^{n-1} = f(\mathrm{Fr}\,Q^n)$.*

Let $K_1, K_2, \cdots, K_r$ be the nondegenerate elements, with $K_i \subset \mathrm{int}\,Q^n$. By the theorem on invariance of domain, f maps $\mathrm{int}\,Q^n \setminus \bigcup K_i$ homeomorphically onto a domain U in S^n, and $f(\mathrm{Fr}\,Q^n) \subset \mathrm{Fr}\,U$. But $f(\mathrm{Fr}\,Q^n)$ is a sphere S^{n-1}, since f is a homeomorphism on $\mathrm{Fr}\,Q^n$; and $f(K_i) \subset \overline{U}$, $i = 1, \cdots, r$, since K_i contains limit points of $Q^n \setminus \bigcup K_i$. Since $f(Q^n)$ is connected, it is contained in one of the domains, say D, complementary to S^{n-1} in S^n. But

$$\mathrm{Fr}\,U = f(\mathrm{Fr}\,Q^n) \cup \bigcup_{i=1}^{r} \{p_i\},$$

where $p_i = f(K_i)$, so that $p_i \in \overline{U}$. Hence $\overline{U} = D$, since the interior of D contains only finitely many points of the boundary of U.

---

1) I.e., taking a fixed homeomorphism between the boundaries and identifying points corresponding to each other under the homeomorphism.

Remark 1.1. If instead of the cube Q^n one considers an arbitrary domain W in S^n, it follows in the same way that $f(\overline{W})$ is a domain in S^n bounded by $f(\operatorname{Fr} \overline{W})$, provided $\overline{W}_f$ has only finitely many nondegenerate point-inverses and these lie in W.

Theorem 1.2 (Brown). *Let f be a continuous mapping of a cube Q^n into the sphere S^n which has only finitely many nondegenerate point-inverses K_i, these lying in the interior of Q^n. Then each K_i is a cellularly imbedded continuum in Q^n.*

We prove the theorem by induction on the number of elements K_i.

1. Suppose there is a single nondegenerate point-inverse K. By Lemma 1.1, $f(Q^n)$ is a closed domain $\overline{U}$ in S^n bounded by the sphere $f(\operatorname{Fr} Q^n)$. Let V be an arbitrary neighborhood of K in $\operatorname{int} Q^n$. We must show that V contains a cellular neighborhood of K.

$f(K) = p$ is a point in $f(V)$. Take two balls δ_1, δ_2 with center p, such that

$$p \in \delta_1 \subset \delta_2 \subset f(V),$$

and pick a point $q \in S^n \setminus \overline{U}$. Next construct a homeomorphism $h\colon S^n \to S^n$ such that

$$h(\overline{U} \setminus \delta_1) \subset \delta_2 \setminus \delta_1, \quad h \mid \delta_1 = 1.$$

To do so, first construct a homeomorphism $h_1\colon S^n \to S^n$ taking q into the point q' diametrically opposite p, stationary outside a small neighborhood of that segment qq' of the meridian not containing p, and such that $h_1(y) = y$ for $y \in \delta_2$. Then construct a central projection h_2 from the point q', taking the boundary of a ball $\Delta \subset h_1(S^n \setminus \overline{U})$ with center q' into the boundary of δ_2, and the domain between the boundaries of Δ and δ_1 onto the domain $\delta_2 \setminus \delta_1$, with $h_2|_{\delta_1} = 1$ (Figure 5.1). Let $h = h_2 h_1$. Since $\Delta \subset h_1(S^n \setminus \overline{U})$, we have $h_1(\overline{U}) \subset S^n \setminus \Delta$, so that $h(\overline{U}) = h_2 h_1(\overline{U}) \subset \delta_2$.

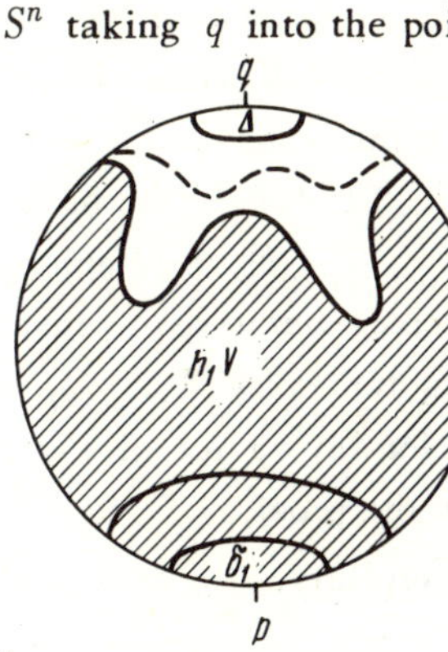

Figure 5.1

Now consider the mapping $g\colon Q^n \to Q^n$ defined as follows:

$$\begin{aligned} g(x) &= x, \quad x \in f^{-1}(\delta_1), \\ g(x) &= f^{-1} h f(x), \quad x \in Q^n \setminus f^{-1}(\delta_1). \end{aligned} \tag{1}$$

g is a homeomorphism. This is obviously the case on $f^{-1}(\delta_1)$; on $Q^n \setminus f^{-1}(\delta_1)$ it is also true, since f is a homeomorphism on $Q^n \setminus K$ (and therefore on $Q^n \setminus f^{-1}(\delta_1)$), and f^{-1} is a homeomorphism on $S^n \setminus \delta_1$, so that

$$f^{-1}(\operatorname{Fr} \delta_1) = \operatorname{Fr} f^{-1}(\delta_1).$$

Since $h(y) = y$ for $y \in \delta_1$, we have $f^{-1} h f(x) = x$ for $x \in f^{-1}(\operatorname{Fr} \delta_1)$. Hence the values of g on $f^{-1}(\delta_1)$ and on $Q^n \setminus f^{-1}(\delta_1)$ agree on $\operatorname{Fr} f^{-1}(\delta_1)$, so that equations (1) define a homeomorphism.

Furthermore, $g(Q^n) \subset V$, since

$$h(\overline{U}) = hf(Q^n) \subset \delta_2 \subset f(V),$$

so that

$$f^{-1}hf(Q^n) \subset f^{-1}f(V) = V.$$

2. Now let the number r of nondegenerate point-inverses be greater than one, and suppose the theorem proved for $k < r$. The cube Q^n can be regarded as a hemisphere in S^n. We show that *any set of* $r - 1$ *elements* K_i has the property that all the K_i contained in it are cellularly imbedded. But then all r elements K_i are cellularly imbedded.

Pick a neighborhood V of the element K_j such that

$$\overline{V} \cap K_i = \Lambda, \quad i \neq j.$$

Let $p = f(K_j)$, and take $p \in \delta_1 \subset \delta_2 \subset f(V)$ as above. Define the homeomorphism $h\colon S^n \to S^n$ and a continuous mapping g by equations (1) in the same way as for the case $r = 1$. Then

$$g(Q^n) \subset V \subset Q^n \subset S^n$$

and g is a continuous mapping of the cube into the sphere.

We assert that g has $r - 1$ nondegenerate point-inverses. Indeed, each element K_i, $i \neq j$, maps via f into a point; since $K_i \cap \overline{V} = \Lambda$ it follows from (1) that also g takes K_i into a point $x_i \in g(Q^n \setminus V) \subset f^{-1}(\delta_2 \setminus \delta_1)$. But $f(K_j) \subset \delta_1$, so that $K_j \subset f^{-1}(\delta_1)$, and hence g is a homeomorphism on K_j by (1). For the points $y \in Q^n \setminus \bigcup_{i=1}^r K_i$ the mapping f is a homeomorphism, and therefore so is g; thus the nondegenerate point-inverses for g are just the K_i, $i \neq j$. By the induction assumption every K_i, $i \neq j$, is cellularly imbedded in Q^n. But since this assertion holds for any choice of j, all the K_i are cellularly imbedded in Q^n. This completes the proof.

Corollary. *If f is a continuous mapping of the sphere S^n into itself with finitely many nondegenerate point-inverses K_i, $i = 1, 2, \cdots, r$, then every K_i is cellularly imbedded in S^n.*[1]

Pick a point $p \in S^n \setminus \bigcup_{i=1}^r K_i$ and a ball δ with center p, such that

$$\overline{\delta} \cap \left(\bigcup_{i=1}^{r} K_i \right) = \Lambda.$$

Then $Q^n = S^n \setminus \operatorname{int} \delta$ is a ball containing all the K_i, and f maps Q^n into S^n. By Theorem 1.2 every K_i is cellularly imbedded in Q^n, and so also in S^n.

Definition 1.3. A sphere $S^{n-1} \subset S^n$ (or E^n) is said to be *cylindrically imbedded* in S^n (or E^n) if there exists a homeomorphism h of the topological product $\Sigma^{n-1} \times [-1, +1]$ of the standard sphere Σ^{n-1} by a line segment into

1) V. P. Kompaniec has recently proved that this remains true, for $n \neq 4$, if the number of nondegenerate point-inverses is countable.

S^n (or E^n) such that

$$h(\Sigma^{n-1} \times 0) = S^{n-1}.$$

Theorem 1.3 (the generalized Schoenflies theorem). *If a sphere* S^{n-1} *is cylindrically imbedded in* S^n, *it separates* S^n *into domains whose closures are topological balls* [32].

Let

$$h : \Sigma^{n-1} \times [-1, +1] \xrightarrow{\text{into}} S^n; \quad h(\Sigma^{n-1} \times 0) = S^{n-1},$$

where h is a homeomorphism.

Set $\Sigma^{n-1} \times [-1, +1] = C$; $h(C) = \pi$. Obviously $S^{n-1} \subset \pi$, and $h(\Sigma^{n-1} \times -1)$ and $h(\Sigma^{n-1} \times +1)$ are topological $(n-1)$-spheres in S^n, so that each of them separates S^n. Let U_- be the domain bounded by $h(\Sigma^{n-1} \times -1)$ that does not contain S^{n-1}, and U_+ the domain bounded by $h(\Sigma^{n-1} \times +1)$ that does not contain S^{n-1}. Then

$$S^n = U_- \cup \pi \cup U_+.$$

Consider a continuous mapping ϕ of the cylinder C onto the sphere Σ^n which is a homeomorphism on the interior points of C, such that

$$\varphi(\Sigma^{n-1} \times -1) = p, \quad \varphi(\Sigma^{n-1} \times +1) = q,$$

where p and q are the poles of Σ^n, and

$$\varphi(\Sigma^{n-1} \times 0) \text{ is the equator of } \Sigma^n.$$

For points $x \in \pi$, set

$$f(x) = \varphi h^{-1}(x).$$

Obviously f maps π onto the sphere Σ^n, with

$$fh(\Sigma^{n-1} \times -1) = p, \quad fh(\Sigma^{n-1} \times +1) = q,$$

$$f(S^{n-1}) = fh(\Sigma^{n-1} \times 0) = \text{equator of } \Sigma^n.$$

Since $h(\Sigma^{n-1} \times -1)$ is the boundary of U_- and $h(\Sigma^{n-1} \times +1)$ is that of U_+, and since f is constant on each of these boundaries, f can be extended to a mapping of S^n onto Σ^n if we set

$$\underset{x \in U_-}{f(x) = p}, \quad \underset{x \in U_+}{f(x) = q}.$$

This is a continuous mapping of the sphere S^n into the sphere Σ^n with two nondegenerate point-inverses, $\bar{U}_-$ and $\bar{U}_+$, under which the sphere S^{n-1} goes into

$\phi(\Sigma^{n-1} \times 0)$, the equator of Σ^n. By the corollary to Theorem 1.2, $\bar{U}_-$ and $\bar{U}_+$ are cellularly imbedded in S^n.

The closures $\bar{V}_1$ and $\bar{V}_2$ of the components of $S^n \setminus S^{n-1}$ map respectively into the two hemispheres $\bar{D}_1$ and $\bar{D}_2$ bounded by the equator of Σ^n. On each $\bar{V}_i$ the mapping has just one nondegenerate point-inverse, and this is cellularly imbedded. But then Corollary 2 to Theorem 5.1 of Chapter I implies that $\bar{V}_i$ is homeomorphic to $\bar{D}_i$, i.e. that $\bar{V}_i$ is a topological n-cube.

Corollary 1. *If S_1^{n-1} and S_2^{n-1} are two spheres cylindrically imbedded in S^n (E^n), there exists a homeomorphism of S^n (E^n) onto itself taking S_1^{n-1} into S_2^{n-1}.*

This follows from Theorems 1.1 and 1.3 in the case that S_1^{n-1} is the standard sphere.

For imbeddings in E^n it suffices to enlarge E^n to S^n by addition of a point p, and to extend to S^n the homeomorphism given on S_1^{n-1} in such a way that the point p goes into itself; this is always possible.

Now let S_1^{n-1} and S_2^{n-1} be two arbitrary spheres cylindrically imbedded in S^n (or E^n), and F_i: $S^n \to S^n$, $i = 1, 2$, homeomorphisms such that F_i takes the standard sphere Σ^{n-1} into S_i^{n-1}. Then $F = F_2 F_1^{-1}$ is the required homeomorphism of S^n taking S_1^{n-1} into S_2^{n-1}.

Observe that for a cylindrically imbedded sphere S^{n-1} the set $h[\Sigma^{n-1} \times (-1, +1)]$ is a neighborhood. Consequently the Schoenflies theorem can be given the alternative formulation of Corollary 2 below.

Definition 1.4. A k-sphere S^k (resp. k-cell Q^k) is said to be imbedded *together with a neighborhood* in S^n (or E^n) if there exists a neighborhood U of the standard sphere Σ^k (resp. the standard cube I^k) in E^n, and a homeomorphism h: $U \to S^n$, such that

$$h(\Sigma^k) = S^k,\ h(I^k) = Q^k.$$

Corollary 2. *If a sphere S^{n-1} is imbedded in S^n together with a neighborhood, then the closures of the complementary domains of S^{n-1} are topological balls.*

Any neighborhood U of the sphere Σ^{n-1} contains a cylindrical neighborhood of Σ^{n-1}, so that S^{n-1} is cylindrically imbedded in S^n and the assertion follows by Theorem 1.3.

Corollary 3. *Let S^n be an arbitrary sphere, and S^{n-1} a sphere cylindrically imbedded in S^n. Then there exists a homeomorphism of S^n onto the boundary of an $(n+1)$-ball, taking S^{n-1} into an equator.*

Since the sphere S^{n-1} is cylindrically imbedded in S^n, there exists a homeomorphism $h: \Sigma^{n-1} \times [-1, +1] \to S^n$, where Σ^{n-1} is the standard sphere, such that $h(\Sigma^{n-1} \times 0) = S^{n-1}$. Let $H: S^n \to \Sigma^n$ be a homeomorphism of S^n onto the standard sphere Σ^n. Then $H(S^{n-1}) = Hh(\Sigma^{n-1} \times 0)$ is cylindrically imbedded in Σ^n, since $Hh(\Sigma^{n-1} \times [-1, +1]) \subset \Sigma^n$ and $H(S^{n-1}) = Hh(\Sigma^{n-1} \times 0)$. Consequently there exists a homeomorphism $\Phi: \Sigma^n \to \Sigma^n$ taking $H(S^{n-1})$ into the equator of Σ^n. Hence ΦH is a homeomorphism of S^n onto the standard sphere Σ^n, taking S^{n-1} into the equator of Σ^n.

Remark 1.2. The assertion remains true if instead of the sphere S^n we consider a space $\mathcal{E}^n$ homeomorphic to E^n, and instead of the equator, the standard sphere in E^n.

§2. Some consequences of the Schoenflies theorem

1. The generalized Schoenflies theorem has as consequence a large number of further theorems, many of which have already received wide application in the topology of manifolds. We present some of them here.

Theorem 2.1 [71] (see also [58]). *A closed n-dimensional manifold which is the union of two open sets, each homeomorphic to Euclidean space, is an n-sphere.*

Let $M^n = \mathcal{E}^n_1 \cup \mathcal{E}^n_2$, where $\mathcal{E}^n_i$, $i = 1, 2$, is homeomorphic to E^n. Since M^n is a compactum and $\mathcal{E}^n_i$ is open in M^n, it follows that $K = M^n \backslash \mathcal{E}^n_2$ is a compactum contained in $\mathcal{E}^n_1$. Since $\mathcal{E}^n_1$ is homeomorphic to E^n, the compactum K is contained in an n-cell $\bar{V}^n_1$, the boundary of $\bar{V}^n_1$ being a sphere S^{n-1} cylindrically imbedded in $\mathcal{E}^n_1$ (homeomorphic to the boundary of a ball in E^n of sufficiently large radius). Thus

$$K \subset \operatorname{int} \bar{V}^n_1 \subset \mathcal{E}^n_1, \quad S^{n-1} = \operatorname{Fr} V^n_1 \subset \mathcal{E}^n_1 \cap \mathcal{E}^n_2.$$

By Remark 1.2, the sphere S^{n-1} bounds, in $\mathcal{E}^n_2$, an n-cell $\bar{V}^n_2 = \mathcal{E}^n_2 \backslash V^n_1$. Hence

$$M^n = \bar{V}^n_1 \cup \bar{V}^n_2, \quad \bar{V}^n_1 \cap \bar{V}^n_2 = S^{n-1},$$

and we are in the situation of Corollary 2 to Theorem 1.1. Consequently M^n is a sphere.

We remark that Theorem 2.1 was used in Stallings' proof [71] of the generalized Poincaré conjecture.

2. Topological imbeddings of k-cells and continua in S^n (or E^n). In the same way as we considered cylindrical imbedding for an $(n-1)$-sphere S^{n-1} in S^n, we may consider an imbedding of a k-sphere S^k in S^n, $k < n-1$, for which there exists a homeomorphism h of the topological product $\Sigma^k \times I^{n-k}$ into S^n such that $h(\Sigma^k \times O) = S^k$, where O is the center of the standard cube I^{n-k}.

For a sphere S^k in S^n such an imbedding is equivalent to an imbedding with neighborhood; for a k-cell Q^k, on the other hand, an imbedding $h: I^k \times I^{n-k} \to S^n$ such that $h(I^k \times O) = Q^k$ is of no interest, since $h(I^k \times I^{n-k})$ is not a neighborhood of Q^k. This condition on an imbedding of a k-cell in E^n is insufficient for the imbedding to be trivial. It is easily verified that for either of the wild simple arcs of Examples 2.1 and 2.4 of Chapter IV there exists a homeomorphism $H: I^1 \times I^2 \to E^3$ such that $h(I^1 \times O)$ is the wild simple arc in question.

Theorem 2.2 (Sosinskiĭ) [9]). *If a k-cell Q^k, $k \leq n$, is imbedded in S^n (or E^n) with neighborhood, it is trivial.*

If $k = n$ the boundary of Q^n is an $(n-1)$-sphere cylindrically imbedded in S^n. Hence for $k = n$ the assertion follows from Corollary 1 to Theorem 1.3. Suppose, then, that $k < n$. It is convenient to consider an imbedding in E^n.

Let Q^k be the given cell, I^k the standard cube, U a neighborhood of I^k in E^n, and h a homeomorphism of U into E^n such that $h(I^k) = Q^k$.

U contains a cylindrical neighborhood V of the cube, which is the interior of the topological product $\tilde{I}^k \times I^{n-k}$, where $\tilde{I}^k$ is a k-cube concentric with I^k and containing I^k in its *interior*. Fr V is obviously an $(n-1)$-sphere cylindrically imbedded in E^n, and hence $\Gamma = h(\mathrm{Fr}\, V)$ is also a cylindrically imbedded $(n-1)$-sphere in E^n. But then the homeomorphism h on $\bar{V}$ can be extended to a homeomorphism $H: E^n \to E^n$. To do so it suffices to apply Corollary 1 of Theorem 1.3, observing that the extension of the homeomorphism can be carried out independently on each complementary domain of the sphere, so that we may suppose that the homeomorphism $H: E^n \to E^n$, which agrees with h on Fr $\bar{V}$, agrees with h on all of $\bar{V}$. Thus $H|_{I^k} \equiv h|_{I^k}$, so that $Q^k = H(I^k)$ is a trivially imbedded k-cell.

For the case that Q^k is imbedded in S^n, it suffices to define H on $S^n \backslash \{p\}$, where p is a point of $S^n \backslash Q^k$, and then set $H(p) = p$.

We remark that for an imbedding of a k-sphere, where $k \leq n-3$, the theorem analogous to 2.2 is also true, but this is a much more subtle result, implying, as it does, that for $k \leq n-3$ a k-sphere S^k imbedded in S^n with neighborhood cannot be knotted. This theorem was proved by Stallings in [73], and the proof is far from simple.

For an imbedding of a sphere S^{n-2} the theorem is false, since in this dimension one can have knots for any $n > 2$. Stallings gives sufficient conditions for a sphere S^{n-2} to be "unknotted" in S^n (or E^n).

Theorem 2.3. *A necessary and sufficient condition for a continuum K to be cellularly imbedded in E^n is that its complement be homeomorphic to that of a point.*

Necessity. By Corollary 2 to Theorem 5.1 of Chapter I, there exists a continuous mapping f of E^n onto itself which has as its only nondegenerate point-inverse the cellularly imbedded continuum K.

Let $p = f(K)$; then $f(E^n \backslash K) = E^n \backslash \{p\}$, and f is a homeomorphism on $E^n \backslash K$, so that $E^n \backslash K$ is homeomorphic to $E^n \backslash \{p\}$.

Sufficiency. Suppose the complement of the continuum $K \subset E^n$ is homeomorphic to that of a point $p \in E^n$. We are to show that K is cellularly imbedded in E^n.

Indeed, let h be the homeomorphism of $E^n \backslash K$ onto the complement of p. We extend it to a mapping of all of E^n, taking every point of K into p.

The mapping so obtained is obviously continuous, and has as its only nondegenerate point-inverse the continuum K. If we compactify E^n by a point ω and set $h(\omega) = \omega$, we obtain a continuous mapping of the sphere $E^n \cup \omega$ onto itself, for which K is again the only nondegenerate point-inverse. We are now in the situation of the corollary to Theorem 1.2. Consequently, K is cellularly imbedded in the sphere $E^n \cup \omega$. Clearly K is then also cellularly imbedded in E^n.

Observe that only the proof of sufficiency is based on Brown's theorem (1.2). Until the proof of the latter Theorem 2.3 was known only for $n \leq 3$, and for $n = 3$ the proof of sufficiency was based on a theorem of Bing on approximating an arbitrary 2-sphere in E^3 by a polyhedral sphere [26], and on Alexander's theorem that a polyhedral 2-sphere in E^3 bounds a 3-cell [17].

§3. Some problems in the topology of E^n and stable homeomorphisms

1. Isotopy. We shall formulate now some important and difficult problems in the topology of E^n (and S^n). We have seen in Chapter II that for an arbitrary planar closed curve S^1 there exists an isotopy of the plane onto itself taking the unit circle into S^1. For a sphere S^2 in E^3 the corresponding theorem is false, since S^2 may be wildly imbedded in E^3. Here, therefore, it makes sense to pose the problem only for trivial imbeddings. And with this understanding the solution of the problem is positive. That is to say, if the sphere S^2 in E^3 is trivial, i.e. if the homeomorphism $h: \Sigma^2 \to S^2$, where Σ^2 is the standard sphere, extends to a homeomorphism $H: E^3 \to E^3$, then there exists an isotopy from the identity mapping of E^3 onto itself which takes the standard sphere Σ^2 into the sphere S^2. This follows from a theorem of Sanderson [68]. For $n > 3$ the question remains open.

Problem 1. Let S^{n-1} be a cylindrically imbedded $(n-1)$-sphere in E^n (or S^n). Does there exist an isotopy $F_t: E^n \to E^n$ from the identity mapping $F_0 \equiv 1$ taking the standard sphere Σ^{n-1} into S^{n-1}?

This problem is closely related to that of the structure of the group of homeomorphisms of Euclidean space or of an n-sphere onto itself. Consider an arbitrary homeomorphism f of S^n onto itself. If f changes the orientation of S^n, then obviously f cannot be joined by an isotopy to the identity mapping. But all known homeomorphisms $f: S^n \to S^n$ that preserve orientation are obtainable by such an isotopy.

Problem 2. Does there exist, for every orientation-preserving homeomorphism f of an n-sphere S^n onto itself (or of E^n onto itself), an isotopy F_t, $0 \le t \le 1$, such that $F_0 \equiv 1$ and $F_1 \equiv f$?

If Problem 2 were to have a positive solution, then any two orientation-preserving homeomorphisms of S^n onto itself would be isotopic. In other words, the subgroup of the group $H(S^n)$ of homeomorphisms of S^n consisting of the orientation-preserving homeomorphisms would be arcwise connected. Up to the present time this conjecture has been proved only for $n \le 3$ [68]. But even in 1924 Alexander pointed out a class of homeomorphisms of S^n (or E^n) joinable by isotopy to the identity mapping.

Theorem 3.1 (Alexander). *Any homeomorphism f of E^n (or S^n) onto itself which is the identity on some domain is isotopic to the identity mapping.*

Consider first E^n. Suppose

$$f : E^n \to E^n \text{ and } f|_U = 1, \tag{2}$$

where f is a homeomorphism and U is a domain in E^n. U contains a ball δ of radius α. Take the origin of coordinates O to be at the center of δ, and consider the family of homeomorphisms $\phi_t(x) = tx$, $0 < t \le 1$.

Set

$$F_t = \varphi_t^{-1} f \varphi_t, \quad 0 < t \leqslant 1.$$

F_t is a homeomorphism for every $t > 0$, and is the identity on the ball Q_t with center O and radius α/t, since $\phi_t(Q_t) = \delta$ and $f(x) = x$ for $x \in \delta$:

$$F_t|_{Q_t} = 1.$$

Since the diameters of the balls Q_t increase to infinity as $t \to 0$, F_t can be extended to an isotopy by setting $F_0 \equiv 1$.

Since ϕ_1 is the identity mapping we have $F_1 = f$, so that F_t is an isotopy joining f to the identity mapping.

The theorem remains true if E^n is replaced by a homeomorphic image $\mathfrak{E}^n = h(E^n)$. Indeed, if $f: \mathfrak{E}^n \to \mathfrak{E}^n$ is a homeomorphism of $\mathfrak{E}^n$ onto itself which is the identity on some domain, then $\psi = h^{-1} f h$ is a similar such homeomorphism of E^n onto itself. Let F_t be an isotopy joining ψ to the identity mapping; then

hF_th^{-1} is an isotopy joining the homeomorphism $f\colon \mathfrak{S}^n \to \mathfrak{S}^n$ to the identity mapping of $\mathfrak{S}^n$ onto itself.

Now let f be a homeomorphism of the sphere S^n onto itself which is the identity on a domain V. Take a point $p \in V$; then $f(p) = p$, so that $f(S^n \backslash p) = S^n \backslash p$; hence, by what we have already proved, there exists an isotopy Φ_t defined on $S^n \backslash p$, joining $f|S^n \backslash p$ to the identity mapping. If we now set $\Phi_t(p) = p$ for all t, we obtain an isotopy joining the given homeomorphism $f\colon S^n \to S^n$ to the identity mapping.

2. **Definition 3.1.** A homeomorphism $f\colon E^n \to E^n$ is called *stable* if it is representable as the composite of a finite number of homeomorphisms:

$$f = f_k f_{k-1} \cdots f_1,$$

such that each homeomorphism f_i, $1 \le i \le k$, is the identity on some domain U_i: $f_i|_{U_i} = 1$.

Corollary. *A stable homeomorphism of E^n onto itself (or of S^n onto itself) is isotopic to the identity mapping.*

By Alexander's theorem each f_i is isotopic to the identity mapping. Let F_t^i be an isotopy such that

$$F_0^i \equiv 1, \quad F_1^i \equiv f_i.$$

Then the isotopy

$$F_t = F_t^k F_t^{k-1} \dots F_t^1$$

takes the identity mapping into f.

A stable homeomorphism obviously preserves orientation, since this is true of any homeomorphism which is the identity on a domain, and the composite of orientation-preserving homeomorphisms is an orientation-preserving homeomorphism.

Problem 3. Is every orientation-preserving homeomorphism of E^n onto itself (or of S^n onto itself) stable?

In view of the corollary, a positive solution of Problem 3 would imply a positive solution of Problem 2.

The annulus problem. An important and difficult problem concerns the imbedding in S^n, or in E^n, of several disjoint $(n-1)$-spheres, each imbedded cylindrically. The annulus problem refers to the case of two disjoint cylindrical $(n-1)$-spheres S_1^{n-1} and S_2^{n-1} in S^n. Since $S_1^{n-1} \cap S_2^{n-1} = \Lambda$, the union $S_1^{n-1} \cup S_2^{n-1}$ separates the sphere S^n into three domains. Two of them—call them U_1 and U_2—are bounded by S_1^{n-1} and S_2^{n-1} respectively; and since the spheres S_i^{n-1} are cylindrical, it follows from the Schoenflies theorem that the closures

$\bar{U}_i$, $i = 1, 2$, of the domains U_i are n-cells. The third domain U_3 is bounded by the pair of spheres $S_1^{n-1} \cup S_2^{n-1}$. It is easy to see that the domain U_3 is an open n-dimensional annulus, i.e. it is homeomorphic to the domain between two concentric standard spheres. Indeed, by the Schoenflies theorem $S^n \backslash U_2$ is an n-cell, and therefore $S^n \backslash \bar{U}_2$ is homeomorphic to E^n. Since S_1^{n-1} is imbedded in it cylindrically, it follows by the Remark after Corollary 3 to the Schoenflies theorem that there exists a homeomorphism h of the domain $S^n \backslash \bar{U}_2$ onto E^n, under which $h(S_1^{n-1})$ is the standard sphere. But then $h(U_3)$ is the complement of the standard ball in E^n, and hence U_3 is an open annulus.

It is natural to expect that the closed domain $\bar{U}_3$ is a closed annulus. But up to the present time this has been proved only for $n \leq 3$.

For $n > 3$ the annulus problem remains open.

Problem 4. Is the *closed* domain in S^n (or E^n) bounded by a pair of disjoint cylindrical $(n-1)$-spheres an annulus?[1]

The problem can also be formulated in the following way:

Problem 4′. Let S^{n-1} be a cylindrical sphere in the interior of an n-ball Q^n. Does there exist a homeomorphism of Q^n onto itself which is the identity on the boundary of Q^n and takes S^{n-1} into a standard sphere in the interior of Q^n?

Let us show that Problems 4 and 4′ are equivalent. The positive solution of 4′ implies that of 4. Indeed, let S_1^{n-1} and S_2^{n-1} be two disjoint cylindrical spheres in S^n. By Corollary 1 to the Schoenflies theorem there exists a homeomorphism $h: S^n \longrightarrow S^n$ taking S_1^{n-1} into the equator of S^n. Let U be the complementary domain of S_1^{n-1} containing S_2^{n-1}. Then $h(\bar{U})$ is a ball containing the cylindrical sphere $h(S_2^{n-1})$. Since we are assuming 4′, there exists a homeomorphism ϕ of the ball $h(\bar{U})$ onto itself which is the identity on the boundary of $h(\bar{U})$ and takes the sphere $h(S_2^{n-1})$ into a standard sphere concentric with $\operatorname{Fr} h(\bar{U})$. Then the homeomorphism ϕh takes the closed domain $\bar{V}$ bounded by the pair of spheres $S_1^{n-1} \cup S_2^{n-1}$ into the spherical annulus bounded by $h(S_1^{n-1}) \cup \phi h(S_2^{n-1})$.

Conversely, the positive solution of 4 implies that of 4′. Let Q^n be a ball in E^n, and S^{n-1} a cylindrical sphere in the interior of Q^n. Since $\operatorname{Fr} Q^n$ and S^{n-1} are disjoint cylindrical spheres, and we are assuming 4, it follows that there exists a homeomorphism h of the closed domain $\bar{V}$ with boundary $S^{n-1} \cup \operatorname{Fr} Q^n$ onto the spherical annulus K between $\operatorname{Fr} Q^n$ and $\operatorname{Fr} \tilde{Q}^n$, where $\tilde{Q}^n$ is a ball concentric with Q^n, $\tilde{Q}^n \subset \operatorname{Int} Q^n$. Thus

1) *Translator's note*. In the case of E^n, of course, one of the spheres must be inside the other.

$$h(\overline{V}) = K, \quad h(\operatorname{Fr} Q^n) = \operatorname{Fr} Q^n, \quad h(S^{n-1}) = \operatorname{Fr} \widetilde{Q}^n.$$

By Corollary 1 to Theorem 1.1 and the Schoenflies theorem, the homeomorphism h defined on S^{n-1} extends to a homeomorphism $H: \overline{W} \to \widetilde{Q}^n$, where W is the domain in Q^n bounded by S^{n-1}. Thus,

$$H: Q^n \to Q^n, \quad H|_{\overline{V}} \equiv h.$$

Now let $\Psi: Q^n \to Q^n$ be the homeomorphism which agrees with h^{-1} on $\operatorname{Fr} Q^n$ and takes the radius of Q^n to any point $x \in \operatorname{Fr} Q^n$ linearly into the radius to the point $h^{-1}(x)$. Then ΨH is again a homeomorphism of Q^n onto itself, with

$$\Psi H|_{\operatorname{Fr} Q^n} = h^{-1}h = 1, \quad \Psi H|_{S^{n-1}} = \Psi h.$$

Hence

$$\Psi H(S^{n-1}) = \Psi h(S^{n-1}) = \Psi(\operatorname{Fr} \widetilde{Q}^n) = \operatorname{Fr} \widetilde{Q}^n.$$

Consequently ΨH is a homeomorphism of Q^n onto itself satisfying the conditions of 4'.

3. The annulus problem has a positive solution when the pair of disjoint spheres is such that the one is obtainable from the other by a stable homeomorphism of S^n.

Lemma 3.1. *The union of two n-dimensional annuli intersecting in a sphere S^{n-1}, belonging to the boundary of each, is an annulus. On the other hand, let $\overline{W}$ be the closure of the difference between two annuli K and K', where $K' \subset K$ and $\operatorname{Fr} K'$ consists of two spheres: $S_1 \subset \operatorname{Fr} K$ and $S_1' \subset \operatorname{Int} K$, with S_1' cylindrically imbedded in K. Then $\overline{W}$ is an annulus.*

Let $K = K_1 \cup K_2$ and $K_1 \cap K_2 = \operatorname{Fr} K_1 \cap \operatorname{Fr} K_2 = S$. There exists a homeomorphism h_1 of the standard annulus T_1 between the spheres of radius 1 and $2/3$ onto K_1, and a homeomorphism h_2 of the standard annulus T_2 between the spheres of radius $2/3$ and $1/3$ onto K_2. $T_1 \cap T_2 = \Sigma$ is the sphere of radius $2/3$. $h_2^{-1}h_1|_\Sigma$ is a homeomorphism of Σ onto itself. As in the proof of Theorem 1.1, construct a homeomorphism $\Psi: T_2 \to T_2$ agreeing on Σ with $h_2^{-1}h_1$, and consider the homeomorphism $h_2\Psi: T_2 \to K_2$; then since $h_2\Psi|_\Sigma = h_1$, we obtain, setting

$$\Phi|_{T_1} = h_1, \quad \Phi|_{T_2} = h_2\Psi,$$

a homeomorphism of the annulus $T_1 \cup T_2$ onto $K_1 \cup K_2$. Consequently $K_1 \cup K_2$ is an annulus.

Now let $\overline{W} = \overline{K \setminus K'}$, $\operatorname{Fr} K = S_1 \cup S_2$, $\operatorname{Fr} K' = S_1 \cup S'$, where S' is a cylindrically imbedded sphere in $\operatorname{Int} K$. Then there exists a homeomorphism $h: \Sigma \times [-1, +1] \to \operatorname{Int} K$ such that $h(\Sigma \times 0) = S'$. Set $S_\alpha' = h(\Sigma \times \alpha)$, $-1 \leq \alpha \leq +1$, and denote by $[S'_\alpha, S'_\beta]$ the closed domain between S'_α and S'_β; obviously $[S'_\alpha, S'_\beta]$ is an annulus. $[S'_{-1}, S'_{+1}]$ will be called the cylinder over S'.

By what we have just proved, $K' \cup [S', S'_{1/2}] = [S_1, S'_{1/2}]$ and $K' \cup [S', S'_1] = [S_1, S'_1]$ are annuli. Hence there exists a homeomorphism H of the annulus $[S_1, S'_1]$ onto the annulus $[S', S'_1]$ such that

$$H(S_1) = S', \quad H|_{S_1'} = 1.$$

Setting $H|_{[S_1', S_2]} = 1$, we obtain a homeomorphism of the annulus K onto $\overline{W}$, so that $\overline{W}$ is likewise an annulus. The homeomorphism H can be constructed as the product of two homeomorphisms, of which the first is the identity on $S'_1 \cup [S_1, S'_{-1/2}]$ and takes S' into $S'_{+1/2}$, and the second is the identity on $[S'_{1/2}, S'_1]$ and takes S_1 into S'.

Theorem 3.2 (Brown and Gluck [36]). *In order that a pair of disjoint cylindrical $(n-1)$-spheres in S^n (resp. E^n) bound an annulus, it is necessary and sufficient that there exist a stable homeomorphism of S^n (E^n) onto itself taking one of the spheres into the other.*

Necessity. Let S and S' be $(n-1)$-spheres cylindrically imbedded in S^n, with $S \cap S' = \Lambda$ and $[S, S']$ an annulus. Let $[S_{-1}, S_1]$ and $[S'_{-1}, S'_1]$ be cylinders around S and S', chosen so that

$$[S_{-1}, S_1] \cap [S'_{-1}, S'_1] = \Lambda,$$

and $Q_{S_{-1}}$ and $Q_{S'_1}$ the balls with boundaries S_{-1} and S'_1 that do not contain these cylinders. We may then construct, in the same way as in the proof of Lemma 3.1, a stable homeomorphism $F\colon S^n \to S^n$ such that

$$F|_{Q_{S_{-1}} \cup Q_{S_1'}} = 1, \quad F(S_{-1/2}) = S, \quad F(S) = S', \quad F(S') = S'_{1/2}.$$

Sufficiency. Let S and $F(S)$ be two cylindrical spheres in S^n, where $S \cap F(S) = \Lambda$ and F is a stable homeomorphism of S^n onto itself. Then F is of the form

$$F = f_k f_{k-1} \cdots f_1, \tag{3}$$

where each homeomorphism f_i, $1 \le i \le k$, is the identity on some domain U_i. We prove the theorem by induction on the number of homeomorphisms f_i.

1) $k = 1$, i.e. $F|_U = 1$. The pair of cylindrical spheres $S \cup F(S)$ separates S^n into three domains: two balls Q_S and Q_F, and a third, W, with boundary $S \cup F(S)$. We must show that $\overline{W}$ is an annulus. The domain U either intersects at least one of the domains Q_S, Q_F, or else is contained in W. We consider the two cases.

a) $U \cap Q_S \neq \Lambda$; the case $U \cap Q_F \neq \Lambda$ reduces to this if we replace F by F^{-1}. In $U \cap Q_S$ we select a topological ball δ with boundary S', such that $[S', S]$ is an annulus. Since $F|_\delta = 1$, we have $\delta \subset F(Q_S)$. Furthermore, since $F|_{S'} = 1$ we have

$$F[S', S] = [S', F(S)],$$

so that $[S', F(S)]$ is an annulus; and since S is a cylindrical sphere in the interior of the annulus $[S', F(S)]$, and $[S', S]$ is an annulus, it follows by the lemma that $[S, F(S)]$ is an annulus.

b) $U \subset W$. We select a ball δ in U with boundary S' such that $[S', S]$ is an annulus, and construct a homeomorphism $H: S^n \longrightarrow S^n$ such that

$$H|_{Q_F} = 1, \ H(S) = S', \ H(S') = S. \tag{4}$$

To do so, i.e. to select S' and construct H, consider a homeomorphism $\phi: S^n \longrightarrow S^n$ taking S into the equator, select in $\phi(U)$ a standard sphere Σ', and set $S' = \phi^{-1}(\Sigma')$; let δ be the complementary domain of S' that does not contain S. In addition, select in $\phi(W \setminus \delta)$ another standard sphere Σ, bounding a ball $\Pi \subset \phi(W \setminus \delta)$. Using Theorem 2.2 of Chapter I, construct three homeomorphisms $h_i: S^n \longrightarrow S^n$, $i = 1, 2, 3$, such that

$$h_1|_{\varphi(Q_F \cup \delta)} = 1, \quad h_1\varphi(S) = \Sigma,$$

$$h_2|_{\varphi(Q_F) \cup \Pi} = 1, \quad h_2(\Sigma') = \varphi(S),$$

$$h_3|_{\varphi(Q_F \cup Q_S)} = 1, \quad h_3(\Sigma) = \Sigma'.$$

Set $H' = h_3 h_2 h_1$ and $H = \phi^{-1} H' \phi$. Then the homeomorphism H satisfies requirement (4).

Since $S' \subset U$, we have $F|_{S'} = 1$; hence

$$F[S', S] = [S', F(S)],$$

and therefore $[S', F(S)]$ is an annulus. By (4),

$$H[S', F(S)] = [S, F(S)],$$

and so $[S, F(S)]$ is an annulus.

Now suppose $k > 1$ and the assertion holds when the number of homeomorphisms f_i is less than k.

2) Take first the case that the spheres S, $F_{k-1}(S)$ and $F(S)$ are disjoint, where $F_{k-1} = f_{k-1} \cdots f_1$. Then by the induction assumption, $[S, F_{k-1}(S)]$ and $[F_{k-1}(S), F(S)]$ are annuli, since $F = f_k F_{k-1}$. If one of the three spheres separates the other two, it follows easily from Lemma 3.1 that $[S, F(S)]$ is an annulus. We suppose, then, that no one of the three separates the other two.

Let Q_S, Q_F and Q_{k-1} be the disjoint topological balls bounded respectively by the spheres S, $F(S)$ and $F_{k-1}(S)$:

$$Q_S \cap Q_F = Q_S \cap Q_{k-1} = Q_{k-1} \cap Q_F = \Lambda.$$

Since $[F_{k-1}(S), F(S)]$ is an annulus, there exists a homeomorphism $\phi: S^n \to S^n$ taking $[F_{k-1}(S), F(S)]$ into the annulus between two standard spheres $[\Sigma_1, \Sigma_2]$. To construct ϕ, it suffices first to map the annulus $[F_{k-1}(S), F(S)]$ homeomorphically onto $[\Sigma_1, \Sigma_2]$, then to extend this map to the cells Q_{k-1} and Q_F. Arguing, therefore, as in case 1b), we may construct a homeomorphism $H: S^n \to S^n$ such that

$$H|_{Q_S} = 1, \; H(F_{k-1}(S)) = F(S).$$

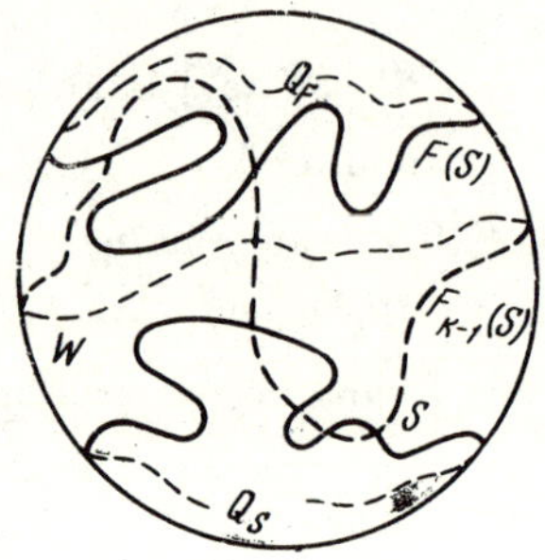

Figure 5.2

Then $H[S, F_{k-1}(S)] = [S, F(S)]$, so that $[S, F(S)]$ is an annulus.

3) Now take the case that $S \cap F_{k-1}(S) \neq \Lambda$ or $F_{k-1}(S) \cap F(S) \neq \Lambda$ (Figure 5.2). We shall construct a homeomorphism $H: S^n \to S^n$ such that

$$S \cap HF_{k-1}(S) = HF_{k-1}(S) \cap F(S) = \Lambda$$

and

$$f_k H^{-1}|_v = 1, \quad Hf_{k-1}|_{v'} = 1, \tag{5}$$

where v and v' are domains in S^n. F is then representable as

$$F = (f_k H^{-1})(Hf_{k-1}) f_{k-2} \cdots f_1 \tag{6}$$

and the proof reduces to case 2). Let $f_k|_U = 1$ and $f_{k-1}|_{U'} = 1$. Consider the various possibilities.

a) $U \cap Q_S \neq \Lambda$.

Take a sphere $S_{-\epsilon} \subset Q_S$ such that

$$Q_{S_{-\varepsilon}} \cap U \neq \Lambda,$$

$$\text{and if } U' \cap Q_S \neq \Lambda, \text{ then } U' \cap Q_{S_{-\varepsilon}} \neq \Lambda. \tag{7}$$

In $S^n \setminus Q_S$ take two topological balls: δ, with boundary S', and $\tilde{\delta} \subset \delta$, with boundary $\tilde{\delta}'$, such that $[\tilde{S}', S']$ and $[S', S]$ are annuli,

$$F_{k-1}(S) \cap \delta = \Lambda, \tag{8}$$
$$\text{and if } U' \subset S^n \setminus Q_S, \text{then } \delta \subset U'.$$

Now construct the homeomorphism $H: S^n \to S^n$ so that

$$H|_{\tilde{\delta} \cup Q_{S_{-\varepsilon}}} = 1, \quad H(S') = S. \tag{9}$$

This can be done because $[S', S]$ is an annulus disjoint from $\tilde{\delta} \cup Q_{S_{-\epsilon}}$. Then

$$H(\delta) = \overline{S^n \setminus Q_S}, \qquad HF_{k-1}(S) \subset \operatorname{Int} Q_S,$$

and it follows from (7)–(9) that requirements (5) are satisfied. Thus F is representable in the form (6), the spheres S, $F(S)$ and $HF_{k-1}(S)$ are disjoint, and $[S, F(S)]$ is an annulus.

The case $U \cap Q_S = \Lambda$, $U \cap Q_F \neq \Lambda$ is treated in the same way, replacing Q_S by Q_F.

b) $U \subset W$, where $\overline{W} = [S, F(S)]$.

In the interior of W, take a domain $\tilde{W}$ bounded by two spheres S_ϵ and $F(S)_\epsilon'$, such that $U \cap \tilde{W} \neq \Lambda$ and, if $U' \cap W \neq \Lambda$, then $U' \cap \tilde{W} \neq \Lambda$. In the interior of the balls Q_S and Q_F, take, as in case a), topological balls $\tilde{\delta}_S$, δ_S, $\tilde{\delta}_F$ and δ_F; make the choice so that δ_F and δ_S do not meet $F_{k-1}(S)$ and, if $U' \subset Q_S$ (Q_F), then $\tilde{\delta}_S \subset U'$ $(\tilde{\delta}_F \subset U')$.

Now construct the homeomorphism H so that

$$H|_{\tilde{\delta}_S \cup \tilde{\delta}_F \cup \tilde{W}} = 1, \quad H(\delta_S) = Q_S, \quad H(\delta_F) = Q_F.$$

Then $HF_{k-1}(S) \subset W$ and requirements (5) are satisfied. Hence F is again representable in the form (6), the spheres S, $HF_{k-1}(S)$ and $F(S)$ are disjoint, and $[S, F(S)]$ is an annulus.

Corollary. *Let S_1^{n-1} and S_2^{n-1} be two disjoint spheres in S^n, such that for each S_i^{n-1} there exists a stable homeomorphism f_i: $S^n \longrightarrow S^n$ taking the equator into S_i^{n-1}. Then $[S_1^{n-1}, S_2^{n-1}]$ is an annulus.*

Obviously S_1^{n-1} and S_2^{n-1} are cylindrical, since each f_i is defined on all of S^n. The homeomorphism $f_2 f_1^{-1}$: $S^n \longrightarrow S^n$ is stable and takes S_1^{n-1} into S_2^{n-1}; consequently the hypothesis of Theorem 3.2 is satisfied and $[S_1^{n-1}, S_2^{n-1}]$ is an annulus.

Observe that the requirement that the spheres be cylindrical is essential in Theorem 3.2. For example, let S_1 and S_2 be two wild (therefore noncylindrical) 2-spheres in E^3, like that in Example 2.1b of Chapter IV, with S_1 in the interior domain of S_2. It is easy to construct a homeomorphism f: $E^3 \longrightarrow E^3$ such that $f(S_1) = S_2$ and $f|_{E^3 \setminus Q} = 1$, where Q is a ball containing S_2. Nevertheless, the domain $[S_1, S_2]$ is not an annulus.

It follows from Theorem 3.2 that a positive solution to Problem 3 would entail a positive solution to the annulus problem (Problem 4). The set of all stable homeomorphisms of the sphere onto itself is a group, since the product (i.e. composite) of stable homeomorphisms is a stable homeomorphism, and if $F = f_k f_{k-1} \cdots f_1$ is a stable homeomorphism, so is $F^{-1} = f_1^{-1} \cdots f_k^{-1}$, f_i and f_i^{-1} being the identity on the same domain U_i.

Let $\mathcal{H}$ be the group of stable homeomorphisms of the sphere. $\mathcal{H}$ is a subgroup of the group Ξ of all homeomorphisms of the sphere onto itself, or, better, $\mathcal{H}$ is a subgroup of the subgroup Ξ' of Ξ consisting of the orientation-preserving homeomorphisms of S^n.

Thus Problem 3 is equivalent to the question: is the group $\mathcal{H}$ the same as the group Ξ' of all orientation-preserving homeomorphisms of the sphere?

4. **Some further consequences.**

Theorem 3.3 (Brown [33]). *Let X be a space which is the union of an increasing sequence of subsets $\mathcal{E}_k$, each homeomorphic to E^n:*

$$X = \bigcup_{k=1}^{\infty} \mathcal{E}_k, \quad \mathcal{E}_k \subset \mathcal{E}_{k+1}, \quad \mathcal{E}_k \sim E^n.$$

Then X is itself homeomorphic to E^n.

It suffices to prove that X is the union of an increasing sequence of topological balls B_k such that $\overline{B_{k+1} \setminus B_k}$ is an annulus. Indeed, suppose

$$X = \bigcup_{k=1}^{\infty} B_k, \tag{10}$$

let Q_k be the ball in E^n with center O and radius k, $T_{k+1} = \overline{Q_{k+1} \setminus Q_k}$, and let ϕ_k be homeomorphisms:

$$\varphi_1 : Q_1 \to B_1, \quad \varphi_k : T_k \to \overline{B_k \setminus B_{k-1}}, \quad k \geqslant 2.$$

Assume that for $i \leq k$ we have constructed homeomorphisms Ψ_i: $T_i \to \overline{B_i \setminus B_{i-1}}$ such that

$$\Psi_1 |_{Q_1} = \varphi_1; \quad \Psi_i |_{\mathrm{Fr}\, Q_i} = \Psi_{i+1} |_{\mathrm{Fr}\, Q_i}, \quad i < k.$$

Let Φ_{k+1} be a homeomorphism of T_{k+1} onto itself agreeing on $\mathrm{Fr}\, Q_k$ with $\phi_{k+1}^{-1} \Psi_k$, and set

$$\Psi_{k+1} = \varphi_{k+1} \Phi_{k+1}.$$

Then

$$\Psi_{k+1} |_{\mathrm{Fr}\, Q_k} = \Psi_k |_{\mathrm{Fr}\, Q_k}, \quad \Psi_{k+1}(T_{k+1}) = \overline{B_{k+1} \setminus B_k}.$$

Thus, setting

$$\Psi |_{Q_1} = \varphi_1, \quad \Psi |_{T_k} = \Psi_k, \quad k > 1,$$

we obtain a homeomorphism Ψ of E^n onto X.

To prove the theorem it remains to represent X in the form (10).

For each k there exists a homeomorphism f_k of E^n onto $\mathcal{E}_k$. Since $\mathcal{E}_1 \subset \mathcal{E}_k$ for $k > 1$, we may without loss of generality suppose that $f_k(O) = f_1(O) = O'$ for every k. Each $\mathcal{E}_k$ can be represented in the form

$$\mathfrak{S}_k = \bigcup_{r=1}^{\infty} f_k(Q_r). \tag{11}$$

Set $B_1 = f_1(Q_1)$, and assume that the B_i have been constructed for $i \leq k$, so that to each B_i is associated a ball $Q_{m_i} \subset E^n$ such that

$$B_i \supset \bigcup_{j=1}^{i} f_j(Q_{m_{i-1}}) \cup f_i(Q_{m_i}), \quad i \leqslant k, \quad m_i \geqslant i, \tag{12}$$

and, furthermore, that each $\overline{B_i \setminus B_{i-1}}$ is an annulus and the boundary of each B_i is a cylindrical sphere.[1)]

Since $\mathfrak{S}_{k+1} \supset \mathfrak{S}_i$, we may pick a ball $Q_{m_{k+1}}$ such that

$$f_{k+1}(Q_{m_{k+1}}) \supset \bigcup_{i=1}^{k} f_i(Q_{m_k}) \cup B_k. \tag{13}$$

By (13), $Q_{m_{k+1}} \supset f_{k+1}^{-1}(B_k)$; on the other hand, by (12), $\operatorname{Int} B_k \ni O'$, so that $\operatorname{Int} f_{k+1}^{-1}(B_k) \ni O$, and we may therefore find balls δ and δ' with center O and of sufficiently small radius, such that

$$\delta' \subset \delta \subset \operatorname{Int} f_{k+1}^{-1}(B_k). \tag{14}$$

Let h be a homeomorphism of E^n onto itself such that

$$h|_{\delta'} = 1, \quad h(\delta) = Q_{m_{k+1}}. \tag{15}$$

Set

$$B_{k+1} = \Phi_{k+1}(B_k), \quad \Phi_{k+1} = f_{k+1} h f_{k+1}^{-1}.$$

Since the boundary of B_k is a cylindrical sphere, and the homeomorphism Φ_{k+1} is defined on all of $\mathfrak{S}_{k+1}$, the boundary of B_{k+1} is again a cylindrical sphere. From (13)–(15) it follows that

$$\operatorname{Int} h f_{k+1}^{-1}(B_k) \supset Q_{m_{k+1}} \supset f_{k+1}^{-1}(B_k)$$

(Figure 5.3). Consequently B_{k+1} satisfies (12) and $\operatorname{Int} B_{k+1} \supset B_k$.

By (15) the homeomorphism Φ_{k+1} is the identity on $f_{k+1}(\delta')$, so that Theorem 3.2 implies that $[B_k, B_{k+1}]$ is an annulus. Finally,

$$X = \bigcup_{k=1}^{\infty} B_k,$$

because, if $x \in X$, then $x \in \mathfrak{S}_r$ for some r, so that $x \in f_r(Q_{m_{k-1}})$ for some $k \geq r$, and consequently, by (12), $x \in B_k$.

Theorem 3.4 (Mazur [50]). *The open star of a vertex in a triangulation of an n-dimensional manifold is homeomorphic to E^n.*

---

1) *Translator's note.* This last is meaningful because, as must in any case be added to the induction assumption, B_i is to be in $\mathfrak{S}_i$.

Let M^n be the manifold, and T a triangulation. Without loss of generality we may suppose that T is realized as a rectilinear polyhedron in Euclidean space and is endowed with the induced metric.

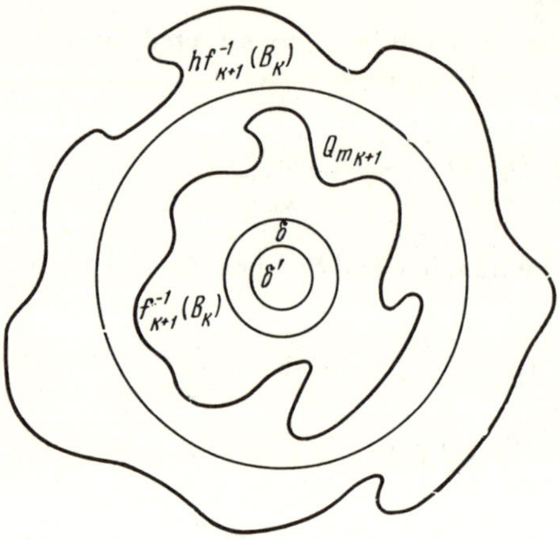

Figure 5.3

Let x be a vertex of T, and $\mathrm{St}_T x$ its open star. Since M^n is a manifold, there exists a neighborhood u of x homeomorphic to E^n, such that $\bar{u} \subset \mathrm{St}_T x$.

Denote by $\rho \mathrm{St}_T x$ the polyhedron obtained by stretching $\mathrm{St}_T x$ from the center x by a factor of ρ. There exists an integer $N > 2$ such that

$$\frac{1}{N} \mathrm{St}_T x \subset u \subset \bar{u} \subset \mathrm{St}_T x. \qquad (16)$$

Let ϕ_k be a homeomorphism of $\mathrm{St}_T x$ onto itself taking every rectilinear segment from x to the boundary of $\mathrm{St}_T x$ semilinearly onto itself, such that

$$\varphi_k \left(\frac{1}{N} \mathrm{St}_T x \right) = \frac{k}{k+1} \mathrm{St}_T x, \quad k = 1, 2, \dots,$$

and

$$\varphi_k (\mathrm{St}_T x) = \mathrm{St}_T x.$$

By (16) we have

$$\varphi_k \left(\frac{1}{N} \mathrm{St}_T x \right) \subset \varphi_k (u) \subset \varphi_k (\bar{u}) \subset \mathrm{St}_T x.$$

It follows that for any $k \geq 1$ there exists a $k' > k$ such that

$$\varphi_k (u) \subset \frac{k'}{k'+1} \mathrm{St}_T x \subset \varphi_{k'} (u).$$

We may therefore find a sequence of integers $k_1 = 1, k_2, \cdots$ such that

$$\varphi_{k_1} (u) \subset \frac{k_2}{k_2+1} \mathrm{St}_T x \subset \varphi_{k_2} (u) \subset \frac{k_3}{k_3+1} \mathrm{St}_T x \subset \dots .$$

Since $\mathrm{St}_T x = \bigcup_{i=1}^{\infty} (k_i/(k_i + 1)) \mathrm{St}_T x$, we have, consequently,

$$\mathrm{St}_T x = \bigcup_{i=1}^{\infty} \varphi_{k_i} (u), \quad \varphi_{k_i} (u) \subset \varphi_{k_{i+1}} (u).$$

Thus, since each $\phi_{k_i}(u)$ is homeomorphic to u, and therefore to E^n, it follows from Theorem 3.3 that $\mathrm{St}_T x$ is homeomorphic to E^n.

It is important to observe that Theorem 3.4 concerns only the open star of a vertex in a triangulation of a manifold. Unsolved, and very difficult, is

Problem 5. Is the *closed* star of a vertex in a triangulation of a manifold a topological ball?

Since the answer to this question is at present unknown, in studying a polyhedral manifold it becomes necessary to impose certain conditions on the triangulation. What is usually studied is the so-called combinatorial manifold; sometimes the formally somewhat more general notion of star manifold is considered. We shall define these notions in §5. Problem 5 can also be formulated as follows.

Problem 5′. Is the boundary of the star of a vertex in a triangulation of a manifold a topological sphere?

Obviously a positive solution of Problem 5 implies a positive solution of 5′. Conversely, a positive solution of 5′ implies a positive solution of 5, since the closed star $\overline{\mathrm{St}_T x}$ is the cone over its boundary Γ, so that a homeomorphism ϕ of the boundary of the standard ball onto Γ extends to the whole ball, the radius to any point y of its boundary going into the segment joining x to $\phi(y)$.

5. Imbedding of a finite number of spheres. We have seen that it is unknown whether every domain between two cylindrical $(n-1)$-spheres is an annulus. So much the more is it unknown whether, given a finite number of disjoint $(n-1)$-spheres $S_1, S_2, \cdots, S_k$ cylindrically imbedded in S^n, there exists a homeomorphism $f: S^n \to S^n$ taking each S_i, $1 \le i \le k$, into a standard sphere, or, equivalently, into the boundary of a spherical n-simplex.

The first proof of the Schoenflies theorem, but with a restriction, was given by Mazur [49,50]. Namely, Mazur assumed that the homeomorphism $f: \Sigma^{n-1} \times [-1, +1] \to S^n$ is linear on some n-simplex $\Delta^n \subset \Sigma^{n-1} \times (-1, +1)$. It was shown that if two disjoint $(n-1)$-spheres S_1 and S_2 are images of $\Sigma^{n-1} \times 0$ under such homeomorphisms, then $[S_1, S_2]$ is an annulus. In fact, there exists an isotopy $F_t: S^n \to S^n$ from the identity mapping, taking the standard annulus between two concentric spheres into $[S_1, S_2]$. In a somewhat different form, a sufficient condition for a "nice" imbedding in S^n of a finite number of disjoint cylindrical $(n-1)$-spheres was given by Černavskiĭ in [12]. It is as follows.

Theorem 3.5. *Let $S_1, \cdots, S_k$ be a finite number of disjoint cylindrical $(n-1)$-spheres in S^n, such that each sphere S_i contains a spherical $(n-1)$-simplex. Then there exists a stable homeomorphism $F: S^n \to S^n$ taking each S_i, $1 \le i \le k$, into the boundary of a spherical n-simplex.*

The spheres S_i can be numbered so that one of the domains, say U_i, bounded by S_i contains no S_j with index $> i$. Indeed, since the spheres S_i are finite in number, there exists among them some which bound domains containing no other spheres; number these. Among the remaining spheres, there exist some which bound domains containing only already numbered spheres; number these; etc. Thus we eventually number all the spheres S_i in the required fashion.

We shall first construct a stable homeomorphism $F_1: S^n \to S^n$ which takes the sphere S_1 into the boundary of a simplex and is the identity on all the S_i, $i \geq 2$. We then construct, in succession, stable homeomorphisms F_r, $r \leq k$, such that

$$F_r(U_i) = d_i, \quad i \leqslant r; \quad F_r | S_j = 1, \quad j > r, \tag{17}$$

where the d_i are n-simplices.

1. *Construction of* F_1. By assumption S_1 is a cylindrical sphere containing an $(n-1)$-simplex D^{n-1} and bounding a domain U_1 that contains no spheres S_i, $i \geq 2$. Take an $(n-1)$-simplex d^{n-1} in D^{n-1} such that $\bar{d}^{n-1} \subset D^{n-1}$. Since S_1 is cylindrical, there exists a homeomorphism f of S^n onto itself taking S_1 into the equator Σ. f can be chosen so that $f(d^{n-1}) = \delta^{n-1}$ is a spherical simplex in Σ. Indeed, let $f': S^n \to S^n$ be an arbitrary homeomorphism such that $f'(S_1) = \Sigma$. The boundary of d^{n-1} is cylindrically imbedded in D^{n-1}, therefore also in S_1, so that $f'(\operatorname{Fr} d^{n-1})$ is a cylindrically imbedded $(n-2)$-sphere in Σ. Hence there exists a homeomorphism $h: \Sigma \to \Sigma$ such that

$$hf'(\operatorname{Fr} d^{n-1}) = \operatorname{Fr} \delta^{n-1}, \quad hf'(d^{n-1}) = \delta^{n-1}$$

where δ^{n-1} is a spherical $(n-1)$-simplex in Σ. The homeomorphism h defined on Σ extends to a homeomorphism $\bar{h}: S^n \to S^n$, the meridian through any point $x \in \Sigma$ going into the meridian through $h(x)$. Setting $f = \bar{h} f'$, we thus obtain a homeomorphism $f: S^n \to S^n$ such that

$$f(S_1) = \Sigma, \; f(d^{n-1}) = \delta^{n-1}. \tag{18}$$

In the domain U_1 we take an n-simplex d^n with base d^{n-1}; and in the domain $f(U_1)$, an n-simplex δ^n with base δ^{n-1} (Figure 5.4). We shall construct the homeomorphism $F_1: S^n \to S^n$ so that it takes S_1 into $\operatorname{Fr} d^n$ and is the identity outside an arbitrarily close neighborhood of $\bar{U}_1$ and on an n-simplex $d' \subset d^n$. Set

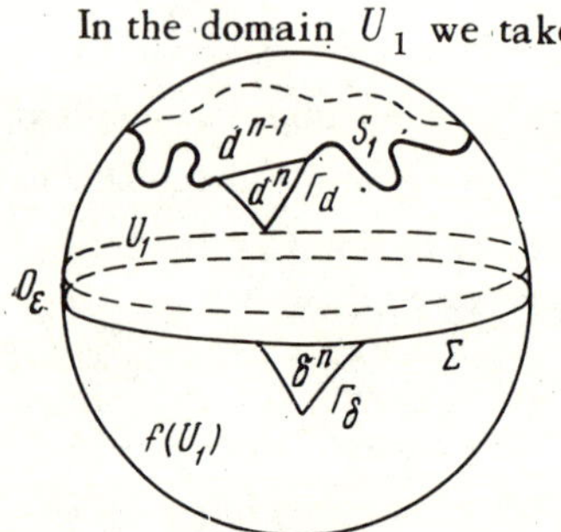

Figure 5.4

$$\Gamma'_d = \operatorname{Fr} d^n \setminus d^{n-1}, \; \Gamma_\delta = \operatorname{Fr} \delta^n \setminus \delta^{n-1}.$$

We first define two auxiliary homeomorphisms $\phi: S^n \to S^n$ and $\psi: S^n \to S^n$ such that

$$\begin{aligned} \varphi(d^{n-1}) &= \Gamma_d, \quad \varphi |_{S_1 \setminus d^{n-1}} = 1, \\ \psi(\delta^{n-1}) &= \Gamma_\delta, \quad \psi |_{\Sigma \setminus \delta^{n-1}} = 1. \end{aligned} \tag{19}$$

To construct ϕ (the construction is similar for ψ) we select two n-simplices D^n and $\tilde{D}^n$, each with base d^{n-1}, such that

$$d^n \subset D^n \subset U_1, \quad \tilde{D}^n \subset S^n \setminus \overline{U}_1;$$

the homeomorphism ϕ, which is to take d^{n-1} into Γ_d and be the identity on $S^n \setminus (D^n \cup \tilde{D}^n)$, is defined by formulas (8) of Chapter I, where a, b, c, d are points lying on a line, with

$$a \in \operatorname{Fr} \tilde{D}^n, \quad b \in d^{n-1}, \quad c \in \Gamma_d, \quad d \in \operatorname{Fr} D^n.$$

Next we construct a homeomorphism $\Phi: S^n \to S^n$ such that

$$\Phi(S_1) = \Sigma, \quad \Phi(d^n) = \delta^n.$$

Define Φ first on $S^n \setminus d^n$, by setting

$$\Phi \big|_{\overline{U_1 \setminus d^n}} = \psi f \varphi^{-1}, \qquad \Phi \big|_{S^n \setminus U_1} = f. \tag{20}$$

Φ is continuous, since

$$\overline{(U_1 \setminus d^n)} \cap (S^n \setminus U_1) = S_1 \setminus d^{n-1}$$

and, by (18) and (19),

$$\psi f \varphi^{-1} \big|_{S_1 \setminus d^{n-1}} = f.$$

From (20) we have

$$\Phi(S_1) = \Sigma, \quad \Phi(d^{n-1}) = \delta^{n-1}, \tag{21}$$

$$\Phi(\Gamma_d) = \Gamma_\delta, \quad \Phi(U_1 \setminus d^n) = f(U_1 \setminus \delta^n).$$

Consequently the homeomorphism Φ can be extended to d^n so that

$$\Phi(d^n) = \delta^n. \tag{22}$$

Since $f(\overline{U}_1)$ is a standard ball, and $\delta^n \subset f(U_1)$ is a spherical simplex, we can define, again by means of formulas (8) of Chapter I, a homeomorphism $H: S^n \to S^n$ which is the identity outside an arbitrary neighborhood O_ϵ of the ball $f(\overline{U}_1)$ and on a neighborhood $O(\delta^{n-1})$, and for which

$$H(\Sigma \setminus \delta^{n-1}) = \Gamma_\delta. \tag{23}$$

Pick the neighborhood O_ϵ of $f(\overline{U}_1)$ so that

$$f^{-1}(O_\varepsilon) \cap S_i = \Lambda, \quad i > 1. \tag{24}$$

Now set

$$F_1 = \Phi^{-1} H \Phi.$$

It follows from (21)–(23) that

$$F_1(U_1) = d^n, \qquad F_1(S_1) = \operatorname{Fr} d^n,$$

$$F_1 \big|_{\Phi^{-1}(O\delta^{n-1})} = 1, \qquad F_1 \big|_{\Phi^{-1}(S^n \setminus O_\varepsilon)} = 1. \tag{25}$$

By (20) we have $\Phi^{-1}(S^n \setminus O_\epsilon) = f^{-1}(S^n \setminus O_\epsilon)$, and $\Phi^{-1}(O\delta^{n-1})$ is a neighborhood of d^{n-1}. Hence $\Phi^{-1}(O\delta^{n-1}) \cap U_1$ contains an n-simplex d' with base d^{n-1}, and, by (25),

$$F_1|_{d'} = 1, \qquad F_1|_{S^n \setminus f^{-1}(O_\varepsilon)} = 1. \tag{26}$$

Thus, in view of (24), we have satisfied (17) for $r = 1$, and

$$F_1(S_i) = S_i, \quad i > 1.$$

2. *Construction of* F_r, $1 < r \leq k$. Suppose we have constructed a stable homeomorphism F_{r-1} so as to satisfy (17) with $r - 1$. We are to construct a stable homeomorphism F_r so as to satisfy (17) for r. We have

$$F_{r-1}(S_j) = S_j, \; j \geqslant r. \tag{17'}$$

Consider the sphere $S_r = F_{r-1}(S_r)$ and the domain U_r bounded by it. Because of the way the spheres have been numbered we have $U_r \cap S_j = \Lambda$ for $j > r$, but U_r may contain some simplices $\delta_i^n = F_{r-1}(U_i)$, $i < r$. Let $\delta_{i_1}^n, \cdots, \delta_{i_s}^n$ be these simplices, and D^{n-1} and $d^{n-1} \subset \bar{d}^{n-1} \subset D^{n-1}$ spherical simplices in S_r. Take an n-simplex d^n in U_r, with base d^{n-1}; we construct a homeomorphism $h\colon S^n \to S^n$ such that

$$h|_{S^n \setminus U_r} = 1, \qquad h(\delta_{i_\rho}^n) = \delta_\rho'^n \subset d^n, \qquad 1 \leqslant \rho \leqslant s,$$

where each $\delta_\rho'^n$ is a spherical simplex.

To do so, it suffices to select s disjoint simplices $\delta_\rho'^n$ in d^n, and construct s domains v_ρ in U_r such that

$$\delta_{i_\rho}^n \cup \delta_\rho'^n \subset v_\rho \subset \bar{v}_\rho \subset U_r, \quad \bar{v}_\rho \cap \bar{v}_{\rho'} = \Lambda, \qquad \rho \neq \rho'.$$

There exists a homeomorphism of a ball Q^n onto itself which is the identity on the boundary and takes a ball $\tilde{Q}^n$, contained in Q^n and concentric with Q^n, into an n-simplex inside $\tilde{Q}^n$. Using this fact and Theorem 2.2 of Chapter I, it is easy to construct, for each $\bar{v}_\rho$, a homeomorphism $\phi_\rho\colon \bar{v}_\rho \to \bar{v}_\rho$ which is the identity on $\operatorname{Fr} v_\rho$ and takes the simplex $\delta_{i_\rho}^n$ into the simplex $\delta_\rho'^n$. Then set

$$h|_{v_\rho} = \varphi_\rho, \quad 1 \leqslant \rho \leqslant s; \quad h|_{S^n \setminus \cup v_\rho} = 1.$$

The composite mapping $F'_{r-1} = hF_{r-1}$ is again a stable homeomorphism, and

$$F'_{r-1}|_{S_j} = 1, \quad j \geqslant r; \quad F'_{r-1}(U_{i_\rho}) = \delta_\rho'^n \subset d^n, \quad 1 \leqslant \rho \leqslant s. \tag{27}$$

Now take an n-simplex d'^n in U_r with base d^{n-1}, such that $d^n \subset d'^n$, and a neighborhood $O(\bar{U}_r)$ such that

$$O(\bar{U}_r) \cap S_j = \Lambda, \; j > r, \tag{28}$$

and, in the same way as F_1 was constructed for S_1, construct a homeomorphism $f_r\colon S_n \to S_n$ such that

$$f_r(S_r) = \operatorname{Fr} d'^n, \quad f_r(U_r) = d'^n, \qquad f_r|_{d^n} = 1, \quad f_r|_{S^n \setminus O(\bar{U}_r)} = 1. \tag{29}$$

Set $F_r = f_r F'_{r-1}$. Then F_r is a stable homeomorphism of S^n onto itself. Furthermore, by (17$'$), (27) and (29), we have

$$F_r(S_r) = \operatorname{Fr} d'^n, \quad F_r(U_r) = d'^n,$$

$$F_r(U_{i_\rho}) = \delta'^n_\rho, \quad F_r(U_i) = F_{r-1}(U_i), \quad i \neq i_\rho, \quad i \neq r.$$

Consequently (17) is satisfied for F_r.

Thus in k steps we construct the required stable homeomorphism F_k.

Corollary 1. *Let $S_1, \cdots, S_k$ be a finite number of disjoint cylindrical $(n-1)$-spheres in S^n, such that each S_i contains a spherical $(n-1)$-simplex. Then there exists an isotopy of S^n onto itself from the identity mapping, taking each S_i into the boundary of a spherical n-simplex:*

$$F_t\colon S^n \to S^n, \quad F_0 \equiv 1, \qquad F_1(S_i) = \operatorname{Fr} d_i^n, \qquad 1 \leqslant i \leqslant k.$$

This follows from Theorem 3.5 and the corollary to Alexander's theorem (3.1).

Corollary 2. *Let S_1 and S_2 be two disjoint cylindrical $(n-1)$-spheres in S^n, such that each contains a spherical $(n-1)$-simplex. Then $[S_1, S_2]$ is an annulus.*

The complement in S^n of a pair of disjoint spherical n-simplices σ_1 and σ_2 is an annulus, since there exists a homeomorphism $h\colon S^n \to S^n$ which is the identity on σ_2 and takes σ_1 into the simplex σ_1' symmetric to σ_2 with respect to the equator of S^n, where the center of σ_2 is a pole. Hence Corollary 2 follows from Corollary 1.

§4. Local conditions

1. **Definitions.** In the examples of Chapter IV, a "bad" imbedding in E^3 of a manifold or of a manifold with boundary is a locally "bad" imbedding at certain points. Proofs are now known, for spheres of dimension $k \neq n-2$ and for k-cells, $k \le n$, that under the assumption of local flatness an imbedding of I^k, $k \le n$, into E^n (or S^n) or of S^k into E^n (or S^n), $k \neq n-2$, is trivial. For a sphere S^{n-1} this is a theorem of Brown, which we shall prove here. For a sphere S^k, $k \le n-3$, it is a theorem of Stallings, and this case is much harder than the case $k = n-1$; Stallings' proof [73] was published in 1963. Finally, the triviality of the imbedding in E^n of a locally flat k-cell, $k \le n$, was proved by Černavskiĭ [14]

by elementary means. Brown and Michael observed that certain notions pertaining to an imbedding are relevant to the study not only of the imbedding of manifolds in E^n or S^n, but, more generally, of that of certain closed sets in a metric space, although the most important results are obtained in the case of imbeddings in a manifold or a manifold with boundary. Let us now state the basic definitions and consider the connections between them.

Definition 4.1. A manifold (with boundary) N^k, contained in an n-dimensional manifold M^n, is called *locally flat* at a point $x \in N^k$ if there exists a neighborhood u_x of x in M^n and a homeomorphism ϕ of u_x into E^n such that

$$\varphi (N^k \cap u_x) \subset E^k, \tag{30}$$

where E^k is a k-dimensional hyperplane in E^n.

Observe that if x is an interior point of N^k (i.e., is not on the boundary dN^k), then (30) is equivalent to the condition

$$\varphi (N^k \cap u_x') = E^k, \quad \varphi (u_x') = E^n, \quad x \in \operatorname{Int} N^k \tag{31}$$

Indeed, $\phi(N^k \cap u_x) \supset q_x^k$, and $\phi(u_x) \supset q_x^n$, where $q_x^n = q_x^k \times q_x^{n-k}$, and q_x^n, q_x^k, q_x^{n-k} are cubes with center at x, with $q_x^k \subset E^k$ and q_x^{n-k} orthogonal to E^k. Setting $u_x' = \phi^{-1}(\operatorname{Int} q_x^n)$ gives (31).

Definition 4.2. By a *pair* (X, Y) we mean a space X and a subset Y of X; we say that the pair (X, Y) is *homeomorphic* to the pair (X', Y') if there exists a homeomorphism H of X onto X' such that $H(Y) = Y'$. Condition (31), when N^k is locally flatly imbedded at the point x, amounts to the statement that the pair $(u_x', u_x' \cap N^k)$ is homeomorphic to the pair (E^n, E^k).

By Corollaries 2 and 3 to Theorem 1.3, to prove Brown's theorem that a locally flat imbedding of a sphere S^{n-1} into S^n is trivial, it suffices to show that S^{n-1} is imbedded in S^n together with a neighborhood.

Stallings showed in [73] that for any $k < n - 2$, $n \geq 5$, if S^k is locally flat in S^n, then S^k is imbedded in S^n together with a neighborhood (the case of a knotted S^{n-2} in S^n, $n > 3$, remains unresolved). Below, we give examples of the imbedding of a sphere in a manifold where this is not the case.

We have already considered a cylindrical imbedding of a sphere S^{n-1} into S^n. This notion can be defined for the imbedding of subsets of an arbitrary space.

Definition 4.2′. A subset Y of a space X is called *cylindrical* in X if it has a neighborhood in X homeomorphic to the topological product $Y \times (-1, +1)$, with $Y = Y \times 0$.

In other words, there exists a homeomorphism $h: Y \times (-1, +1) \to U(Y)$ onto a neighborhood of Y in X, with $h(Y \times 0) = Y$.

Definition 4.3. A subset Y of a space X is called *locally cylindrical* at a point $y \in Y$ if there exists a neighborhood u_y of y in X such that the pair $(u_y, Y \cap u_y)$ is homeomorphic to the pair

$$[(Y \cap u_y) \times (-1, +1), (Y \cap u_y) \times 0]. \tag{32}$$

It is easily seen that in the case of an imbedding of an $(n-1)$-manifold M^{n-1} in an n-manifold M^n the notions of locally flat imbedding and of locally cylindrical imbedding are equivalent.

Indeed, if M^{n-1} is locally flat at a point x there exists a neighborhood u_x' of x such that (31) holds; consequently $M^{n-1} \cap u_x'$ is homeomorphic to E^{n-1} and u_x' is homeomorphic to E^n, i.e. to $(M^{n-1} \cap u_x') \times (-1, +1)$. Conversely, if M^{n-1} is locally cylindrical at a point y, then since M^{n-1} is a manifold, the neighborhood u_y can be chosen so that $u_y \cap M^{n-1}$ is homeomorphic to E^{n-1}, so that the homeomorphism of the pair $(u_y, M^{n-1} \cap u_y)$ to the pair (32), with $Y = M^{n-1}$, implies (31).

In the general case of an $(n-1)$-manifold imbedded in an n-manifold or in an n-manifold with boundary, and even in the case of a sphere S^{n-1} imbedded in a manifold M^n, a locally cylindrical imbedding need not be cylindrical.

Example 4.1. The circle C which is the equator of the projective plane P^2 (or of the Moebius strip) is imbedded locally cylindrically in P^2 (so also locally flatly), but not cylindrically; the imbedding of C in P^2 is not an imbedding with neighborhood (Definition 1.4), since no neighborhood of C in P^2 is separated by C, in contrast to the case of a circle in the sphere S^2.

Example 4.2. A locally flat imbedding of a circle in a three-dimensional manifold, which is not an imbedding with neighborhood. Let M^3 be the 3-manifold obtained from the solid torus by identification of diametrically opposite points on the boundary of every meridional cross section, and C the circle in M^3 obtained as a result of this identification from the boundary of a particular meridional cross section θ. The imbedding of C in M^3 is locally flat at every point; the neighborhood u_x can be taken as the topological product of a line segment orthogonal to θ by a square which is a neighborhood of x in the projective plane P^2 obtained from θ as a result of the identification (C is the equator of P^2). But this imbedding of C in M^3, like the imbedding of C in P^2, is not an imbedding with neighborhood. Note, however, that the projective plane P^2 is imbedded cylindrically in M^3 (just as is the disk θ in the solid torus).

2. **Locally semicylindrical imbedding.** Brown and Michael observed that in

studying imbeddings there is another useful property which enables one, under broad assumptions, to pass from a local condition to a global one.

Definition 4.4. A subset Y of a space X is called *semicylindrical* in X if it has a neighborhood in X which is homeomorphic to the topological product $Y \times [0, 1)$, where $Y = Y \times 0$.

Thus, for example, the boundary of a ball is semicylindrically imbedded in the ball.

Definition 4.5. A subset Y of a space X is called *locally semicylindrical* at a point $y \in Y$ if there exists a neighborhood u_y of y in X such that the pair $(u_y, u_y \cap Y)$ is homeomorphic to the pair

$$[(Y \cap u_y) \times [0, 1), (Y \cap u_y) \times 0].$$

Brown and Michael showed that for metric spaces a locally semicylindrical imbedding is semicylindrical [35]. This theorem has important applications in the imbedding of manifolds. The proof we present here was suggested by A. V. Černavskiĭ for the imbedding of a metric space with countable base into a metric space. The proof in this case is almost no different from that for the imbedding of a manifold into a manifold.

If Y is a metric space with countable base, then so are $Z = Y \times [0, 1)$ and $\bar{Z} = Y \times [0, 1]$. We denote a point of the space $\bar{Z}$ by (y, t), where $y \in Y$ and $0 \leq t \leq 1$.

A metric in $\bar{Z}$ is given by

$$\rho[(y, t), (y', t')] = \sqrt{\rho^2(y, y') + (t - t')^2},$$

where $\rho(y, y')$ is the distance between y and y' in Y.

Lemma 4.1. *Let h: $Y \times [0, 1] \to X$ be a homeomorphism into X, let Y be a metric space, and T a subset of $Y \times [0, 1)$ such that for any point $(y, 0) \in Y$, $\rho((y, 0), T) > 0$. Then there exists a homeomorphism h': $Y \times [0, 1] \to X$ such that*

$$h' |_{Y \times 0} \equiv h, \; h'(Y \times [0, 1]) \cap h(T) = \Lambda.$$

Define on Y the function $\mu(y) = \frac{1}{2} \min\{\rho((y, 0), T), 1\}$. The function $\mu(y)$ is continuous and does not exceed half the length of the line segment $[(y, 0), (y, t_0)]$ from the point $(y, 0)$ to the nearest point (y, t_0) in T, and $0 < \mu(y) \leq \frac{1}{2}$. Hence the set of all points (y, t) such that $t < \mu(y)$ is an open subset W of the space $Y \times [0, 1]$ containing $Y \times 0$, and $\bar{W} \cap T = \Lambda$. Define the homeomorphism π: $Y \times [0, 1] \to \bar{W}$ by $\pi(y, t) = (y, t\mu(y))$.

Setting $h' = h\pi$, we obtain the required homeomorphism.

Theorem 4.1 (**Brown and Michael** [34, 35]). *Let Y be a metric locally compact space with countable base, and $H\colon Y \to X$ a topological imbedding of Y into an arbitrary space X. If $H(Y)$ is locally semicylindrical at every point, then $H(Y)$ is semicylindrical in X.*

Let $Z = Y \times [0, 1]$, and let $H\colon Y \times 0 \to X$ be the given imbedding. Since the imbedding is semicylindrical at each point of $H(Y)$ and Y is locally compact, there exists for each point $y \in Y$ a neighborhood u_y' such that $\bar{u}_y'$ is compact and a homeomorphism $h_y\colon u_y' \times [0, 1] \to X$ such that

$$h_y|_{u_y' \times 0} \equiv H \tag{33}$$

and $h_y(u_y' \times [0, 1))$ is open in X.

Let $\{u_i'\}$ be a countable covering of Y refining the covering $\{u_y'\}$ and let $h_i\colon u_i' \times [0, 1] \to X$ be a homeomorphism satisfying condition (33). Assume that $\mathcal{E}^x = \mathcal{E} \times [0, 1]$, where $\mathcal{E} \subset Y$. We shall show that if $h_i\colon u^x \to X$ defines a semicylindrical imbedding in X of the set u^x, open in Y, and if $K \subset u^x$, where K is compact, then there exists a semicylindrical imbedding $h'\colon u^x \to X$ such that $h'(u^x) \subset h(u^x)$, $h'|_{u\times 0} = Hu$:

$$h'(K^x) \cap h'(Y \setminus K) = \Lambda. \tag{34}$$

Since h is defined on u^x, we have

$$\mathcal{P} = h(K^x) \cap H(Y \setminus K) = h(K^x) \cap H(Y \setminus u).$$

But $\mathcal{P}$ is a closed subset of $H(Y)$, so that its closure $\bar{\mathcal{P}}$ in X does not intersect $H(u)$. Then $T = h^{-1}(\bar{\mathcal{P}})$ is closed in u^x and $T \cap u = \Lambda$, so that by Lemma 4.1 there exists a homeomorphism $h'\colon u^x \to X$ such that $h'(u^x) \cap \bar{\mathcal{P}} = \Lambda$. If h' is the homeomorphism obtained from $h\colon u^x \to X$ by means of Lemma 4.1, then for an arbitrary set $\mathcal{E} \subset u$ we have

$$h'(\mathcal{E}^x) = h(\mathcal{E}^x) \cap h'(u^x) \subset h(\mathcal{E}^x), \tag{35}$$

since $h'(y \times [0, 1]) \subset h(y \times [0, 1])$ for $y \in u$. For $\mathcal{E} = K$ this result implies (34). The condition (34) holds for an *arbitrary* set $R \subset K$, since $h(R^x) \cap h(K \setminus R)^x = \Lambda$, and for an arbitrary homeomorphism $\tilde{h}$, if $\tilde{h}(R^x) \subset h(R^x)$.

Let $\{u_n\}$ be a covering of Y with $u_n \subset \bar{u}_n \subset u_n'$. If the $\bar{u}_n'$ are compact, then the $\bar{u}_n$ are also compact, and thus there exist homeomorphisms $h_n'\colon u_n'^x \to X$ satisfying (34) with $K = \bar{u}_n$. We shall show that for an arbitrary u_n and a finite number of sets u_i, $i \le N$ there exists a homeomorphism $h_n\colon u_n^x \to X$ such that

$$h_n(u_n^x) \cap h_i'(\bar{u}_i \setminus u_n)^x = \Lambda \tag{36}$$

$$i \le N.$$

Let $\mathcal{P} = h_n'(u_n^x) \cap [\bigcup_{i=1}^N h_i'(\bar{u}_i \setminus u_n)^x]$. Since $\bar{u}_i \setminus u_n$ is compact, $h_i'(\bar{u}_i \setminus u_n)^x$ is also compact, so that $\mathcal{P}$ is closed in $h_n'(u_n^x)$. Moreover, by (34) we have

$\mathcal{P} \cap H(u_n) = \Lambda$ for $h' = h_i'$, $K = \bar{u}_i \backslash u_n \subset \bar{u}_i$. Setting $T = h_n'^{-1}(\mathcal{P})$, we can construct, by Lemma 4.1, a homeomorphism $h_n: u_n^x \to X$ such that $h_n(u_n^x) \cap \mathcal{P} = \Lambda$, which implies (36), since $h_n(u_n^x) \subset h_n'(u_n^x)$. Then h_n satisfies (34) with $K = u_n$. If the homeomorphisms h_n are defined for all $n \leq N$, then by (35) and by the compactness of $h_i'(\bar{u}_i \backslash u_n)^x$ we have $\overline{h_i(u_i \backslash u_n)^x} \subset h_i'(\bar{u}_i \backslash u_n)^x$ and from (36) it follows that

$$h_n(u_n^x) \cap \overline{h_i(u_i \backslash u_n)^x} = \Lambda \tag{37}$$
$$i, n \leq N.$$

2. We shall show that if $A \subset Y$ is compact, $H(A)$ has a neighborhood of X which is cylindrically imbedded in $H(Y)$. Then A is contained in a finite number of elements u_i: $A \subset \bigcup_{i=1}^N u_i = w$ (for convenience we renumber the u_i). Let $h_i: u_i^x \to X$ be a homeomorphism satisfying (33) and (34) with $K = u_i^x \subset \bar{u}_i^x$ and (37) for an arbitrary pair (n, i), $n, i \leq N$. We shall show that $H(w)$ is semicylindrical in X.

Let

$$V = \bigcup_{i=1}^{N} h_i(u_i^x) \quad \text{and} \quad V' = \{H(w) \times [-1, 0]\} \bigcup_H V,$$

where V' is obtained by identification of the points $H(y) \times 0$ in $H(w) \times [-1, 0]$ and $H(y)$ in V. We construct a homeomorphic mapping $F: V' \xrightarrow{\text{onto}} V$ satisfying the condition

$$F(H(w) \times -1) = H(w) \times 0. \tag{38}$$

The homeomorphisms $h_i: u_i^x \to \bar{V}$ can be continued to a homeomorphic mapping $h_i: u_i^x \times [-1, 1] \to \bar{V}'$ by setting

$$h_i(y, t) = [H(y), t], \quad -1 \leq t \leq 0, \quad y \in u_i.$$

For every i, $1 \leq i \leq N$, let us define on w a continuous function f_i equal to the distance from a given point to $w \backslash u_i$. Then

$$f_i\big|_{u_i} = 0; \quad f_i\big|_{w \backslash u_i} > 0.$$

Since the covering $w = \bigcup_{i=1}^N u_i$ is finite, at each point w the following function is defined and continuous:

$$f(y) = \sum_{i=1}^{N} f_i(y);$$

and at an arbitrary point $y \in w$ we have $f_i(y) > 0$. The functions $\phi_i = f_i/f$ form a partition of unity corresponding to the covering $\{u_i\}$, since

$$0 \leq \phi_i(y) \leq 1; \qquad \phi_i|_{w \setminus u_i} = 0; \quad \sum_{i=1}^{N} \phi_i(y) = 1. \tag{39}$$

Let θ_i be the homeomorphic mapping of $(w \cap \overline{u}_i) \times [-1, +1]$ onto itself defined by the formulas:

$$\theta_i(y, t) = (y, t + \phi_i(y)), \text{ if } -1 \leq t \leq -\phi_i(y);$$

$$\theta_i(y, t) = \left[y, \frac{t + \phi_i(y)}{1 + \phi_i(y)}\right], \text{ if } -\phi_i(y) \leq t \leq 1. \tag{40}$$

In the homeomorphism θ_i every interval $y \times [-1, -\phi_i(y)]$ is shifted by an amount $\phi_i(y) \leq 1$ and θ_i is the identity on $u_i \times 1$ and $[(\overline{u}_i \setminus u_i) \cap w] \times [-1, +1]$, since $\phi_i|_{(\overline{u}_i \setminus u_i) \cap w} = 0$. Here, since $\theta_i(y, -1) = (y, t)$, $t \leq 0$, we have

$$\theta_i(u_i \times [-1, 1]) \supset u_i^x \supset \theta_i(u_i^x). \tag{41}$$

For each i, $1 \leq i \leq N$, we define the homeomorphic mapping Φ_i: $V' \to V'$ by the formulas

$$\Phi_i(x) = x, \text{ if } x \in V' \setminus h_i(u_i \times [-1, 1))$$

$$\Phi_i(x) = h_i \theta_i h_i^{-1}(x), \text{ if } x \in h_i(u_i \times [-1, 0)). \tag{42}$$

We shall show that this definition is correct, i.e. for a sequence of points $x_n \in h_i(u_i \times [-1, 0))$ tending to the point $x \in V' \setminus h_i(u_i \times [-1, 0))$ we have $\lim_{x_n \to x} \Phi_i(x_n) = x$. This follows from the formulas (40) if $x \in h_i(u_i \times 1)$ or $x = (H(y), t)$, $t < 0$, $y \in w \cap (\overline{u}_i \setminus u_i)$, since then $\phi_i(y) = 0$ and $\lim_{x_n \to x} \theta_i h_i^{-1}(x_n) = \lim h_i^{-1}(x_n)$. If $x \in V \cap [\overline{h_i(u_i^x)} \setminus h_i(u_i^x)]$, then for some $n \leq N$ we have

$$x \in h_n(u_n^x) \cap [\overline{h_i(u_i^x)} \setminus h_i(u_i^x)] \subset h_n(u_n^x) \cap h_i'(\overline{u}_i \setminus u_i)^x,$$

and by (36) we also have $x = h_i'(y, t)$, where $y \in u_n \cap (\overline{u}_i \setminus u_i)$. Consequently, $\theta_i(y, t) = (y, t)$ and $\lim_{x_n \to x} \Phi_n(x_n) = x$.

By (41) and (42) we see that

$$V' \supset \Phi_i(V') \supset V \supset \Phi_i(V). \tag{43}$$

Let us consider the sequence of homeomorphic mappings of V' onto itself: $F_1 = \Phi_1$, $F_2 = \Phi_2\Phi_1, \cdots, F_N = \Phi_N \Phi_{N-1} \cdots \Phi_1$. For arbitrary $i \leq N$ we have

$$V' \supset F_i(V') \supset V \supset F_i(V). \tag{44}$$

This is true for $i = 1$. We shall show that (44) for i implies (44) for $i + 1$. The homeomorphism Φ_{i+1} takes the point $z = h_{i+1}(y, t)$, $y \in u_{i+1}$ into the point $h_{i+1}(y, t')$, $t' > t$ and $\Phi_{i+1}(z) = z$, if $z \in V' \setminus h_{i+1}(u_{i+1}^x)$, from which it follows that $F_i(V') \supset F_{i+1}(V')$ and $F_i(V) \supset F_{i+1}(V)$. By (39) and (42):

$$F_{i+1}(H(y), -1) = \left[H(y), -1 + \sum_{j=1}^{i+1} \phi_j(y)\right]. \tag{45}$$

Consequently, $F_{i+1}(H(y), -1) = (H(y), t)$, $t \leq 0$, from which it follows that $F_{i+1}(V') \supset V$ and the proof of (44) is complete. From (45), (34) and (39) we have $F_N(H(y), -1) = (H(y), 0)$, which means that the homeomorphism F_N satisfies (38) and $F_N(V') = V$. The set $H(w) \times [-1, 0)$ is open in V', so that $F_N[H(w) \times [-1, 0)]$ is open in V, and therefore also in X, so that $H(w)$ is semicylindrical in X.

3. Since Y is locally compact, we have

$$Y = \bigcup_{n=1}^{\infty} u_n, \quad \bar{u}_n \subset u_{n+1},$$

where u_n is open in Y and $\bar{u}_n$ is compact. Then $K_n = \bar{u}_n \backslash u_{n-1}$ is compact, and by what has been proved above this K_n has a neighborhood v_n' semicylindrically imbedded in X. Since $K_i \cap K_j = \Lambda$ if $|i - j| > 1$, v_n' can be so chosen that

$$v_i' \cap v_j' = \Lambda, \quad |i - j| > 1. \tag{46}$$

In each v_n' let us choose a neighborhood v_n such that $K_n \subset v_n \subset \bar{v}_n \subset v_n'$. By (46) we have $v_n' \subset u_{n+1}$, so that $\bar{v}_n$ is compact. Let us choose homeomorphisms h_n': $v_n'^x \to X$ such that (34) is satisfied for $K = \bar{v}_n$.

For each n let us construct a homeomorphism h_n: $v_n^x \to X$ such that (36) is satisfied for $i \leq n + 1$. We shall show that an arbitrary pair of homeomorphisms h_n, h_i satisfies (37). For $i \leq n + 1$ this follows from the fact that (36) is satisfied by construction and $\overline{h_i(u_i \backslash u_n)^x} \subset h_i'(\bar{u}_i \backslash u_n)^x$. If $i \geq n + 2$, then $\bar{v}_n \cap \bar{v}_i = \Lambda$, and therefore $h_n'(\bar{v}_n^x) \cap h_i'(\bar{v}_i^x)$ is compact and does not intersect $\bar{v}_n \cup \bar{v}_i$, and by the construction of h_i we obtain from Lemma 4.1:

$$\overline{h_n'(v_n)} \cap \overline{h_n(v_i^x)} = \Lambda,$$

from which (37) follows for $i \geq n + 2$.

Since $\{v_n\}$ is a covering of Y of order 2, we can define homeomorphisms Φ_n: $V' \to V$ such that for a neighborhood of the compact with $V = \bigcup_{n=1}^{\infty} h_n(v_n^x)$ the homeomorphisms $F_n = \Phi_n \Phi_{n-1} \cdots \Phi_1$ also satisfy the conditions (44) and (45). We shall show that the sequence $\{F_n\}$ converges to the homeomorphic mapping F: $V' \to V$ satisfying (38) for $w = Y$. Let $z \in V'$, and let n be the smallest number for which $z = h_n(y, t)$, $-1 \leq t < 1$ and $y \in v_n$. Then $F_{n-1}(z) = z$. If $y \in v_n \backslash v_{n-1} \backslash v_{n+1}$, then from $h_i(y, t) = [H(y), t]$, $t \leq 0$ and (37) if follows that

$$h_n[(v_n \backslash v_{n+k}) \times [-1, 1)] \cap h_{n+k}(v_{n+k} \times [-1, 1)) = \Lambda, \quad k \geq 1.$$

Thus $F_{n+k}(z) = F_n(z)$, $k \geq 1$.

If $y \in v_n \cap v_{n+1}$, then $F_{n+k}(z) = F_{n+1}(z)$, $k \geq 2$, since by (37) and (45) we have $F_{n+1}(z) \in h_n(v^x_n \cap v^x_{n+1}) \cup h_{n+1}(v^x_n \cap v^x_{n+1})$ and $F_{n+1}(z) \notin h_{n+k}(v^x_{n+k})$, $k \geq 2$. Thus the sequence $\{F_n\}$ converges to the mapping F defined by the equality

$$F\Big|_{\bigcup_{i=1}^{n} h_i(v_i \times [-1, 1))} = F_{n+1}. \tag{47}$$

Here F is continuous and one-to-one. The continuity of F^{-1} follows from the fact that in the neighborhood of an arbitrary point $z \in V'$ the mapping F^{-1} coincides with some F_n, so that F is a homeomorphism. Since $H(Y \times [-1, 0))$ is open in V', therefore $F(H(Y) \times [-1, 0))$ is open in V, and thus also in X, so that $H(Y)$ is semicylindrical in X.

3. Consequences of Theorem 4.1. This theorem has important applications to the imbedding of manifolds.

Corollary 1. *In a manifold with boundary, the boundary is imbedded semicylindrically.*

It suffices to show that the boundary ∂N^n of an n-manifold with boundary, N^n, is imbedded semicylindrically at every point. Recall that the boundary is the set of points of N^n which do not have neighborhoods in N^n homeomorphic to E^n, but have neighborhoods whose closures are homeomorphic to an n-cube.

Take $x \in \partial N^n$ and let u_x be a neighborhood such that $\bar{u}_x$ is an n-cell. Since x has no neighborhood homeomorphic to E^n, x must be on the boundary of the n-cell $\bar{u}_x$. There exists a neighborhood u'_x of x such that $\bar{u}'_x \subset u_x$; u'_x intersects $\partial \bar{u}_x$ in a set v_x which is open in the latter, and which obviously lies in ∂N^n and is a neighborhood of x in ∂N^n.

Then v_x is semicylindrically imbedded in u_x, therefore also in N^n. Consequently ∂N^n is an $(n-1)$-manifold, whose imbedding in N^n is locally cylindrical at every point, so that by Theorem 4.1 ∂N^n is cylindrically imbedded in N^n.

Now consider the imbedding of an $(n-1)$-manifold M^{n-1} into an n-manifold M^n. M^{n-1} is said to be two-sided in M^n if it separates every sufficiently close neighborhood of itself in M^n. Otherwise M^{n-1} is said to be one-sided. Observe that two-sidedness is not an intrinsic property of the manifold M^{n-1}, but only characterizes the imbedding of M^{n-1} in M^n (see Example 4.2).

If M^{n-1} is imbedded in M^n cylindrically, then M^{n-1} is two-sided in M^n, since $M^{n-1} \times 0$ separates the neighborhood $M^{n-1} \times (-1, +1)$. Thus, for one-sided manifolds M^{n-1} in M^n, local cylindricalness cannot imply cylindricalness.

Corollary 2. *If the imbedding of a manifold M^{n-1} in a manifold M^n is two-sided and locally cylindrical at every point, then the imbedding of M^{n-1} in M^n*

is cylindrical [35].

Since M^{n-1} is two-sided, it separates some neighborhood U in M^n:

$$U \setminus M^{n-1} = U_{-1} \cup U_{+1};$$

U and U_i, $i = -1, +1$, are n-manifolds. We assert that each $\bar{U}_i = U_i \cup M^{n-1}$ is an n-manifold with boundary M^{n-1}. It suffices to show that every point $x \in M^{n-1}$ has a neighborhood in $\bar{U}_i$ whose closure is an n-cell. By assumption, M^{n-1} is locally cylindrical at x. This means that there exists a neighborhood u_x of x in U which is homeomorphic to $(u_x \cap M^{n-1}) \times (-1, +1)$, with $u_x \cap M^{n-1} = (u_x \cap M^{n-1}) \times 0$. Since M^{n-1} is a manifold, $u_x \cap M^{n-1}$ contains a neighborhood v_x^{n-1} of x in M^{n-1} which is homeomorphic to E^{n-1} and whose closure is an $(n-1)$-cell.

$\bar{U}_i$ contains a closed neighborhood of x in $\bar{U}_i$ which is homeomorphic to $\bar{v}_x \times [i/2, 0]$, $i = \pm 1$. Since $\bar{v}_x$ is an $(n-1)$-cell, $\bar{v}_x \times [-\frac{1}{2}, 0]$ and $\bar{v}_x \times [0, \frac{1}{2}]$ are n-cells; consequently $\bar{U}_i$, $i = \pm 1$, is a manifold with boundary M^{n-1}. By Corollary 1 the boundary M^{n-1} is semicylindrically imbedded in each $\bar{U}_i$. Consequently there exists a neighborhood V_i of the boundary M^{n-1} in $\bar{U}_i$ which is homeomorphic to $M^{n-1} \times [0, i]$. Then $V = V_{-1} \cup V_{+1}$ is a neighborhood of M^{n-1} in M^n which is homeomorphic to $M^{n-1} \times (-1, +1)$, with $M^{n-1} = M^{n-1} \times 0$; that is to say, M^{n-1} is cylindrically imbedded in M^n.

Corollary 3. *If a sphere S^{n-1} is locally cylindrically imbedded in a sphere S^n (or in E^n), it is trivial.*

Since S^{n-1} is a two-sided manifold in S^n (or E^n), it follows from Corollary 2 that local cylindricalness of S^{n-1} implies cylindricalness. By Corollary 3 to the Schoenflies theorem (1.3), S^{n-1} is therefore trivially imbedded in S^n.

§5. Imbedding of combinatorial and star $(n-1)$-spheres in S^n

1. As we have already indicated, it is unknown up to now whether for $n > 3$ an arbitrary polyhedral $(n-1)$-sphere in E^n is trivially imbedded. In general, for $n > 3$, arbitrary polyhedral manifolds, or equivalently, arbitrary triangulable manifolds, lend themselves poorly to study. To investigate them one must impose restrictions on the triangulation.

The most important class is that of combinatorial manifolds, already introduced in the twenties by Newman and Alexander [19]. The theory of combinatorial manifolds has in recent years been subject to a renewed and intensive development. A number of important results have been obtained. In part they are based on the methods of differential topology. We present here some basic definitions in the topology of combinatorial manifolds.

Recall that a continuous mapping f of a polyhedron P_1 onto a polyhedron P_2 is called semilinear if there exist triangulations T_1 and T_2 of P_1 and P_2 with respect to which f maps every simplex of T_1 linearly onto a simplex of T_2. If f is a homeomorphism it is called a semilinear homeomorphism, and P_2 is said to be a semilinearly homeomorphic image of P_1.

Definition 5.1. A polyhedral sphere Σ^n is called *combinatorial* if it is the semilinear homeomorphic image of the boundary of an $(n+1)$-simplex. A polyhedral n-cell is called *combinatorial* if it is the semilinear homeomorphic image of a closed n-simplex.

In other words, this says that a combinatorial sphere (cell) has a rectilinear triangulation which is isomorphic to some triangulation of the boundary of an $(n+1)$-simplex (some triangulation of an n-simplex).

Definition 5.2. A polyhedral n-mainfold M^n with or without boundary is called *combinatorial* if it has a triangulation in which the boundary of the star of any vertex is a combinatorial $(n-1)$-sphere or a combinatorial $(n-1)$-cell. Any such triangulation of M^n is called combinatorial.

For $n \leq 3$ every polyhedral sphere is combinatorial [57] and every triangulation of a polyhedral manifold with or without boundary is combinatorial. For $n > 3$ the following problems are very hard and very important.

Problem 6. Is every polyhedral n-sphere (or n-cell) combinatorial, for $n > 3$?

Problem 7. Is every polyhedral n-dimensional manifold, or manifold with boundary, combinatorial for $n > 3$?

We show now that Euclidean space E^n, a combinatorial sphere, and a combinatorial n-cell are combinatorial manifolds.

Lemma 5.1. *If the closure of a neighborhood O_x of a point x in E^n is the cone over the boundary of O_x, and the boundary is a rectilinear polyhedron P, then P is a combinatorial sphere.*

Firstly, P is a sphere, since every ray from x to P intersects the boundary S^{n-1} of a sufficiently small ball in O_x with center x, so that P is homeomorphic to S^{n-1}. We are to show that P is a combinatorial sphere, i.e. that P has a triangulation isomorphic to some triangulation of the boundary of an n-simplex. Inside S^{n-1} we take an n-simplex σ^n such that $x \in \sigma^n$ and consider the $(n-1)$-dimensional hyperplanes π_r passing through the point x and all the $(n-2)$-dimensional faces of σ^n. They divide the simplices of the given triangulation T of the polyhedron P into convex sets G_i. We construct a new triangulation T' of P. The vertices of T' are to include, to begin with, all the vertices of T, as well as the points which are intersections of the form $\sigma^p \cap \pi_{r_1} \cap \cdots \cap \pi_{r_p}$, where

$\sigma^p \in T$. Suppose we have constructed the simplices of T' which lie in the sets G_i of dimension $i \leq k-1 < n-1$; and let G_k be a k-dimensional convex set cut off by the hyperplanes π_r from a simplex $\sigma^k \in T$, or a k-dimensional intersection $\sigma^{k+p} \cap \pi_{r_1} \cdots \cap \pi_{r_p}$, where $\sigma^{k+p} \in T$. The boundary of G_k is a $(k-1)$-dimensional polyhedron in the triangulation T'. Select a point x_k in G_k, adding it to the vertices of T'. In addition, adjoin to T' all simplices in G_k of dimension $\leq k$ which are cones with vertex x_k over the simplices of the boundary of G_k. Continuing this process for all $k \leq n-1$, we arrive at a triangulation $T' = \{t_\rho\}$ of the polyhedral sphere P. The cone with vertex x over any simplex $t_\rho \in T'$ intersects only one face of the simplex σ^n, and the intersection is a simplex τ_ρ of the same dimension as t_ρ. Thus the triangulation T' of P is isomorphic to the triangulation $\{\tau_\rho\}$ of the boundary of σ^n, so that P is a combinatorial $(n-1)$-sphere.

Corollary. *An arbitrary rectilinear triangulation of Euclidean space is combinatorial.*

This follows from the lemma, since in any triangulation of E^n the closed star of a vertex x is the cone, with vertex x, over its own boundary.

Theorem 5.1. *Any rectilinear triangulation of a closed n-simplex, or of its boundary, is combinatorial.*

Let Σ^n be the boundary of an $(n+1)$-simplex σ^{n+1}, let T be a triangulation of Σ^n, let x be a vertex of the triangulation, and let $\operatorname{St} x$ be its star. The closure of $\operatorname{St} x$ fails to meet at least one n-dimensional face of σ^{n+1}, so that there exists a semilinear homeomorphism h (projection into the hyperplane of this face) taking $\overline{\operatorname{St}}\, x$ into E^n. Then [1] $h(\operatorname{St} x)$ is a neighborhood of $h(x)$ in E^n, and its boundary is a polyhedron $P = h(\partial \operatorname{St} x)$.

Since $\overline{\operatorname{St}}\, x$ is the cone over its own boundary with vertex x, and the homeomorphism h is semilinear, $h(\overline{\operatorname{St}}\, x)$ is a cone over $h(\partial \operatorname{St} x)$. Applying Lemma 5.1, we conclude that P is a combinatorial $(n-1)$-sphere. Hence $\partial \operatorname{St} x = h^{-1}(P)$ is likewise a combinatorial $(n-1)$-sphere.

Now consider a triangulation of a closed simplex σ^n. We may suppose σ^n is contained in E^n. The assertion then follows immediately from the lemma if the vertex x is in the interior of σ^n. Suppose, then, that $x \in \partial\sigma^n$. As in the proof of the lemma, using the $(n-1)$-dimensional hyperplanes that pass through x and the $(n-2)$-dimensional faces of σ^n not containing x, construct a triangulation T of the polyhedron $\partial \operatorname{St}_{\sigma^n} x$ such that the cone from x over any simplex in T intersects only one face of σ^n. $\operatorname{St} x$ fails to meet at least one $(n-1)$-dimensional face of σ^n. By central projection from x, the boundary of $\operatorname{St} x$ maps

1) *Translator's note.* This seems inaccurate. Better: take h to be projection parallel to the line joining x and a point inside Σ^n.

semilinearly onto the closure K of the union of those $(n-1)$-dimensional faces of σ^n that fail to meet $\operatorname{St} x$; and K in turn maps linearly (by projection) onto the closure $\bar{\sigma}^{n-1}$ of one of the $(n-1)$-dimensional faces of σ^n not in K. Every simplex in T then goes into a simplex of the same dimension in a certain triangulation of $\bar{\sigma}^{n-1}$. Consequently the triangulation T is isomorphic to a triangulation of $\bar{\sigma}^{n-1}$, so that $\partial \operatorname{St} x$ is a combinatorial $(n-1)$-cell.

Thus the boundary of the star of a vertex is in every case either a combinatorial sphere or a combinatorial $(n-1)$-cell.

Corollary. *A combinatorial sphere is a combinatorial manifold, and a combinatorial n-cell is a combinatorial manifold with boundary.*

Let Σ^n be a combinatorial sphere, and T_Σ a triangulation isomorphic to a triangulation of the boundary $\partial\sigma^{n+1}$ of an $(n+1)$-simplex. By Theorem 5.1 T_Σ is combinatorial, and therefore Σ^n is a combinatorial manifold. Similarly for a combinatorial n-cell.

The converse assertion is, up to now, entirely unproved. Until recently it was known only for $n \leq 3$. Smale has recently proved, using methods of differential topology, that a combinatorial manifold homeomorphic to an n-sphere is a combinatorial sphere, for $n \geq 6$ [70]. This has been proved now also for $n = 5$, and so remains unknown only for $n = 4$.

Problems 6 and 7 are part of the so-called Principal Conjecture of combinatorial topology (Hauptvermutung), that triangulations of two homeomorphic polyhedra have isomorphic subdivisions. In its general form the problem has been solved in the negative: Milnor has constructed two homeomorphic seven-dimensional polyhedra which have no isomorphic triangulations [52]. But Milnor's polyhedra are not manifolds. For manifolds the question remains open. The problem is unsolved even for spheres: it is unknown whether every polyhedron homeomorphic to an n-sphere, $n > 3$, is a combinatorial sphere, or whether it has a combinatorial triangulation, i.e. is a combinatorial manifold.

The theory of combinatorial manifolds occupies an important position in the topology of manifolds. But up to the present time, not only is it unknown, for $n > 3$, whether every polyhedral manifold is combinatorial, but the following very important problem is unsolved.

Problem 8 (triangulation of manifolds). Is every n-manifold of dimension $n > 3$ homeomorphic to a rectilinear polyhedron?

For $n \leq 3$ the problem has a positive solution. For $n \leq 2$ this is a classical result, and all such manifolds have been described. The triangulation theorem for three-dimensional manifolds was proved by Moise, and also by Bing [57, 29]. And since every triangulation of a three-dimensional manifold is combinatorial, the study of three-dimensional manifolds reduces to that of combinatorial manifolds.

For $n \leq 3$ all the problems we have formulated are presently known to have positive solutions [53–60, 66–68]. For $n = 2$ the solutions of a majority of them have in fact been given in Chapter II of this monograph. The solution of Problem 1 is given by Antoine's theorem and its corollaries. Furthermore, since the homeomorphism constructed for the proof of that theorem was stable, it follows by Theorem 3.2 of Chapter V that Problem 4 (the annulus problem) has a positive solution for $n = 2$. In addition, it is easily proved that the boundary of the star of a vertex in a triangulation of a two-dimensional manifold is a polyhedral closed curve (cf. the proof of Lemma 3.1 of Chapter II). From this follows the positive solution of Problems 5, 6 and 7 for $n = 2$.

2. As we have previously indicated, it was already shown in 1920 by Alexander that every polyhedral 2-sphere is imbedded trivially in E^3. The corresponding question concerning the imbedding of a polyhedral $(n-1)$-sphere in E^n or in the boundary of an $(n+1)$-simplex, for $n > 3$, is up to the present time unsolved.

For a combinatorial sphere, on the other hand, it was solved by Newman [61]. For this purpose it was convenient to consider a more general notion, that of a star sphere, which is defined by induction.

Definition 5.3. A pair of points is a star 0-sphere. A polyhedral n-sphere is called a star sphere if the boundary of the star of any vertex is a star $(n-1)$-sphere.

Definition 5.4. A point is a star 0-cell. A polyhedral n-cell is called a star cell if the boundary of the star of any vertex is either a star $(n-1)$-sphere or a star $(n-1)$-cell.

Definition 5.5. A polyhedral n-manifold with or without boundary is called a star manifold if the boundary of the star of any vertex is either a star $(n-1)$-sphere or a star $(n-1)$-cell.

Theorem 5.2. *A combinatorial sphere (cell) is a star sphere (cell).*

This follows easily from Theorem 5.1. A pair of points is both a combinatorial and a star 0-sphere, and a single point is a combinatorial and a star 0-cell. Suppose the statement of the theorem true for $k \leq n-1$, and let Σ^n be a combinatorial n-sphere and Q^n a combinatorial n-cell. By the corollary to Theorem 5.1, the boundary $\partial \mathrm{St}\, x$ of the star of a vertex of Σ^n (or Q^n) in some triangulation of Σ^n (or Q^n) is a combinatorial $(n-1)$-sphere or $(n-1)$-cell. Then Σ^n (or Q^n) is a star sphere (or star cell) by the induction assumption.

Corollary. *Every combinatorial manifold with or without boundary is a star manifold.*

For, the star of any vertex in some triangulation is a combinatorial, and therefore a star, $(n-1)$-sphere or $(n-1)$-cell.

Theorem 5.3 (Newman [61]). *A star $(n-1)$-sphere S^{n-1} which is a subpolyhedron of a star n-sphere S^n separates S^n into domains whose closures are star n-cells.*

Newman gave a direct proof of the theorem by showing that S^{n-1} is imbedded in S^n cylindrically. We present here a considerably simpler proof due to Brown, which is based on Theorem 4.1.

We prove the theorem by induction on n. A star 0-sphere, i.e. a pair of points, separates a polyhedral 1-sphere containing it into two polygonal lines, i.e. into two star 1-cells. Suppose the theorem proved for all $k < n$; we must prove that it holds for n.

Let S^n be a star n-sphere, with a given triangulation, and S^{n-1} a star $(n-1)$-sphere which is a subpolyhedron of S^n. By Corollary 3 to Theorem 4.1, to prove that the closures of the complementary domains of S^{n-1} are topological cubes, it suffices to show that the imbedding of S^{n-1} in S^n is locally cylindrical. And since S^{n-1} is the union of the open stars of its vertices, if suffices to show that the open star of any vertex of S^{n-1} is cylindrically imbedded in S^n. We denote by $\mathrm{St}_P x$ the star of vertex x in polyhedron P.

Let x be a vertex of S^{n-1}. Consider $\mathrm{St}_{S^n} x$ and $\mathrm{St}_{S^{n-1}} x$. Since S^n and S^{n-1} are star spheres, $\partial \mathrm{St}_{S^n} x$ is a star $(n-2)$-sphere which is a subpolyhedron of $\partial \mathrm{St}_{S^n} x$.

By the induction assumption $\partial \mathrm{St}_{S^{n-1}} x$ bounds domains in $\partial \mathrm{St}_{S^n} x$ whose closures, D_{-1} and D_{+1}, are star $(n-1)$-cells. Therefore the cone K_i, $i = \pm 1$, over D_i with x as vertex is an n-cell with boundary $D_i \cup \mathrm{St}_{S^{n-1}} x$. This means that $\overline{\mathrm{St}}_{S^{n-1}} x$ is cylindrically imbedded in the closure $\overline{V}_i$ of each complementary domain of S^{n-1} in S^n, i.e. K_i is a closed neighborhood of x in $\overline{V}_i$, homeomorphic to $\overline{\mathrm{St}}_{S^{n-1}} x \times [0, i]$, $i = \pm 1$. Hence x has a neighborhood $\mathrm{Int}\,(K_{-1} \cup K_{+1})$ in $\overline{V}_{-1} \cup \overline{V}_{+1}$, i.e. in S^n, which is homeomorphic to the topological product $\mathrm{St}_{S^{n-1}} x \times (-1, +1)$, with $\mathrm{St}_{S^{n-1}} x = \mathrm{St}_{S^{n-1}} x \times 0$; that is to say, $\mathrm{St}_{S^{n-1}} x$ is cylindrically imbedded in S^n. Thus S^{n-1} is cylindrically imbedded in S^n, and by the Schoenflies theorem $\overline{V}_i$, $i = \pm 1$, is an n-cell. We must still show that $\overline{V}_i$ is a star n-cell. Since S^n is a star sphere, for any vertex of $\overline{V}_i$ in the interior of $\overline{V}_i$, the boundary of the star is obviously a star $(n-1)$-sphere. On the other hand, if $x \in \mathrm{Fr}\,\overline{V}_i$, then $\partial \mathrm{St}_{\overline{V}_i} x$ is the closed domain D_i in $\partial \mathrm{St}_{S^n} x$, which, as we have seen, is a star $(n-1)$-cell.

Observe that Newman's theorem holds not only for rectilinear polyhedra, but for any topological sphere S^n which has a star triangulation T and a sphere

$S^{n-1} \subset S^n$ which is a topological subpolyhedron of S^n and is a star sphere in the triangulation $T|_{S^{n-1}}$.

Corollary 1. *A combinatorial sphere* Σ^{n-1} *which is a subpolyhedron of a combinatorial sphere* Σ^n *separates* Σ^n *into two domains* V_i *whose closures* $\bar{V}_i$ *are star n-cells.*

This follows from Newman's theorem and the fact that a combinatorial sphere is a star sphere. Observe that we are not able to assert that the n-cells $\bar{V}_i$ are combinatorial, since it is unknown whether a star n-cell or a star sphere is combinatorial. The methods of Newman and Brown do not give the means of answering this question.

Corollary 2. *Every two-sided star* $(n-1)$*-manifold* M^{n-1} *which is a subpolyhedron of a star manifold* M^n *is imbedded cylindrically in* M^n.

Since M^{n-1} is the union of the stars of its vertices, it suffices to prove that the star of any vertex of M^{n-1} is cylindrically imbedded in M^n. This follows from Newman's theorem and the fact that $\partial \mathrm{St}_{M^n} x$ and $\partial \mathrm{St}_{M^{n-1}} x$ are star spheres.

Corollary 3. *For any rectilinear combinatorial* $(n-1)$*-sphere* Σ^{n-1} *in* E^n, *there exists a homeomorphism* $h: E^n \longrightarrow E^n$ *taking* Σ^{n-1} *into the boundary of an n-simplex.*

By Corollary 1 to the Schoenflies theorem (1.3) it suffices to show that Σ^{n-1} is cylindrically imbedded in E^n.

Take an n-simplex σ^n such that $\Sigma^{n-1} \subset \sigma^n$. Let T_Σ be a combinatorial triangulation of Σ^{n-1}. Using all the $(n-1)$-dimensional hyperplanes containing the $(n-1)$-simplices of T_Σ, subdivide σ^n, as well as every simplex $\sigma^r \in T_\Sigma$, into a finite number of convex sets and, as in the proof of Theorem 5.1, construct a triangulation T of σ^n, whose restriction to Σ^{n-1} is a subdivision of the triangulation T_Σ. Since σ^n is a simplex and the sphere Σ^{n-1} is combinatorial, it follows by Theorem 5.1 that the triangulation T is combinatorial and that the induced subdivision of T_Σ is also combinatorial. Consequently both of these are star triangulations, so that, by Corollary 2, Σ^{n-1} is cylindrically imbedded in σ^n, and so also in E^n.

We are not able to assert that the homeomorphism $h: E^n \longrightarrow E^n$ satisfying the conditions of the theorem can be taken to be semilinear as well, since it is unknown whether the n-cell bounded by the sphere Σ^{n-1} is combinatorial.

§6. Conclusion

1. We state now without proof some results of Stallings [73], Zeeman [77] and Černavskiĭ [14,15], concerning imbeddings of k-spheres and k-cells in E^n and S^n, $k < n-2$. Stallings studies the imbedding of locally flat topological

spheres, and Zeeman the imbedding of combinatorial spheres, for $k < n - 2$.

Quite recently, Gluck and Greathouse [42,43] and Černavskiĭ [14] have concerned themselves with the question of existence, for trivially imbedded spheres S^k and cells Q^k, of an isotopy of E^n (or S^n) from the identity mapping, taking S^k (or Q^k) into the standard sphere (or cube). In this direction we have at present only partial results: Gluck [42] and Greathouse [43] have answered the question for $k \leq n/2 - 1$, and Černavskiĭ for $k < 2n/3 - 1$ [14].[1] We remark that the increase in the bound involves grave difficulties. As we have seen in §3, the isotopy problem is closely related to that of the structure of the homeomorphisms of S^n (or of E^n) onto itself.

Theorem I (Stallings). *Every locally flat sphere S^k in S^n is trivially imbedded, if $k < n - 2$ and $h \geq 5$* [73].

The corresponding theorem for a sphere S^{n-2} in S^n is false, since for any $n \geq 3$ a locally flat sphere can be "knotted" in S^n, i.e. imbedded in such a way that there is no homeomorphism $h: S^n \to S^n$ taking S^{n-2} into the standard $(n-2)$-sphere.

Theorem II (Stallings). *In order that a locally flat $(n-2)$-sphere S^{n-2} in S^n, $n \geq 5$, be trivial, it is necessary and sufficient that $S^n \setminus S^{n-2}$ be homotopically equivalent to a circle* [73].

Thus Stallings' result has a double content. On the one hand, it shows that, assuming local conditions satisfied, knots can exist only in dimension $n - 2$; and on the other hand, it gives a convenient criterion for a topological imbedding of a k-sphere into an n-sphere to be trivial, if $k \leq n - 2$, $n \geq 5$. To obtain these results the author needed to apply a new method, of his own construction, based on his so-called engulfing lemma [72].

The question of the imbedding of k-cells in S^n is, in contrast to spheres, much simpler, and was answered by Černavskiĭ [14] by elementary means: every locally flat k-cell in S^n, $k \leq n$, is trivially imbedded in S^n. Specifically, it is shown that a locally flat k-cell is imbedded in S^n together with a neighborhood. It follows by Theorem 2.2 that Q^k is trivial in S^n. Observe that for an imbedding of a sphere S^{n-2} it does not follow from the fact that S^{n-2} is imbedded with neighborhood that it is imbedded trivially.

2. Zeeman studied the imbedding of combinatorial k-spheres in E^n, or in the boundary of an $(n+1)$-simplex, for $k < n - 2$. The formulation itself of the problem is here different from that of Stallings, and it is not a simple matter to carry the latter over to the case of a combinatorial sphere. First of all, Zeeman imposes no local conditions. But a combinatorial sphere may not be locally flat

1) And very recently for $k \leq n - 3$ (see [81]).

at some points. Consider an arbitrary one-dimensional polyhedral knot C^1 in a hyperplane E^3 of 4-space E^4, and let C^2 be a suspension over C^1 in E^4, i.e. a double cone over C^1, whose vertices p_1 and p_2 lie in different complementary domains of E^3 in E^4; C^2 is a 2-sphere in E^4. It is easily seen that the points p_1 and p_2 are points of "local knottedness," i.e., points at which C^2 is not locally flat.[1)]

In general, if we take a one-dimensional knot C^1 in a three-dimensional hyperplane E^3 of E^n, and construct over C^1 an $(n-3)$-fold suspension, we obtain a combinatorial $(n-2)$-sphere in E^n whose "locally knotted" subset is an $(n-4)$-sphere. It follows from Zeeman's results that such points of "local knottedness" cannot appear on a combinatorial sphere S^k if $k \leq n-3$.

Furthermore, Zeeman proves a stronger fact for combinatorial spheres than triviality of the imbedding. Namely, he proves the existence of a semilinear isotopy of S^n taking a combinatorial sphere S^k into the boundary of a $(k+1)$-simplex.

Definition. By a *semilinear isotopy* of a polyhedron P onto itself is meant a semilinear homeomorphism $H\colon P \times [0, 1] \to P \times [0, 1]$, such that $H|_{P\times 0} = 1$, and $H|_{P\times t} = [H_t, t]$ is a semilinear homeomorphism of P onto itself for every t, $0 < t \leq 1$.

Zeeman's theorem. *For any combinatorial sphere S^k, $k \leq n-3$, which is a subpolyhedron of a combinatorial sphere S^n (or of E^n), there exists a semilinear isotopy of S^n onto itself taking S^k into the boundary of a $(k+1)$-simplex* [77].

3. For locally flat spheres S^k and k-cells in S^n, the existence of an isotopy was proved by Černavskiĭ for dimensions $k < n-2$.

The solution of the problem was first given for $k \leq n/2 - 1$ by Gluck [42] and Greathouse [43], who used the fact that in these dimensions a sphere S^k (or a k-cell Q^k) can be approximated arbitrarily closely by a rectilinear combinatorial sphere Σ^k. To do so, it suffices to take a sufficiently fine combinatorial triangulation T of the sphere S^k, approximate the set of all vertices of T by a set K of points $\{x_r\}$ in general position, and then span every set of $r+1$ points of K approximating the vertices of an r-simplex in T by a rectilinear r-simplex. The proof is based on a theorem of Homma on isotopy for topological polyhedra [45]. The better results were achieved by Černavskiĭ by means of an entirely different method [14, 15].

The isotopy problem would receive a positive solution if Problem 3 did, i.e. if it were proved that every orientation-preserving homeomorphism of S^n onto

---

1) In fact, the local fundamental group of the complement at either of these points is isomorphic to the fundamental group of the knot.

itself is stable. Indeed, suppose the homeomorphism $H\colon S^n \to S^n$ taking a locally flat sphere into the standard sphere Σ^k changes orientation. The sphere Σ^k is contained in a standard sphere Σ^{n-1}. Let $h\colon S^n \to S^n$ be the homeomorphism which is the identity on Σ^{n-1} and takes any point $x \in S^n \setminus \Sigma^{n-1}$ into the point $h(x)$ symmetric to x with respect to Σ^{n-1}. Then the homeomorphism $hH\colon S^n \to S^n$ preserves orientation, and $hH(S^k) = H(S^k) = \Sigma^k$. Thus, if a sphere S^k, $k \le n-1$, is trivial in S^n, there always exists a homeomorphism $H\colon S^n \to S^n$ which preserves orientation and takes S^k into the standard sphere.

Černavskiĭ studied the structure of the homeomorphisms of S^n onto itself, and proved his theorem on isotopy of S^n, for locally flat spheres or cells of dimension $k < 2n/3 - 1$, by basing it on the following structure theorem for homeomorphisms of S^n.

Černavskiĭ's theorem [14]. *Every homeomorphism h of the sphere S^n onto itself is representable in the form*

$$h = h_k g, \tag{48}$$

where g is a stable homeomorphism and h_k is a homeomorphism which is the identity on some k-dimensional ball B^k of dimension $k < 2n/3 - 1$.

$$h_k|_{B^k} = 1,\ k < \frac{2}{3}n - 1.^{1)} \tag{49}$$

Corollary. *For every locally flat k-cell Q^k or sphere S^k in S^n (or E^n), of dimension $k < 2n/3 - 1$, there exists an isotopy F_t from the identity mapping of S^n (or E^n) onto itself, taking Q^k into a standard k-cell (or S^k into a standard k-sphere).*

The argument is especially simple in the case of a k-cell in E^n. We give it here. Let Q^k be a locally flat k-cell in E^n, of dimension $k < 2n/3 - 1$.

By the theorem cited above Q^k is trivial in E^n; consequently there exists a homeomorphism $h\colon E^n \to E^n$ such that $h(Q^k) = I^k$ is a standard k-dimensional cube in E^n. By Černavskiĭ's theorem h is representable in the form (48) so as to satisfy (49), where B^k is a second standard cube in E^n. Our assertion will be proved for Q^k if we show that there exists a stable homeomorphism $H\colon S^n \to S^n$ taking Q^k into I^k. Set $g(Q^k) = \tilde{Q}^k$. Then $h_k(\tilde{Q}^k) = I^k$, and since g is stable, it suffices to construct a stable homeomorphism $\tilde{h}\colon E^n \to E^n$ taking $\tilde{Q}^k$ into I^k, and to set $H = \tilde{h}g$. By Theorem 2.2 of Chapter I there exists a homeomorphism $\phi\colon E^n \to E^n$ which is the identity on some domain V not containing $B^k \cup I^k$ and which takes I^k into B^k:

---

1) *Translator's note.* The integer k can be specified beforehand; thus it can be taken as the greatest integer satisfying the inequality.

$$\varphi|_V = 1, \ \varphi(I^k) = B^k.$$

Set

$$\tilde{h} = \varphi^{-1} h_k^{-1} \varphi h_k.$$

The homeomorphism $\tilde{h}$ is stable, since

$$\varphi|_V = 1, \ (h_k^{-1} \varphi h_k)|_{h_k^{-1}(V)} = 1.$$

Furthermore, by assumption, $h_k(\tilde{Q}^k) = I^k$ and $h_k|_{B^k} = 1$. Hence

$$h_k^{-1} \varphi h_k(\tilde{Q}) = h_k^{-1} \varphi(I^k) = h_k^{-1}(B^k) = B^k,$$

so that $\tilde{h}(\tilde{Q}^k) = \phi^{-1}(B^k) = I^k$. Thus $\tilde{h}$ is a stable homeomorphism and takes $\tilde{Q}^k$ into I^k. Consequently the homeomorphism $H = \tilde{h}g$ is also stable and takes Q^k into I^k.

By the corollary to Theorem 3.1, therefore, there exists an isotopy taking H into the identity mapping.

Very recently, new results have been obtained in the topology of imbeddings. Homma has proved that any topological imbedding of a combinatorial k-manifold, $k \leq n-3$, into a sphere can be approximated by a semilinear imbedding. Using this result and his theorem on the union of locally flat cells [15], Černavskiĭ has now extended his above-stated results up to dimension $k = n-3$ (see [81]).

BIBLIOGRAPHY

[1] P. Aleksandrov, *Sur les ensembles de la première classe et les espaces abstraits*, C. R. Acad. Sci. Paris **178** (1924), 185–187.

[2] G. Ja. Zuev and A. B. Sosinskiĭ, *On spheres containing wild zero-dimensional sets*, Vestnik Moskov. Univ. Ser. I Mat. Meh. 22 (1967), no. 3, 5–9. (Russian) MR 35 #7321.

[3] A. A. Ivanov, *Isotopy of compacts in Euclidean spaces*, Dokl. Akad. Nauk SSSR 71 (1950), 1021–1022. (Russian) MR 11, 675.

[4] L. V. Keldyš, *The imbedding in E^4 of some monotone images of E^3*, Dokl. Akad. Nauk SSSR 136 (1961), 18–21 = Soviet Math. Dokl. 2 (1961), 9–12. MR 23 #A2871.

[5] ______, "Some theorems on topological imbedding" (Russian) in *General topology and its relations to modern analysis and algebra*, Proc. Sympos., Prague 1961, Academic Press, New York and Publ. House Czech. Acad. Sci., Prague, 1962, pp. 230–234. MR 26 #3025.

[6] ______, *On imbedding certain monotone images of E^n in E^n and E^{n+1}*, Mat. Sb. 57 (99) (1962), 95–104; English transl., Amer. Math. Soc. Transl. (2) 52 (1966), 32–41. MR 25 #1542.

[7] ______, *Imbedding of zero-dimensional compact sets in E^n*, Dokl. Akad. Nauk SSSR **147** (1962), 772–775 = Soviet Math. Dokl. 3 (1962), 1701–1704. MR 32 #8323.

[8] L. S. Pontrjagin, *Continuous groups*, 2nd ed., GITTL, Moscow, 1954; English transl., *Topological groups*, Gordon and Breach, New York, 1966. MR 17, 171; MR 34 #1439.

[9] A. B. Sosinskiĭ, *Imbedding of a k-dimensional element in E^n*, Dokl. Akad. Nauk SSSR 139 (1961), 1311–1313 = Soviet Math. Dokl. 2 (1961), 1096–1098. MR 24 #A1716.

[10] ______, *Multidimensional topological knots*, Dokl. Akad. Nauk SSSR 163 (1965), 1326–1329 = Soviet Math. Dokl. **6** (1965), 1119–1123.

[11] P. S. Uryson, *Works on topology and other fields of mathematics*. Vol. II, GITTL, Moscow, 1951. (Russian) MR 14, 122.

[12] A. V. Černavskiĭ, *The configuration of $(n-1)$-spheres in the n-sphere*, Dokl. Akad. Nauk SSSR **140** (1961), 540–542 = Soviet Math. Dokl. 2 (1961), 1244–1246. MR 24 #A2376.

[13] ______, *Homeomorphisms of euclidean space and topological imbeddings of polyhedra into euclidean space*. I, Mat. Sb. 68 (110) (1965), 581–613; English transl., Amer. Math. Soc. Transl. (2) 76 (1968) (to appear). MR 33 #1842.

[14] A. V. Černavskiĭ, *Homeomorphisms of euclidean space and topological imbeddings of polyhedra into euclidean space.* II, Mat. Sb. 72 (114) (1967), 573–601 = Math. USSR Sbornik 1 (1967), no. 4 (to appear). MR 34 #8391.

[15] ______, *Singular points of topological imbeddings of manifolds and the union of locally flat cells*, Dokl. Akad. Nauk SSSR 167 (1966), 528–530 = Soviet Math. Dokl. 7 (1966), 433–436. MR **33** #6605.

[16] J. W. Alexander, *On the deformation of an n-cell*, Proc. Nat. Acad. Sci. U.S.A. 9 (1923), 406–407.

[17] ______, *On the subdivision of 3-space by a polyhedron*, Proc. Nat. Acad. Sci. U.S.A. 10 (1924), 6–8.

[18] ______, *An example of a simply connected surface bounding a region which is not simply connected*, Proc. Nat. Acad. Sci. U.S.A. 10 (1924), 8–10.

[19] ______, *The combinatorial theory of complexes*, Ann. of Math. (2) 31 (1930), 292–320.

[20] L. Antoine, *Sur l'homéomorphie de deux figures et de leurs voisinages*, J. Math. Pures Appl. (8) 4 (1921), 221–325.

[21] E. Artin and R. Fox, *Some wild cells and spheres in three-dimensional space*, Ann. of Math. (2) 49 (1948), 979–990. MR **10**, 317.

[22] R. H. Bing, *A characterization of 3-space by partitionings*, Trans. Amer. Math. Soc. **70** (1951), 15–27. MR 13, 484.

[23] ______, *Locally tame sets are tame*, Ann. of Math. (2) **59** (1954), 145–158. MR 15, 816.

[24] ______, *Upper semicontinuous decompositions of E^3*, Ann. of Math. (2) 65 (1957), 363–374. MR 19, 1187.

[25] ______, *A decomposition of E^3 into points and tame arcs such that the decomposition space is topologically different from E^3*, Ann. of Math. (2) 65 (1957), 484–500. MR 19, 1187.

[26] ______, *Approximating surfaces with polyhedral ones*, Ann. of Math. (2) 65 (1957), 456–483. MR **19**, 300.

[27] ______, *Necessary and sufficient conditions that a 3-manifold be S^3*, Ann. of Math. (2) 68 (1958), 17–37. MR **20** #1973.

[28] ______, *Conditions under which a surface in E^3 is tame*, Fund. Math. 47 (1959), 105–139. MR 21 #5954.

[29] ______, *An alternative proof that 3-manifolds can be triangulated*, Ann. of Math. (2) 69 (1959), 37–65. MR 20 #7269.

[30] ______, *Tame Cantor sets in E^3*, Pacific J. Math. 11 (1961), 435–446. MR 24 #A539.

[31] W. A. Blankinship, *Generalization of a construction by Antoine*, Ann. of Math. (2) 53 (1951), 276–297. MR 12, 730.

[32] M. Brown, *A proof of the generalized Schoenflies theorem*, Bull. Amer. Math. Soc. 66 (1960), 74–76. MR 22 #8470b.

[33] ______, *The monotone union of open n-cells is an open n-cell*, Proc. Amer. Math. Soc. 12 (1961), 812–814. MR 23 #A4129.

[34] ______, "Locally flat embeddings of topological manifolds" in *Topology of 3-manifolds and related topics*, Proc. Univ. Georgia Inst. 1961, Prentice-Hall, Englewood Cliffs, N. J., 1962, pp. 83–91. MR 28 #1598.

[35] ______, *Locally flat imbeddings of topological manifolds*, Ann. of Math. (2) 75 (1962), 331–341. MR 24 #A3637.

[36] M. Brown and H. Gluck, *Stable structures on manifolds*. I, II, III, Ann. of Math. (2) 79 (1964), 1–58. MR 28 #1608a, b, c.

[37] J. C. Cantrell, *Separation of the n-sphere by an* $(n-1)$*-sphere*, Trans. Amer. Math. Soc. 108 (1963), 185–194. MR 27 #5242.

[38] J. C. Cantrell and C. H. Edwards, Jr., *Almost locally polyhedral curves in Euclidean n-space*, Trans. Amer. Math. Soc. 107 (1963), 451–457. MR 26 #6941.

[39] M. K. Fort, Jr., *A note concerning a decomposition space defined by Bing*, Ann. of Math. (2) 65 (1957), 501–504. MR 19, 1188.

[40] H. Gluck, *The embedding of two-spheres in the four-sphere*, Trans. Amer. Math. Soc. 104 (1962), 308–333. MR 26 #4327.

[41] ______, *Unknotting* S^1 *in* S^4, Bull. Amer. Math. Soc. 69 (1963), 91–94. MR 25 #5507.

[42] ______, *Embeddings in the trivial range*, Bull. Amer. Math. Soc. 69 (1963), 824–831. MR 27 #5241b.

[43] C. Greathouse, *Locally flat, locally tame, and tame embeddings*, Bull. Amer. Math. Soc. 69 (1963), 820–823. MR 27 #5241a.

[44] V. K. A. M. Gugenheim, *Piecewise linear isotopy and embedding of elements and spheres*. I, Proc. London Math. Soc. (3) 3 (1953), 29–53. MR 15, 336.

[45] T. Homma, *On the imbedding of polyhedra in manifolds*, Yokohama Math. J. 10 (1962), 5–10. MR 27 #4236.

[46] A. Kirkor, *Wild 0-dimensional sets and the fundamental group*, Fund. Math. 45 (1958), 228–236. MR 21 #1569.

[47] B. Knaster, *Un continu dont tout sous-continu est indécomposable*, Fund. Math. 3 (1922), 247–286.

[48] K. W. Kwun, *Upper semicontinuous decompositions of the n-sphere*, Proc. Amer. Math. Soc. 13 (1962), 284–290. MR 25 #3512.

[49] B. Mazur, *On embeddings of spheres*, Bull. Amer. Math. Soc. 65 (1959), 59–65. MR 22 #8469.

[50] ______, *On embeddings of spheres*, Acta Math. 105 (1961), 1–17. MR 23 #A2870.

[51] D. R. McMillan, Jr., *The singular points of a topological embedding*, Duke Math. J. 31 (1964), 711–716. MR 29 #6478.

[52] J. Milnor, *Two complexes which are homeomorphic but combinatorially distinct*, Ann. of Math. (2) 74 (1961), 575–590. MR 24 #A2961.

[53] E. E. Moise, *Affine structures in 3-manifolds. I: Polyhedral approximations of solids*, Ann. of Math. (2) 54 (1951), 506–533. MR 13, 484.

[54] ______, ibid. II: *Positional properties of 2-spheres*, Ann. of Math. (2) 55 (1952), 172–176. MR 13, 574.

[55] ______, ibid. III: *Tubular neighborhoods of linear graphs*, Ann. of Math. (2) 55 (1952), 203–214. MR 13, 765.

[56] ______, ibid. IV: *Piecewise linear approximations of homeomorphisms*, Ann. of Math. (2) 55 (1952), 215–222. MR 13, 765.

[57] ______, ibid. V: *The triangulation theorem and Hauptvermutung*, Ann. of Math. (2) 56 (1952), 96–114. MR 14, 72.

[58] ______, ibid. VI: *Compact spaces covered by two Euclidean neighborhoods*, Ann. of Math. (2) 58 (1953), 107. MR 14, 1003.

[59] ______, ibid. VII: *Disks which are pierced by intervals*, Ann. of Math. (2) 58 (1953), 403–408. MR 15, 337.

[60] ______, ibid. VIII: *Invariance of the knot-types; local tame imbedding*, Ann. of Math. (2) 59 (1954), 159–170. MR 15, 889.

[61] M. H. A. Newman, *On the division of Euclidean n-space by topological $(n-1)$-spheres*, Proc. Roy. Soc. London Ser. A 257 (1960), 1–12. MR 23 #A633.

[62] C. D. Papakyriakopoulos, *On Dehn's lemma and the asphericity of knots*, Ann. of Math. (2) 66 (1957), 1–26. MR 19, 761.

[63] R. Penrose, J. H. C. Whitehead and E. C. Zeeman, *Imbedding of manifolds in euclidean space*, Ann. of Math. (2) 73 (1961), 613–623. MR 23 #A2218.

[64] K. Reidemeister, *Knotentheorie*, Chelsea, New York, 1948.

[65] R. H. Rosen, *The five dimensional polyhedral Schoenflies theorem*, Bull. Amer. Math. Soc. 70 (1964), 511–516. MR 29 #601.

[66] D. E. Sanderson, *Isotopy in 3-manifolds. I: Isotopic deformations of 2-cells and 3-cells*, Proc. Amer. Math. Soc. 8 (1957), 912–922. MR 19, 760.

[67] ______, ibid. II: *Fitting homeomorphisms by isotopy*, Duke Math. J. 26 (1959), 387–396. MR 21 #5956.

[68] ______, ibid. III: *Connectivity of spaces of homeomorphisms*, Proc. Amer. Math. Soc. 11 (1960), 171–176. MR 22 #2986.

[69] A. Schoenflies, *Beiträge zur Theorie der Punktmengen*. III, Math. Ann. 62 (1906), 286–328.

[70] S. Smale, *Differentiable and combinatorial structures on manifolds*, Ann. of Math. (2) **74** (1961), 498–502. MR 24 #A2967.

[71] J. R. Stallings, *Polyhedral homotopy-spheres*, Bull. Amer. Math. Soc. 66 (1960), 485–488. MR 23 #A2214.

[72] ______, *The piecewise-linear structure of Euclidean space*, Proc. Cambridge Philos. Soc. 58 (1962), 481–488. MR 26 #6945.

[73] ______, *On topologically unknotted spheres*, Ann. of Math. 77 (1963), 490–503. MR 26 #6946.

[74] E. R. van Kampen, *On the connection between the fundamental groups of some related spaces*, Amer. J. Math. 55 (1933), 261–267.

[75] E. C. Zeeman, "Isotopies and knots in manifolds" in *Topology of 3-manifolds and related topics*, Proc. Univ. Georgia Inst., 1961, Prentice-Hall, Englewood Cliffs, N. J., 1962, pp. 187–193. MR 25 #3520.

[76] ______, "Polyhedral n-manifolds. II: Embeddings" in *Topology of 3-manifolds and related topics*, Proc. Univ. Georgia Inst., 1961, Prentice-Hall, Englewood Cliffs, N. J., 1962, pp. 64–70. MR 28 #1596.

[77] ______, *Unknotting combinatorial balls*, Ann. of Math. (2) 78 (1963), 501–526. MR 28 #3432.

[78] E. C. Zeeman and J. F. P. Hudson, *On combinatorial isotopy*, Inst. Hautes Études Sci. Publ. Math. No. 19 (1964), 69–94. MR 29 #4063.

[79] L. V. Keldyš, *Topological imbeddings in manifolds and pseudo-isotopy*, Mat. Sb. 71 (113) (1966), 433–453; English transl., Amer. Math. Soc. Transl. (2) 76 (1968) (to appear). MR 34 #6784.

[80] H. Seifert and W. Threfall, *Lehrbuch der Topologie*, Teubner, Leipzig, 1934.

[81] A. V. Černavskiĭ, *Homeomorphisms of R^n are k-stable for $k \leq n - 3$*, Mat. Sb. 70 (112) (1966), 605–606; English transl., Amer. Math. Soc. Transl. (2) 76 (1968) (to appear). MR 33 #6642.